A Royal Road to Algebraic Geometry

The Greek historian, Herodotus of Halicarnassus in the fifth Century BC, describes the Persian Royal Road in detail. Part of his description is given in the introduction to this book. But there are other such Royal Roads in other countries, build for the kings to use in keeping control of their kingdoms. One such road follows the south western part of the Norwegian coastline, for example.

Audun Holme

Audun Holme

A Royal Road to Algebraic Geometry

 Springer

Audun Holme
Department of Mathematics
University of Bergen
Johs. Brunsgt. 12
5008 Bergen
Norway
audun.holme@gmail.com

ISBN 978-3-642-42921-7   ISBN 978-3-642-19225-8 (eBook)
DOI 10.1007/978-3-642-19225-8
Springer Heidelberg Dordrecht London New York

Mathematics Subject Classification (2010): 14-01

Cover illustration: Herodotus statue © Istockphoto
Cover design: deblik

Printed on acid-free paper

Springer is part of Springer Science+Business Media (www.springer.com)

Preface

Why the "royal road"? The great geometer who today is known as *Euclid of Alexandria* was invited to come to that city by Ptolemy 2., possibly from Athens.[1] Ptolemy was a son of Ptolemy 1., one of Alexander's generals who assumed the title *King of Egypt* and founded the great museum and library in Alexander's city by the delta of the Nile. Euclid and his collaborators assembled practically the entire body of geometry known to them in thirteen books, the fruit of the towering and impressive effort of Greek mathematics. This work is known today as *Euclid's Elements*.

There is an historical anecdote to the effect that when the king wanted to learn some geometry, he found the Elements too long and too hard to read and understand, and asked Euclid if there really existed no simpler way of learning about geometry. Euclid is said to have answered, somewhat arrogantly perhaps, that *"there is no Royal Road to geometry"*.

The Persian Royal Road is described by the *Father of History* Herodotus of Halicarnassus in the fifth Century BC. His text can be found in English translation in [19], 5.52–53, page 129 in the book listed in the present reference. He writes as follows:

> *Here is what the road is like. There are royal way stations and fine inns all along the way, and the whole road runs through safe, inhabited territory. There are twenty way stations on the three-hundred-seventy-seven-and-a-half-mile stretch from Lydia to Phrygia. The Halys river is on the Phrygian border. Gates stand at the river crossing, and it is absolutely necessary to pass through them to pass the well-guarded river.* [...]

---

[1] However, some believe that Euclid of Alexandria never existed, but that the name "Euclid" was taken from Plato's dialogue *Theaetetus*. In the dialogue there appears a minor and insignificant character, totally fictitious, called *Euclid*. As the events related are supposed to find place in the city of Megara, this character is known today as *Euclid of Megara*. Now one theory goes that this name was used as a pseudonym for a group of geometers working in Alexandria. Much as the name *N. Bourbaki* was used by a group of French mathematicians in modern times.

Herodotus then continues to describe the route in great detail, including the number of rest houses and inns along the way, and information on where one needs to cross rivers by boat. Then he concludes by explaining that The Royal Road has been measured to 14400 furlongs. A furlong, later called a *stade* by the Romans, was about 200 meters long, so the total length of the Road should be 2880 kilometers. He estimates that it is reasonable to travel 164 furlongs a day, that is about 33 kilometers. So the whole journey should take roughly 90 days.

Thus the term *A Royal Road* was intended to mean *a simple way* by Euclid himself, perhaps too simple. In that sense one might argue that there is not, or should not be, a Royal Road to Algebraic Geometry either. King Ptolemy was a competent ruler, who probably understood that geometry as such can not be made easy. But he might have wanted to learn the important *ideas* behind the proofs, without going into what we would call the *axiomatic formal deductions*. This is what Archimedes called the *Analysis*. A proof can not be obtained in this way, Archimedes says. But the analysis of the problem provides information on the problem, so it may be easier to find an actual proof.

When you tread the road of algebraic geometry, you tread in the footsteps of queens and kings of mathematics. And if you manage to study four or five pages of this book each day, stopping at some of the resting places, you complete the journey in ninety days by a comfortable margin. But that means you will be travelling along, viewing the landscape and learning the geography. You will, however, not pitch camp to go fishing or hunting from time to time. In more serious terms, I present a number of important theorems in algebraic geometry with only *comments* on how the basic idea of the real proofs work. But references to where the reader may find such complete proofs are always provided, and I try to limit myself to using relatively few, well known and readily available texts. Thus for instance the important result of *Serre duality* is stated but not proved here, the reference I have chosen being to Hartshorne's book [18]. The same applies to intersection theory. Intersection theory, especially in the singular case, is just shown the traveller in the distance, from one of the smaller hills we have climbed.

Thus it should be possible to teach a rapid course from the present text in one term. A course of this type could be useful for students or other young (or older) mathematicians who have specialized in slightly different subjects, and would like to read up on modern algebraic geometry after first gaining some knowledge about the subject.

A more in-depth treatment, with complete proofs taken from the references, with exercises and with more examples, would at least be a full one year graduate course. The term *algebraic geometry* is used in the traditional sense, founded in the fundamental and important works by *André Weil* [41] and *Alexander Grothendieck* [15]. Thus I make no mention of such themes as "*tropical*" geometry, for example, nor do I deal with the so called "*non commutative algebraic geometry*". Some experts of these new areas would therefore have preferred the term *classical algebraic geometry* in the title of the present treatment. But for several reasons the author finds that term to be somewhat misleading. The term *"classical"* is ambiguous, and in the present context it should be stretched, at the very most, to cover the subject roughly up to the first quarter of the 20th century. A suitable title for a treatment including the above mentioned new developments, might be simply *"Algebraic Geometry in the 21th century"*. Perhaps followed by a question mark for the time being.

The book is divided into two parts. Part I, on *Curves*, introduces the basic concepts of algebraic geometry in the context of projective curves. The treatment here is quite simple, and leads up to the intersection theory in a simplified setting, as well as to the statement of the classical Riemann Roch Theorem and the concept of the *dual curve*. Part I is also intended as a preparation and motivation for an introduction to the Grothendieck theory of *schemes*, given in Part II. But Part I could well be used as a text for an undergraduate course on curves, less ambitious but providing more of an overview than the standard texts available. A good supplement would be parts of Fulton's very nice book [10]. For example, our treatment of intersection theory for curves in $\mathbb{P}^2_k$ in Chap. 4 follows this source. Hartshorne's book [18] is another excellent text, more advanced and perfectly suited for a follow up course to one based on the present text. Both of [10] and [18] have a number of very good exercises.

Some of the material in this book was surveyed several years ago, partly in Spanish, in publications of the UNAM in Mexico, [21] and [22]. The same applies to the papers [23] and [24], as well as [25], [29] and [30]. Chapter 22 contains the main part of [26].

Also, parts of the material presented in Sect. 21.4, were developed during work on the joint paper with *Joel Roberts* [31], while I was visiting the University of Minneapolis about twelve years ago. Some of this work has not been published before, and as it now appears here I take the opportunity to thank Joel for so many stimulating conversations on this and related mathematics, back then.

Acknowledgment   The author was supported by a grant from the Norwegian non-fiction Literary Fund under the NFF.

Bergen, Norway   Audun Holme

Contents

Part I
Curves

Chapter 1
Affine and Projective Space

The historical roots of algebraic geometry lie in the study of curved lines in the plane, or as we would prefer to say today, *planar curves*. The treatment of modern algebraic geometry offered in the present book takes a starting point which is more general and at the same time more restricted: More general in the sense that we will study geometric objects such as curves and surfaces, say, in spaces of any dimensions, and where the points, including those at infinity, are described by coordinates which are elements of a general field, not just real numbers. More special in the sense that we consider geometric objects defined by *polynomial equations* in the coordinates of the space in which they lie. Historically the necessity of dealing with points at infinity is one of the reasons why ordinary space had to be completed to *projective space* by adding points at infinity.

In this first chapter we establish these foundations for our subject. The theorem of Desargues is treated in some detail, since it illustrates in a beautiful way the role played by points at infinity, the concept of duality and the interplay between different projective coordinate systems.

1.1 Definitions

Let k denote a field. For simplicity we frequently assume that k be algebraically closed.

Following most standard references such as [18], we introduce the following notation: The set k^n of all n-tuples of elements from k is denoted by $\mathbb{A}_k^n$, and referred to it as the *affine n-space* over the field k. An element $P = (a_1, \ldots, a_n)$ is referred to as a point, and a_i is called the ith coordinate of P. In particular we get the affine line $\mathbb{A}_k^1 = k$ and the affine plane $\mathbb{A}_k^2 = k^2$. If $k = \mathbb{R}$, the field of real numbers, we get the *"usual"* affine (real) spaces. In general k is referred to as the *ground field*.

A. Holme, *A Rōyal Road to Algebraic Geometry*,
DOI 10.1007/978-3-642-19225-8_1, © Springer-Verlag Berlin Heidelberg 2012

We next define the *n-dimensional* projective space over the field k, $\mathbb{P}_k^n$. First consider the set

$$\mathcal{M} = k^{n+1} - \{(0,\ldots,0)\}$$

and define a relation by

$$(a_0, a_1, a_2, \ldots, a_n) \sim (b_0, b_1, b_2, \ldots, b_n)$$

whenever there exists an element, necessarily non-zero, $r \in k$ such that

$$a_i = r b_i \quad \text{for all } i = 0, 1, 2, \ldots, n.$$

One easily verifies that this is an equivalence relation. The set $\mathcal{M}/\sim$ is denoted by $\mathbb{P}_k^n$ and referred to as the *projective n-space over k*. In particular we have the affine and projective n-spaces over the rational numbers $\mathbb{Q}$, the real numbers $\mathbb{R}$, the complex numbers $\mathbb{C}$, over the binary field $\mathbb{Z}_2$, and so on.

The following notation for the equivalence classes will be used:

$$[(a_0, a_1, a_2, \ldots, a_n)] = (a_0 : a_1 : a_2 : \ldots : a_n).$$

Whenever $a_0 \neq 0$, we may assume that $a_0 = 1$, and with this assumption the other coordinates are uniquely determined. Thus we may identify the subset

$$\{(a_0 : a_1 : a_2 : \ldots : a_n) \mid a_0 \neq 0\} \subset \mathbb{P}_k^n$$

with $\mathbb{A}_k^n$ by letting $(a_0 : a_1 : a_2 : \ldots : a_n)$ correspond to $(\frac{a_1}{a_0}, \frac{a_2}{a_0}, \ldots, \frac{a_n}{a_0})$. The set

$$\{(a_0 : a_1 : a_2 : \ldots : a_n) \mid a_0 = 0\}$$

is referred to as *the points at infinity* in $\mathbb{P}_k^n$. This subset may in turn be identified with $\mathbb{P}_k^{n-1}$ in the obvious manner by ignoring the first coordinate, which is zero. We obtain the following description of $\mathbb{P}_k^n$:

$$\mathbb{P}_k^n = \mathbb{A}_k^n \cup \mathbb{P}_k^{n-1}$$

and we say that $\mathbb{P}_k^n$ *is obtained by adjoining to* $\mathbb{A}_k^n$ *a space* $\mathbb{P}_k^{n-1}$ *of points at infinity.*

By dividing up $\mathbb{P}_k^{n-1}$ similarly, and repeating the process all the way down to $\mathbb{P}_k^0$, we get

$$\mathbb{P}_k^n = \mathbb{A}_k^n \cup \mathbb{A}_k^{n-1} \cup \cdots \cup \mathbb{A}_k^1 \cup \mathbb{P}_k^0.$$

But $\mathbb{P}_k^0$ consist of only one point: In fact, for a and $b \in k^*$, i.e. they are non-zero elements of k, we have

$$(a) \sim (b)$$

since

$$b = \frac{b}{a} a.$$

Thus $\mathbb{P}_k^0 = \{pt\}$.

The points of $\mathbb{P}_k^n$ may be viewed as the collection of all lines in $\mathbb{A}_k^{n+1}$ passing through the origin $(0, 0, \ldots, 0)$. In particular the points in $\mathbb{P}_k^2$ are the lines through $(0, 0, 0)$ in $\mathbb{A}_k^3$. Such a line α is uniquely determined by a *vector* $(a, b, c) \neq (0, 0, 0)$. This vector gives the direction of the line, and is given as follows:

$$\alpha = \{(x, y, z) \mid x = at, y = bt, z = ct, \text{ where } t \in k\}.$$

We say that the line is given on parametric form: To every value of the parameter t there corresponds a unique point $P(t) \in \alpha$, and conversely such that $P = P(t_P)$.

We may also describe some curves in $C \subset \mathbb{A}_k^3$ in this way: Then C is given by

$$C = \{(x, y, z) \mid x = f(t), y = g(t), z = g(t), \text{ where } t \in k\}.$$

Thus a line which does not necessarily pass through the origin, will have the following parametric form:

$$\alpha = \{(x, y, z) \mid x = x_0 + at, y = y_0 + bt, z = z_0 + ct, t \in k\}$$

where (x_0, y_0, z_0) is a point on the line which we may chose arbitrarily. Thus clearly the same curve C may be given on different parametric forms.

1.2 Algebraic Subsets and Coordinates

The interplay between affine and projective n-space is important.

Let $P(X_1, \ldots, X_n)$ be a polynomial with coefficients from the ground field k. Then we put

$$V(P) = \{(a_1, \ldots, a_n) \in \mathbb{A}_k^n \mid P(a_1, \ldots, a_n) = 0]\}$$

and

$$D(P) = \{(a_1, \ldots, a_n) \in \mathbb{A}_k^n \mid P(a_1, \ldots, a_n) \neq 0\}.$$

More generally, let $P_1, \ldots, P_m$ be polynomials as above. We then put

$$V(P_1, \ldots, P_m) = \left\{ (a_1, \ldots, a_n) \in \mathbb{A}_k^n \ \middle| \ \begin{array}{l} P_i(a_1, \ldots, a_n) = 0 \\ \text{for all } i = 1, \ldots, m \end{array} \right\}.$$

Such subsets are referred to as *affine subsets* of $\mathbb{A}_k^n$. Affine subsets given by a single polynomial are referred to as affine *hypersurfaces*, and if the polynomial is of degree 1 as an affine *hyperplane*.

For projective n-space we have to work with polynomials in the variables $X_0, X_1, \ldots, X_n$, with coefficient from the ground field k, say $\mathbb{R}$ or $\mathbb{C}$ as the case may be. But here we encounter the difficulty that a polynomial $Q(X_0, X_1, \ldots, X_n)$ may vanish at the point $(a_0, a_1, \ldots, a_n)$, yet have a non-zero value at some $(b_0, b_1, \ldots, b_n)$, even though $(a_0, a_1, \ldots, a_n) \sim (b_0, b_1, \ldots, b_n)$. But if *all the monomials occurring in Q have the same total degree*, then this will not happen. Such a polynomial is called *a homogeneous polynomial*. For a homogeneous polynomial $Q(X_0, X_1, \ldots, X_n)$ we therefore are able to make the following definitions:

$$V_+(Q) = \{(a_1 : \ldots : a_n) \in \mathbb{P}_k^n \mid Q(a_0, \ldots, a_n) = 0\}$$
$$D_+(Q) = \{(a_0 : \ldots : a_n) \in \mathbb{P}_k^n \mid Q(a_0, \ldots, a_n) \neq 0\}$$

and

$$V_+(Q_1, \ldots, Q_m) = \left\{(a_0, \ldots, a_n) \in \mathbb{A}_k^n \;\middle|\; \begin{array}{l} Q_i(a_0, \ldots, a_n) = 0 \\ \text{for all } i = 1, \ldots, m \end{array}\right\}.$$

Subsets of the last type are referred to as *projective subsets*, and in analogy to the affine case as projective hypersurfaces if there is only one polynomial, and finally, as a projective hyperplane if the polynomial is a linear form, that is, of the type $L = b_0 X_0 + \cdots + b_n X_n$, with $b_i \in k$. A common name for affine and projective subsets of $\mathbb{A}_k^n$ and $\mathbb{P}_k^n$, respectively, is *algebraic subsets*.

1.3 Affine and Projective Coordinate Systems

From a mathematical point of view it is more satisfactory to proceed in a slightly different manner. In fact, consider a k-vector space V, of finite dimension. We then define the set $\mathbb{P}(V)$ as the set of all 1-dimensional k-subspaces of $U \subset V$. We shall not pursue this *coordinate free* approach very far right now, but postpone it to Part 2 of this book. However, when V is of dimension $n < \infty$, we note that by fixing a k-basis $\{v_1, \ldots, v_n\}$ for V, V is identified with $\mathbb{A}_k^n = k^n$ by letting the element $v = \alpha_1 v_1 + \cdots + \alpha_n v_n$ correspond to the point $(\alpha_1, \ldots, \alpha_n) \in k^n$.

If W is an $n + 1$-dimensional k-space, then similarly $\mathbb{P}(W)$ is identified with $\mathbb{P}_k^n$ by fixing a basis $w_0, \ldots, w_n$ for W and letting the point $(\alpha_0 : \ldots : \alpha_n)$ correspond to the subspace spanned by $\alpha_0 w_0 + \cdots + \alpha_n w_n$.

It is clear that switching to a different basis for V, say

$$\alpha_{1,1} v_1 + \cdots + \alpha_{1,n} v_n$$
$$\alpha_{2,1} v_1 + \cdots + \alpha_{2,n} v_n$$
$$\vdots$$
$$\alpha_{n,1} v_1 + \cdots + \alpha_{n,n} v_n$$

corresponds to the change of coordinate system in $\mathbb{A}_k^n$ given by

$$\overline{X}_1 = \alpha_{1,1}X_1 + \cdots + \alpha_{1,n}X_n$$
$$\overline{X}_2 = \alpha_{2,1}X_1 + \cdots + \alpha_{2,n}vX_n$$
$$\vdots$$
$$\overline{X}_n = \alpha_{n,1}X_1 + \cdots + \alpha_{n,n}X_n.$$

Similarly, switching to the new basis for W,

$$\alpha_{0,0}w_0 + \cdots + \alpha_{0,n}w_n$$
$$\alpha_{1,0}w_0 + \cdots + \alpha_{1,n}w_n$$
$$\vdots$$
$$\alpha_{n,0}w_0 + \cdots + \alpha_{n,n}w_n$$

corresponds to the change of projective coordinate system in $\mathbb{P}_k^n$ given by

$$\overline{X}_0 = \alpha_{0,0}X_0 + \cdots + \alpha_{0,n}X_n$$
$$\overline{X}_1 = \alpha_{1,0}X_0 + \cdots + \alpha_{1,n}vX_n$$
$$\vdots$$
$$\overline{X}_n = \alpha_{n,0}X_0 + \cdots + \alpha_{n,n}X_n.$$

Now let $L = a_0X_0 + \cdots + X_n$, then by switching to a new projective co-ordinate system, $D_+(L)$ can be made to correspond to $D_+(\overline{X}_0)$ and thus be identified with $\mathbb{A}_k^n$, while $V_+(L)$ is identified with $V_+(\overline{X}_0)$ and thus with the projective $n-1$-space $\mathbb{P}_k^{n-1}$. In general $V_+(X_0)$ is referred to as the *hyperplane at infinity*.

It is clear that a hypersurface in $\mathbb{A}_k^n$, in particular a curve in $\mathbb{A}_k^2$, given by a polynomial of degree d in the coordinates $x_1,\ldots,x_n$ will be given by an equation of the same degree d in the coordinates $\overline{x}_1,\ldots,\overline{x}_n$. The same observation holds in the projective case.

Rather than to view this as moving from one coordinate system to a new coordinate system with a new origin in general, it may regarded as a mapping from $\mathbb{A}_k^n$ to itself, a so called affine transformation:

$$\mathbb{A}_k^n \longrightarrow \mathbb{A}_k^n$$
$$(x_1,\ldots,x_n) \mapsto (\overline{x}_1,\ldots,\overline{x}_n)$$

where

$$\begin{bmatrix} \overline{x}_1 \\ \vdots \\ \overline{x}_n \end{bmatrix} = \begin{bmatrix} \alpha_{1,1} & \ldots & \alpha_{1,n} \\ & \vdots & \\ \alpha_{n,1} & \ldots & \alpha_{n,n} \end{bmatrix} \begin{bmatrix} x_1 \\ \vdots \\ x_n \end{bmatrix} + \begin{bmatrix} b_1 \\ \vdots \\ b_n \end{bmatrix}$$

and where $(b_1,\ldots,b_n)$ is the new origin. The properties of affine subsets or configurations of affine subsets which are preserved by the affine transformations are referred to as *affine properties*. Affine properties include incidence

Fig. 1.1 Gérard Desargues.
Illustration by the author

(that a point lies on a line, or a line passes through a point), collinearity (that several points lie on a common line), concurrency (that several lines pass through a common point). The properties which are not affine include being a circle, while being an ellipse is an affine property. Over $\mathbb{R}$ the property for two lines in $\mathbb{A}^2_{\mathbb{R}}$ forming a right angle is not affine, while the property for a line to bisect in equal parts the angle formed by two given lines, is affine. The property of tangency is affine.

Projective properties are defined analogously: all projective properties are preserved by projective transformations. As for the affine case we may also express this by saying that the projective properties are independent of choice of projective coordinate system. Obviously the points at infinity are not preserved, however.

We now need the following useful observation, the proof of which is of course well known, and is explained in detail when $n = 2$ and $k = \mathbb{R}$ in Chap. 11 of [27], Chap. 12 of [28]. The proof given there carries over to the general case with obvious modifications. We shall not repeat it here.

Proposition 1.1 *Given $n + 2$ points $P_1, P_2, \ldots, P_{n+2} \in \mathbb{P}^n_k$ no three of which are collinear, as well as another set of points $P'_1, P'_2, \ldots, P'_{n+2} \in \mathbb{P}^n_k$ with the same property. Then there exists a projective transformation G of $\mathbb{P}^n_k$ onto itself, mapping P_i to P'_i, $i = 1, 2, \ldots, n$.*

1.4 The Theorem of Desargues

The *Theorem of Desargues* was important in the development of projective geometry. Gérard Desargues came from a wealthy family, and had many influential mathematical friends such as Rene Descartes and the father and son Pascal. He belonged to the famous and influential circle around the secular monk *Marin Mersenne*.

In [27] and [28] the theorem is proved only for $\mathbb{P}^2_k$ and it is assumed that $k = \mathbb{R}$ for this result. But as we shall see below, that assumption is not needed.

Fig. 1.2 Desargues'
theorem, picture with
$k = \mathbb{R}$ in the affine piece
$D_+(X_0) \subset \mathbb{P}_k^3$

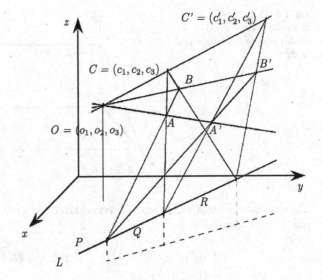

Theorem 1.2 *Let two triangles ABC and $A'B'C'$ be given in $\mathbb{P}_k^3$, such that $A \neq A'$, $B \neq B'$ and $C \neq C'$. Then if the lines through corresponding vertices pass through the same point O, the intersections of the prolongations of corresponding sides will intersect in points lying on the same line L.*

Remark When the triangles have the property that the lines through corresponding vertices pass through O, then O is referred to as the center of perspective, and the triangles are said to be perspective from the point O. If the intersections of the prolongations of corresponding sides intersect in points lying on L, then L is called the *line of perspective*, and the triangles are said to be perspective from the line L.

Proof We introduce notation as in Fig. 1.2. The coordinates of the points are labelled analogously to those of O, C and C'.

   Evidently, no three of the points

$$(0:0:1:0), \quad (0:1:0:0), \quad (0:0:0:1) \quad \text{and} \quad (1:0:0:0)$$

are collinear. It is easily seen that the same is true for the points O, P, Q and A. Thus by Proposition 1.1 there is a linear transformations which makes these four points correspond, and hence, after a change of coordinate system, we may assume that

$$O = (0:0:1:0), \quad P = (0:1:0:0),$$
$$Q = (0:0:0:1) \quad \text{and} \quad A = (1:0:0:0:0).$$

Fig. 1.3 Desargues'
theorem, after a good
choice of coordinate system
in $\mathbb{P}^3_\mathbb{R}$

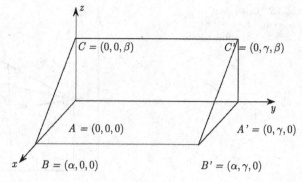

Then a line through O, P and Q, respectively, is given in $D_+(X_0) = \mathbb{A}^2_k$ on parametric form as

$$L_O = \{(x_1, t, z_1) \mid t \in k\}, \qquad L_P = \{(t, y_2, z_2) \mid t \in k\},$$
$$L_Q = \{(x_3, y_3, t) \mid t \in k\}.$$

Since moreover $A = (0,0,0)$, we therefore get in addition that

$$B = (\alpha, 0, 0), \quad C = (0, \beta, 0), \quad A' = (0, \gamma, 0),$$
$$B' = (\alpha, \gamma, 0) \quad \text{and} \quad C' = (0, \beta, \gamma).$$

The situation is illustrated in Fig. 1.3.

We identify the point $(x, y, z) \in \mathbb{A}^3_\mathbb{R}$ with the point $(1 : x : y : z) \in D_+(X_0) \subset \mathbb{P}^3_\mathbb{R}$ as usual, and get the following parametric forms for the lines BC, $B'C'$ and PQ, where we write $(ua : ub : uc : ud)$ as $u(a : b : c : d)$ to simplify the notation:

$$BC = \{u_1(1 : 0 : 0 : \beta) + v_1(1 : \alpha : 0 : 0) \mid (u_1, v_1) \neq (0, 0)\}$$
$$B'C' = \{(u_2(1 : 0 : \gamma : \beta) + v_2(1 : \alpha; \gamma : 0) \mid (u_2, v_2) \neq (0, 0)\}$$
$$PC = \{u_3(0 : 1 : 0 : 0) + v_3(0 : 0 : 0 : 1) \mid (u_3, v_3) \neq (0, 0)\}.$$

It is now easily verified that the two lines BC and $B'C'$ intersect in the point $(0 : \alpha : 0 : -\beta)$, which lies on the line PQ. Thus the claim follows. $\square$

This theorem has the following corollary, which is the converse of the theorem in the plane:

Corollary 1.3 *Let two triangles ABC and $A'B'C'$ be given in $\mathbb{P}^2_\mathbb{R}$, and assume that $A \neq A'$, $B \neq B'$ and $C \neq C'$. If the triangles are perspective from a line, then they are perspective from a point.*

Proof The proof is immediate by the *Principle of Duality* for $\mathbb{P}^2_\mathbb{R}$, which we shall prove in the next section. Here we find that the dual result is actually a *converse* to the assertion of the theorem. The figure in the proof above is actually *self dual*. □

1.5 Duality for $\mathbb{P}^2_k$

We now consider $\mathbb{P}^2_k$, and let P denote some statement involving *points, lines incidence*.

Then the dual statement $P^\vee$ is defined as the statement obtained by interchanging the words "point" and "line", and keeping "incidence", with the adjustments of language which may be necessary to obtain natural geometrical statements.

For example, consider the statement below:

Statement P: Through two distinct points Q_1 and Q_2 there always pass a uniquely determined line ℓ.

Then the dual statement is the following:

Statement $P^\vee$: Two distinct lines q_1 and q_2 always intersect in a uniquely determined point L.

For $\mathbb{P}^2_k$ we have the so called *principle of duality*:

Theorem 1.4 (Duality for $\mathbb{P}^2_k$) *If P denotes a true statement about $\mathbb{P}^2_k$ dealing with points, lines and incidence, then the dual statement $P^\vee$ is also a true statement.*

Proof The statement P may be translated into a possibly infinite set of equations involving the projective coordinates of the points and the coefficients of the equations of the lines occurring in the statement. The equations will all be of the form

$$A_0\alpha_0 + A_1\alpha_1 + A_2\alpha_2 = 0,$$

where $A = (A_0 : A_1 : A_2)$ is a point and α_0, α_1 and α_2 are the coefficients in the equation for a line in $\mathbb{P}^2_k$, so the line is given by

$$\alpha_0 X_0 + \alpha_1 X_1 + \alpha_2 X_2 = 0.$$

If we denote this line by ℓ, then the statement "A lies on ℓ" is equivalent to the relation above. For every point in $\mathbb{P}^2_k$ we now let correspond a *line* in $\mathbb{P}^2_k$ given by the equation whose coefficients are the coordinates of the point, and to every line we let correspond the point whose projective coordinates are the coefficients of the equation giving the line.

Then the set of algebraic equations between the coordinates of points in $\mathbb{P}^2_k$ and coefficients of lines in $\mathbb{P}^2_k$ which expresses the truth of the statement P is the same as the collection of relations which expresses the truth of $P^\vee$.

This completes the proof. Note that the proof remains valid even if the collections of lines and/or points are infinite.   □

Example 1.1 There is an abundance of examples of this principle. Perhaps the simplest is the following: One of the basic axioms of axiomatic plane projective geometry, is that *through two different points in the plane there passes one and only one line.* The dual statement is: *two different lines in the projective plane meet in one and only one point.* Which is also true.

Chapter 2
Curves in $\mathbb{A}^2_k$ and in $\mathbb{P}^2_k$

In this chapter we introduce the first interesting class of planar curves, namely the conic sections. This leads to a first discussion of singular and non singular points. Closely tied to these concepts is the notion of the tangent at a point on a curve. We then move on to a discussion of curves of higher degrees, and introduce the concepts of tangent lines, the tangent cone and the multiplicity of a point on a curve which may have singularities. A number of important examples of higher order curves are discussed. Elliptic curves are briefly discussed, this class of curves (which are *certainly not* ellipses) played an important role for the fruitful interplay between geometry and function theory, so central in the pathbreaking work of *Niels Henrik Abel*.

2.1 Conic Sections

A *conic section*[1] is a curve in the affine plane $\mathbb{A}^2_k$ of degree 2, over a field which we shall assume to be of characteristic $\neq 2$. The general form of the equation is usually written as

$$q(x,y) = Ax^2 + 2Bxy + Cy^2 + 2Dx + 2Ey + F = 0.$$

In [27] these curves are treated with $k = \mathbb{R}$ in some detail, and we refer the reader who is unfamiliar to the basics of this subject to the treatment there, as a suitable basis for reading the present chapter. In particular, this reference gives the proof that all such curves can be obtained as the intersection between a fixed double circular cone and a varying plane, in the case when $k = \mathbb{R}$.

The non-degenerate conics are the ellipses, the parabolas and the hyperbolas. In addition to these, we have the *degenerate cases*. In the three non-degenerate cases the equation can be brought on one of the three canonical forms. We refer to [27] for more information on this.

---

[1] We also say *conic curve* or just *a conic*.

A. Holme, *A Royal Road to Algebraic Geometry*,
DOI 10.1007/978-3-642-19225-8_2, © Springer-Verlag Berlin Heidelberg 2012

In [27] we consider the following problem: There is given 5 distinct points in $\mathbb{A}^2_k$, $(x_1, y_1), (x_2, y_2), \ldots, (x_5, y_5)$. If these points are in sufficiently general position, then there is a unique, non-degenerate conic curve, in other words a non-degenerate curve of degree 2, passing through them. The condition that the points be in sufficiently general position, here amounts to the requirement that *no three of them be collinear*. The equation for the conic in question is

$$\begin{vmatrix} x^2 & xy & y^2 & x & y & 1 \\ x_1^2 & x_1 y_1 & y_1^2 & x_1 & y_1 & 1 \\ x_2^2 & x_2 y_2 & y_2^2 & x_2 & y_2 & 1 \\ x_3^2 & x_3 y_3 & y_3^2 & x_3 & y_3 & 1 \\ x_4^2 & x_4 y_4 & y_4^2 & x_4 & y_4 & 1 \\ x_5^2 & x_5 y_5 & y_5^2 & x_5 & y_5 & 1 \end{vmatrix} = 0.$$

2.2 Singular and Non-singular Points

We need the notion of a *non-singular* point of a plane curve, we return to a refined treatment of this important concept in Sect. 2.8.

We define the derivative of a polynomial with coefficients form a field k formally as follows:

Definition 2.1 Let $P(x) = a_n x^n + a_{n-1} x^{n-1} + \cdots + a_1 x + a_0$ be a polynomial in x with coefficients from k. We define the derivative of $P(x)$ with respect to x as

$$P'(x) = \frac{\mathrm{d}P}{\mathrm{d}x} = n a_n x^{n-1} + (n-1) a_{n-1} x^{n-2} + \cdots + a_1.$$

If $F(x, y, u, \ldots) \in k[x, y, u, \ldots]$, then the partial derivatives are defined analogously in a formal manner.

The basic properties of derivatives still hold: The derivative of a constant is zero, the formulas for the derivatives of sums and products of polynomials hold, as does the chain rule.

In characteristic 0 we still have the Taylor formula in its usual form in one and several variables, and may proceed as for $k = \mathbb{R}$. In characteristic $p > 0$ the procedure is somewhat modified, this is omitted here. All these observations only apply to polynomials, of course.

We note the following result, which was shown in Sect. 12.7 in [27] for $k = \mathbb{R}$. It will be given a different proof in Sect. 2.3.

Theorem 2.1 *The equation*

$$q(x, y) = Ax^2 + 2Bxy + Cy^2 + 2Dx + 2Ey + F = 0$$

yields a non-degenerate conic curve if and only if the following determinantal criterion is satisfied:

$$\begin{vmatrix} A & B & D \\ B & C & E \\ D & E & F \end{vmatrix} \neq 0.$$

We now make the following important definition:

Definition 2.2 Let Z be an affine plane curve given by the equation

$$f(x,y) = 0.$$

Let (x_0, y_0) be a point on the curve such that the two partial derivatives do not both vanish,

$$\left(\frac{\partial f}{\partial x}(x_0, y_0), \frac{\partial f}{\partial y}(x_0, y_0) \right) \neq (0,0).$$

Such a point is called a non-singular point on the curve. At all non-singular points we define the tangent line[2] by the equation

$$\frac{\partial f}{\partial x}(x_0, y_0)(x - x_0) + \frac{\partial f}{\partial y}(x_0, y_0)(y - y_0) = 0.$$

A point which is not non-singular is called a singular point.

We next turn to the tangents of non-degenerate conics in $\mathbb{A}_k^2$, as well as the related concepts of *pole and polar line.*

Let $P = (x_0, y_0)$ be a point on the non-singular conic curve given by the equation

$$q(x,y) = Ax^2 + 2Bxy + Cy^2 + 2Dx + 2Ey + F = 0,$$

the tangent at P is given by

$$(Ax_0 + By_0 + D)(x - x_0) + (Bx_0 + Cy_0 + E)(y - y_0) = 0,$$

which after a short calculation takes the form

$$Ax_0x + B(y_0x + x_0y) + Cy_0y + D(x + x_0) + E(y + y_0) + F = 0.$$

This equation is also of interest when the point is not on $\mathcal{C}$. We have the following result:

---

[2]This concept will be explained in more detail in Sect. 2.9.

Fig. 2.1 The line joining
the two points of tangency

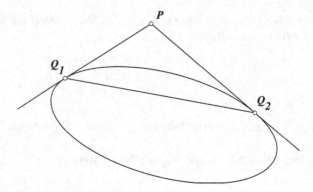

Proposition 2.2 *Let $P = (x_0, y_0)$ be a point and let $\mathcal{C}$ be the conic given by the equation*

$$q(x, y) = Ax^2 + 2Bxy + Cy^2 + 2Dx + 2Ey + F = 0.$$

There are two tangent lines to $\mathcal{C}$ passing through P, coinciding if P is on $\mathcal{C}$. Let the two points of tangency be Q_1 and Q_2. Then the line p passing through Q_1 and Q_2 is given by the equation

$$Ax_0 x + B(y_0 x + x_0 y) + Cy_0 y + D(x + x_0) + E(y + y_0) + F = 0.$$

Proof The situation is shown in Fig. 2.1.

Let $Q_1 = (x_1, y_1)$ and $Q_2 = (x_2, y_2)$, then the two tangents in question will have equations

$$Ax_1 x + B(y_1 x + x_1 y) + Cy_1 y + D(x + x_1) + E(y + y_1) + F = 0,$$

$$Ax_2 x + B(y_2 x + x_2 y) + Cy_2 y + D(x + x_2) + E(y + y_2) + F = 0.$$

These lines pass through $P = (x_0, y_0)$, thus

$$Ax_1 x_0 + B(y_1 x_0 + x_1 y_0) + Cy_1 y_0 + D(x_0 + x_1) + E(y_0 + y_1) + F = 0,$$

$$Ax_2 x_0 + B(y_2 x_0 + x_2 y_0) + Cy_2 y_0 + D(x_0 + x_2) + E(y_0 + y_2) + F = 0.$$

But this demonstrates that the line whose equation is given in the assertion of the proposition, does indeed pass through the two points Q_1 and Q_2. Hence the claim follows.   □

Definition 2.3 The point P and the line p in Proposition 2.2 are called the pole and the polar line corresponding to each other.

In the case when the field k is not algebraically closed, such as when $k = \mathbb{R}$, we encounter some apparently puzzling phenomena. For example, are the "conic sections" given by the equations $x^2 + y^2 + 1 = 0$ and $x^2 + y^2 = 0$

really *curves*? The former has no points in $\mathbb{A}_{\mathbb{R}}^2$, while the latter has only the origin as a point on it. According to Theorem 2.1 the former is non-degenerate while the latter is degenerate. An explanation for this apparent paradox is that we need to consider not only points over k, but also points over the algebraic closure $\overline{k}$ in order to understand an algebro-geometric object such as an affine curve.

Moreover, if we take a point inside an ellipse, then there will be no real points of tangency, even though we get a well defined polar line using the equation we have derived in Proposition 2.2. But if we compute the *complex* points of tangency, we find that corresponding coordinates are complex conjugates, and we get a real line joining them.

Finally, if we choose the center of the unit circle, say, then the "line" given by the formula is just $0 = 1$, which has no points on it. The explanation for this is that the polar of the center is the *line at infinity*. Thus we see here both the need for computing complex points as well as for considering points at infinity in order to understand algebraic curves over $\mathbb{R}$.

2.3 Conics in the Projective Plane

We shall now consider the so-called *projective closure* of the conics in $\mathbb{A}_k^2$. We substitute

$$x = \frac{X_1}{X_0}, \qquad y = \frac{X_2}{X_0}$$

into the equation of the conic $\mathcal{C}$,

$$q(x, y) = Ax^2 + 2Bxy + Cy^2 + 2Dx + 2Ey + F = 0,$$

which yields the following homogeneous equation

$$Q(X_0, X_1, X_2)$$
$$= AX_1^2 + 2BX_1X_2 + CX_2^2 + 2DX_0X_1 + 2EX_0X_2 + FX_0^2 = 0.$$

Hence we get the equation of a curve $\overline{\mathcal{C}}$ in $\mathbb{P}_k^2$, which we refer to as the *projective closure* of $\mathcal{C}$. When intersected with $D_+(X_0) = \mathbb{A}_k^2$ it gives back the original curve.

We first wish to determine its points at infinity. Those are the points $\overline{\mathcal{C}} \cap V_+(X_0)$. The point $(u : v : 0)$ is in $\overline{\mathcal{C}}$ if

$$Au^2 + 2Buv + Cv^2 = 0,$$

and we immediately get the following information:

Proposition 2.3 1. $\mathcal{C}$ *has no real points at infinity if* $B^2 - AC < 0$.
  2. $\mathcal{C}$ *has one real point at infinity if* $B^2 - AC = 0$.
  3. $\mathcal{C}$ *has two points at infinity if* $B^2 - AC > 0$.
  Thus 1. *corresponds to a possibly degenerate ellipse,* 2. *to a possibly degenerate parabola and* 3. *to a possibly degenerate hyperbola.*

In general, let C be the curve in $\mathbb{P}^2_k$ defined by

$$F(X_0, X_1, X_2) = 0.$$

Let $P = (a_0 : a_1 : a_2)$ be a point on it. In Chap. 3, Sect. 3.4 we show that the equation

$$\frac{\partial F}{\partial X_0}(a_0, a_1, a_2)X_0 + \frac{\partial F}{\partial X_1}(a_0, a_1, a_2)X_1 + \frac{\partial F}{\partial X_2}(a_0, a_1, a_2)X_2 = 0$$

yields the tangent line to C at P, provided that the coefficients involved do not all vanish.

Definition 2.4 If the partial derivatives involved in the equation above all vanish at some point on the curve, then the point is said to be a singular point. If they do not all vanish, the point is called non-singular.

The equation for the tangent to the conic curve in $\mathbb{P}^2_k$ given by the equation $Q(X_0, X_1, X_2) = 0$ at the point $P = (x_0, x_1, x_2)$ is

$$(Ax_1 + Bx_2 + Dx_0)X_1 + (Bx_1 + Cx_2 + Ex_0)X_2$$
$$+ (Dx_1 + Ex_2 + Fx_0)X_0 = 0$$

or written on a more appealing form

$$Ax_1X_1 + B(x_1X_2 + x_2X_1) + Cx_2X_2$$
$$+ D(x_0X_1 + x_1X_0) + E(x_0X_2 + x_2X_0) + Fx_0X_0 = 0.$$

This is similar to what we found in the affine case.
  If the point P is singular, then its projective coordinates constitute a non-trivial solution of the following homogeneous system of equations:

$$Au + Bv + Dw = 0$$
$$Bu + Cv + Ew = 0$$
$$Du + Ev + Fw = 0$$

and thus we have in this case

$$\begin{vmatrix} A & B & D \\ B & C & E \\ D & E & F \end{vmatrix} = 0.$$

But the argument works both ways, thus the determinant above vanishes if and only if the conic section has a singular point, at least over $\overline{k}$. On the other hand, if the conic section has such a singular point, then passing to $\overline{k}$ and switching to a suitable projective coordinate system, we may assume that the singular point is $(1:0:0)$. But then $D = E = F = 0$, thus the equation of the conic curve is $Ax^2 + Bxy + Cy^2$ in $D_+(X_0) = \mathbb{A}_{\overline{k}}^2$. Since this polynomial splits as a product of linear forms in x and y, we have proved the theorem stated below:

Theorem 2.4 *Assume that $k = \overline{k}$. The following are equivalent:*

1. *The equation*

$$q(x,y) = Ax^2 + 2Bxy + Cy^2 + 2Dx + 2Ey + F = 0$$

   yields a non-degenerate conic section.
2. *The projective closure in $\mathbb{P}_k^2$ of the curve in $\mathbb{A}_k^2$ given by $q(x,y) = 0$ is non-singular.*
3.
$$\begin{vmatrix} A & B & D \\ B & C & E \\ D & E & F \end{vmatrix} \neq 0.$$

We finally note the following result:

Theorem 2.5 *Let $k = \overline{k}$. Assume that $AX_1^2 + 2BX_1X_2 + CX_2^2 + 2DX_0X_1 + 2EX_0X_2 + FX_0^2 = 0$ is the equation of a non singular conic in $\mathbb{P}_k^2$. Then we may choose the projective coordinate system such that $B = D = E = 0$ and $A = C = E = 1$.*

Proof By Proposition 1.1 we may assume that the following two points lie on the conic:

$$(0:i:1) \quad \text{and} \quad (0:-i:1)$$

where $i = \sqrt{-1} \in k$, as $k = \overline{k}$. Thus

$$-A + 2Bi + C = 0 \quad \text{and} \quad -A - 2Bi + C = 0$$

from which it follows that $B = 0$, hence $A = C$. Since the conic is non-degenerate, we must have $A = C \neq 0$ so we may assume $A = C = 1$, and the equation becomes

$$X_1^2 + X_2^2 + 2DX_0X_1 + 2EX_0X_2 + FX_0^2 = 0.$$

Evidently this is transformed as follows by completing two squares:

$$(X_1 + DX_0)^2 + (X_2 + EX_0)^2 + (F - D^2 - E^2)X_0^2 = 0.$$

Changing projective coordinate system again if necessary we obtain

$$X_1^2 + X_2^2 + GX_0^2 = 0$$

where $G \neq 0$ since otherwise the conic would be degenerate. A final change
of projective coordinate system yields

$$X_1^2 + X_2^2 + X_0^2 = 0$$

and the proof is complete.   □

For more on conics, including elementary proofs of the theorems of Pappus
and Pascal, which we will not include here, we refer to Sects. 12.8 and 12.9
in [27] or Sects. 13.8 and 13.9 in [28].

2.4 The Cubic Curves in $\mathbb{A}_k^2$

The simplest curve of *higher degree*, by which we mean degree higher than 2,
is the curve known as the *cubic parabola*. The parabola has the equation
$y = x^2$, after a suitable change of coordinate system in $\mathbb{A}_k^2$. Classically the
term *parabola* was used in a wider sense, as the name of a curve whose graph
would lie "parallel" to the y-axis.

Thus curves with an equation of the form $y = x^m$, m being a positive
integer *or a rational* number, would be called parabolas as well. Accordingly,
a curve which may be brought on the from $y = x^3$ is referred to as a *cubic
parabola*.

The next step in complexity is a curve which may be brought on the
form $y^2 = x^3$. It is called a *semi-cubic parabola*. It has the graph displayed in
Fig. 2.2.

The concept of *degenerate curves* and the related process of *degeneration
of a family of curves* are important.

A curve is said to be *degenerate* if it decomposes into the union of two or
more curves of lower degrees. For a cubic curve this means that it is a union
of a conic curve and a line, or of three lines (some possibly coinciding).

Planar curves of degree 3 already constitute a much richer and interesting
group of geometrical objects than the ones of degree 2.

The simplest example of a degenerate cubic curve would be the y-axis with
multiplicity 3. Its equation is $x^3 = 0$. We have not yet made the notion of
curves with multiplicity precise, this comes in Sect. 2.7. But we may already
at this point consider a family of semi-cubic parabolas, *degenerating to* the
triple y-axis. Namely, consider the curves depending on the parameter t, as
$t \to 0$: $ty^2 = x^3$.

We show some members of this family in Fig. 2.2. The values of t in the
plots are $t = 10, 4, 1, 0.1$.

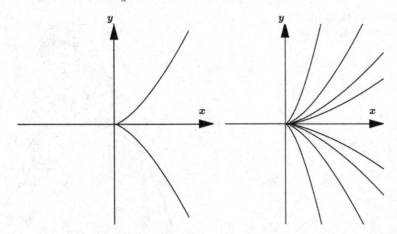

Fig. 2.2 The semi-cubic parabola given by $y^2 = x^3$ to the *left*, to the *right* we show the degeneration $ty^2 = x^3$ of the semi-cubic parabola to the triple y-axis

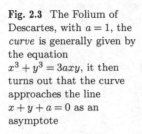

Fig. 2.3 The Folium of Descartes, with $a = 1$, the *curve* is generally given by the equation $x^3 + y^3 = 3axy$, it then turns out that the curve approaches the line $x + y + a = 0$ as an asymptote

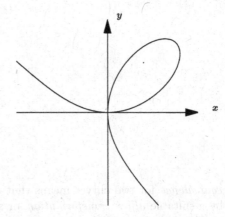

Note also that when $t \to \infty$, then the limit is the x-axis with multiplicity 2, since this degeneration is equivalent to letting u tend to 0 for the family given by the equation $y^2 = ux^3$.

The term degeneration is used rather loosely, without a formal definition. The idea we intend to convey by this, is to have one curve, say the semi-cubic parabola $y^2 = x^3$, be a member of a family of curves depending on a parameter, all but a finite number of which are of the same type. Then the exceptional members are understood as degenerate cases. This is, of course, the way we may view two intersecting lines as a degenerate hyperbola, or a double line as a degenerate hyperbola or a degenerate parabola, and so on.

Two more types of non-degenerate curves of degree three exist, up to a *projective change of coordinate system*. We will explain this *projective equivalence* for curves in $\mathbb{P}_{\mathbb{R}}^2$ (and in $\mathbb{A}_{\mathbb{R}}^2$) later, in Sect. 3.5. The simple *affine*

Fig. 2.4 Réné Descartes.
Illustration by the author

Fig. 2.5 Pierre de Fermat.
Illustration by the author

equivalence for two curves means that one may be obtained from the other by a suitable *affine transformation*, or a change of coordinate system in $\mathbb{A}^2_\mathbb{R}$.

This kind of equivalence is more complicated than the projective equivalence, there are more equivalence classes of affine cubic curves under this affine equivalence. But from our point of view, the projective equivalence is more interesting than the affine one.

The first of the remaining classes of cubic curves is represented by the *Folium of Descartes*. The French mathematician *René Descartes*, 1596–1650, is credited by some historians of mathematics as being the founder of algebraic geometry. However, this is disputed by others.

Descartes was the first to systematically introduce coordinates and equations into geometry, and our usual coordinate system in the plane is named after him, a *Cartesian* coordinate system. His name was originally *Cartes*, and when he was knighted it changed into *Des Cartes*. Descartes was for some time engaged in a bitter feud with another great French mathematician, *Pierre de Fermat*, 1601–1665. One of the issues they could not agree on was the proper way to define the tangent to a curve at a given point.

Fig. 2.6 The usual nodal
cubic

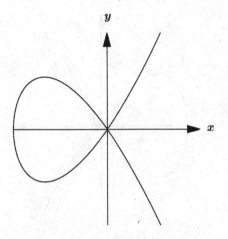

We also give another curve, belonging to the same class as the Folium under projective equivalence, but to a separate class under the affine equivalence. It is often referred to as the *usual nodal cubic*. It is given by the equation $y^2 - x^3 - x^2 = 0$. It looks somewhat similar to the semi-cubic parabola. In fact, the latter may be obtained by deforming the former. At this time the tools from calculus needed for what we regard as the proper solution to this question had not yet been sufficiently developed, and to us the methods of both Descartes and Fermat would look strange and clumsy. The curve was given as an example by Descartes in this argument with Fermat.

We also give another curve, belonging to the same class as the Folium under projective equivalence, but to a separate class under the affine equivalence. It looks somewhat similar to the semi-cubic parabola. In fact, the latter may be obtained by deforming the former. This is the simplest and most used example of a *nodal cubic curve* in $\mathbb{A}_{\mathbb{R}}^2$. It is shown in Fig. 2.6. The deformation referred to is obtained from the family $y^2 - x^3 - tx^2 = 0$.

To the left in Fig. 2.7 we see some of the corresponding plots, for $t = 0, 0.5, 2$, degenerating the usual nodal cubic given by $y^2 - x^3 - x^2 = 0$ to *the semi-cubic parabola* with equation $y^2 - x^3 = 0$. To the right an unusual "nodal cubic" given by $y^2 - x^3 + x^2 = 0$. Actually, the origin is on the curve, but that point appears to be isolated from the main part of it.

But there are complex points, invisible in $\mathbb{A}_{\mathbb{R}}^2$, which establish the connection.

We have now come to a very interesting class of curves. These curves are tied to a real leap forward in mathematics which occurred in the 19th century, and is tied to such mathematical giants as Niels Henrik Abel and Carl Gustav Jacob Jacobi. The ground had been prepared by mathematicians like Leonhard Euler and Adrien-Marie Legendre, who had studied the mysterious so called *elliptic integrals*, occurring when one wanted to compute arc lengths of segments of ellipses and of the *lemniscate*. We have arrived at the concept of an *elliptic curve*.

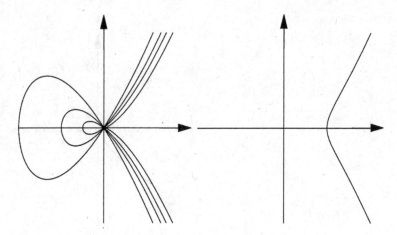

Fig. 2.7 A family of nodal cubics and an unusual one

Fig. 2.8 Gustav Jacob Jacobi to the *left*, Niels Henrik Abel to the *right*. Illustration by the author

2.5 Elliptic Integrals and the Elliptic Transcendentals

The reason for the name *elliptic curve* is that such curves come up when one attempts to compute arc length for ellipses. The corresponding problem for a circle is quite simple: We represent the circle by the equation $x^2 + y^2 = R^2$. We then have to compute the integral $L = \int_\alpha^\beta \sqrt{1 + y'^2}\mathrm{d}x$.

Then as $y = \sqrt{R^2 - x^2}$ we find $y' = -\frac{x}{\sqrt{R^2 - x^2}}$ and thus to find the arc length of the circle between two given points corresponding to x_1 and x_2 in the first or second quadrant, say, we have to compute the integral

$$L = \int_{x_1}^{x_2} \sqrt{1 + y'^2}dx = \int_{x_1}^{x_2} \frac{Rdx}{\sqrt{R^2 - x^2}}.$$

In this case we may introduce polar coordinates, $x = R\cos(\varphi)$ and $y = R\sin(\varphi)$. Then the integral becomes

$$L = R\int_{\varphi_1}^{\varphi_2} \frac{-\sin(\varphi)}{\sqrt{1 - \cos^2(\varphi)}}d\varphi = -R\int_{\varphi_1}^{\varphi_2} d\varphi = R(\varphi_1 - \varphi_2).$$

However, consider the corresponding problem for the *ellipse*

$$\left(\frac{x}{a}\right)^2 + \left(\frac{y}{b}\right)^2 = 1$$

where $a > b$. Then the same method applied to $x = a\cos(\varphi)$, $y = b\sin(\varphi)$ leads to the integral

$$L = a\int \sqrt{1 - k^2\cos^2(\varphi)}d\varphi,$$

where $k = \frac{\sqrt{a^2 - b^2}}{a}$ is the eccentricity of the ellipse. Putting $t = \cos(\varphi)$, this integral is reduced to

$$I_2 = \int_0^x \frac{\sqrt{1 - k^2t^2}}{\sqrt{1 - t^2}}dt,$$

referred to as an *elliptic integral of the second kind.* An elliptic integral of the *first kind* is of the form

$$I_1 = \int_0^x \frac{1}{\sqrt{(1 - t^2)(1 - k^2t^2)}}dt,$$

while an elliptic integral of the *third kind* is

$$I_3 = \int_0^x \frac{1}{(1 + nt^2)\sqrt{(1 - xt^2)(1 - k^2t^2)}}dt.$$

These three forms are referred to as *Legendre's standard forms* for elliptic integrals. Before Abel's (and Jacobi's) time these integrals, as functions of the upper limit x, were considered as the *elliptic functions,* the so-called *elliptic transcendentals.* Abel, and later on Jacobi, turned this around and defined the elliptic functions as the *inverse* of these integral-functions. Thus, as for instance

$$I_2 = \int_0^x \frac{dt}{\sqrt{1 - t^2}} = \operatorname{Arcsin}(x),$$

Fig. 2.9 Elliptic cubic,
given by $y^2 - x^3 + x = 0$

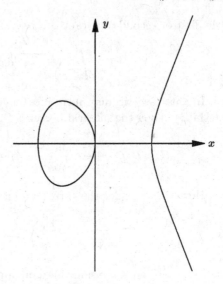

we find the elliptic function associated to an elliptic integral of the second kind with $k = 0$ to be the function $y = \sin(x)$. So elliptic functions are vast generalizations of the trigonometric functions.

In general an elliptic integral is an integral of the form $\int_a^x \frac{dt}{f(t,\sqrt{R})}$ where f is a rational expression in two variables and R is a cubic or biquadratic expression in t. Legendre succeeded in expressing all such integrals in terms of his normal forms above.

Today one uses the *Weierstrass Normal Form*,

$$u = \int_a^x \frac{dt}{\sqrt{4t^3 - g_1 t - g_2}}$$

and we note that the denominator with $g_1 = 4$ and $g_2 = 0$ gives rise to the equation

$$y^2 = 4x(x^2 - 1),$$

giving a curve which is equivalent to the elliptic curve displayed in Fig. 2.9. We explain this in more detail in Sect. 4.12.

2.6 More Curves in $\mathbb{A}^2_\mathbb{R}$

Before proceeding with the general theory, we shall look at some other examples of curves in $\mathbb{A}^2_\mathbb{R}$. Some of them have interesting histories, here we shall just present a curve which is due to *Colin Maclaurin*.

The Trisectrix of Maclaurin is given by the equation

$$x^3 + xy^2 + y^2 - 3x^2 = 0.$$

Fig. 2.10 Colin Maclaurin.
Illustration by the author

Fig. 2.11 The Trisectrix of
Maclaurin

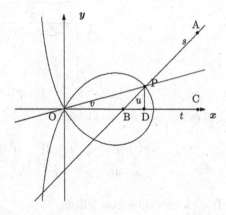

Suppose there is given an angle $u = \angle ABC$. Then the two lines AB and
BC are extended to s and t, respectively, as shown in Fig. 2.11. A Cartesian
coordinate system is introduced, so that t becomes the x-axis and the origin
is located on t to the left of B at a distance of 2. We then plot the curve given
by $x^3 + xy^2 + y^2 - 3x^2 = 0$. This curve intersects the line s in the point P,
and we draw the line OP between the origin O and P. *We claim that if*
$v = \angle POC$, *then* $u = 3v$. Indeed, it suffices to show that $\sin(u) = \sin(3v)$, in
other words that $\sin(u) = 3\sin(v) - 4\sin^3(v)$. Now we have $\sin(u) = \frac{PD}{PB}$ and
$\sin(v) = \frac{PD}{PO}$. Moreover, $PD = y$, $PO = \sqrt{x^2 + y^2}$ and $PB = \sqrt{(x-2)^2 + y^2}$.
Thus we need to verify the following identity:

$$\frac{y}{\sqrt{(x-2)^2 + y^2}} = 3\frac{y}{\sqrt{x^2 + y^2}} - 4\left(\frac{y}{\sqrt{x^2 + y^2}}\right)^3$$

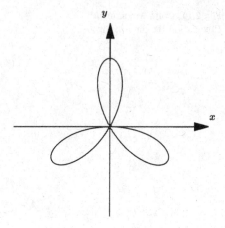

in the presence of the relation $x^3 + xy^2 + y^2 - 3x^2 = 0$, or equivalently

$$\sqrt{\frac{x^2 + y^2}{(x-2)^2 + y^2}} = 3 - 4\frac{y^2}{x^2 + y^2}$$

i.e.

$$(x^2 + y^2)^3 = ((x-2)^2 + y^2)(3x^2 - y^2)^2.$$

An evaluation finally yields

$$(x^2 + y^2)^3 - ((x-2)^2 + y^2)(3x^2 - y^2)^2$$
$$= -4(-2y^2x + y^2 - 3x^2 + 2x^3)(y^2x + y^2 - 3x^2 + x^3)$$

from which the claim follows.

Another curve looks like a *clover leaf*. It has equation $(x^2 + y^2)^2 + 3x^2y - y^3 = 0$ and is shown in Fig. 2.12. According to the picture, the curve is smooth except at the origin. There this curve displays a more complicated behavior, and gives the appearance of being the shadow, or the *projection* of a knot-like space curve. We shall make these features precise later.

Our first curve of degree higher than three was the Clover Leaf Curve above. Another such interesting curve is the famous *Airplane Wing Curve*. Amazingly it looks very similar to a section through the wing of an airplane.

We now have a sufficient base of *examples* to appreciate some more general theory. We already mentioned the need to incorporate complex points in connection with elliptic cubic curves above. In addition to this, a curve may consist of several *components*. That is to say, it may consist of several curves taken together. And some of these components may also occur with a *multiplicity*. Thus for instance, a *double line* is a different curve from a *single line*. We now take a closer look.

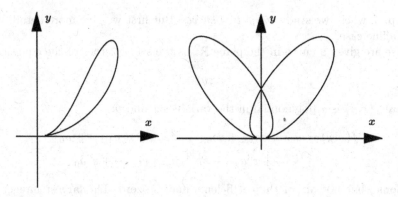

Fig. 2.13 The Airplane Wing Curve to the *left*, to the *right* the curve with equation $2x^4 - 3x^2y + y^4 - 2y^3 + y^2 = 0$. It has two singular points

2.7 General Affine Algebraic Curves

Let the curve C be given by the equation $f(x,y) = 0$. We then study the set of pairs (u,v) of *complex numbers* such that $f(u,v) = 0$. So we consider the zero locus of $f(x,y) = 0$ in $\mathbb{A}_{\mathbb{C}}^2$. We may denote this set by $C(\mathbb{C})$, and the curve considered as a subset of $\mathbb{A}_{\mathbb{R}}^2$ we may denote by $C(\mathbb{R})$. If we identify $\mathbb{A}_{\mathbb{C}}^2$ with $\mathbb{A}_{\mathbb{R}}^4$, this locus is identified with a *surface* defined by two equations. Namely, writing

$$u = x_1 + ix_2, \qquad v = x_3 + ix_4$$

and

$$f(u,v) = f_1(x_1, x_2, x_3, x_4) + if_2(x_1, x_2, x_3, x_4)$$

then f_1 and f_2 are polynomials with real coefficients in four variables, and the set of all complex points on the curve is given as

$$C(\mathbb{C}) = \left\{ (a_1, a_2, a_3, a_4) \in \mathbb{A}_{\mathbb{R}}^4 \,\middle|\, \begin{array}{l} f_1(a_1, a_2, a_3, a_4) = 0 \\ f_2(a_1, a_2, a_3, a_4) = 0 \end{array} \right\}.$$

This is a surface in four-space, in $\mathbb{A}_{\mathbb{R}}^4$, defined by two polynomials. In many situations we really need to include all complex points of a curve, although we usually still confine ourselves to sketch the real points only. And even if the complex points form a surface in $\mathbb{A}_{\mathbb{R}}^4$, it is important to keep in mind that we really are studying a *curve* in the plane, and not a surface in four space. Indeed, of we switch to regard our object under study as a surface in $\mathbb{A}_{\mathbb{R}}^4$, then *it will also have complex points*, thus yielding a *fourfold in* $\mathbb{A}_{\mathbb{R}}^8$, and so on. Thus we have to remember that we are studying complex points on a curve in the plane, rather than the real points of a surface in four space.

The further important extension is to include the *points at infinity* of a curve. This is a somewhat more technical matter, which we come to in

Chap. 3, where we study *projective curves*. But first we give more details on the affine case.

We are given a curve in the plane $\mathbb{R}^2$ as the set of zeroes of the equation

$$f(x,y) = 0$$

where $f(x,y)$ is a polynomial in the variables x and y:

$$f(x,y) = a_{0,0} + xa_{1,0} + ya_{0,1} + x^2 a_{2,0} + xya_{1,1} + y^2 a_{0,2}$$
$$+ \cdots + x^d a_{d,0} + x^{d-1} ya_{d-1,1} + \cdots + y^d a_{0,d}.$$

Some, but not all, of the coefficients may be zero. The largest integer d such that not all $a_{d-i,i}$ are zero is the *degree* of the polynomial, and this is by definition the *degree of the curve*. But here we have a problem, best elucidated by an example.

The equation

$$y = 0$$

defines the x-axis. But so does the equation

$$y^2 = 0$$

at least as a *point-set*. But algebraically we need to distinguish between these two cases. The former equation defines the x-axis as a line, whereas the latter defines a *double line* along the x-axis: Informally speaking, it defines twice the x-axis.

The situation becomes even more difficult when we consider complicated polynomials. Thus for example we may consider the curve defined by the equation

$$(y^2 - x^3 - x^2)(y^2 - x^2) = 0.$$

When we are given the equation on this partly factored form, it is not difficult to see what we get: It is the nodal cubic curve displayed in Fig. 2.6 together with the two lines defined by $y = \pm x$. But suppose that we are given the following equation, on expanded form

$$3y^2 x^4 - 3y^4 x^2 + y^6 - x^7 + 2x^5 y^2 - x^3 y^4 - x^6 = 0$$

then it is not so easy to understand the situation. Using some PC-program to plot this curve, we should get the same picture as above. But this result is quite deceptive. Indeed, if we *factor* the left hand side of the equation, say again by some PC-program, we find that the equation becomes

$$(x^3 + x^2 - y^2)(x + y)^2(x - y)^2 = 0$$

which certainly defines the same point set, but reveals that this time the two lines occurring should *be counted with multiplicity 2*.

Recall that an *irreducible polynomial* in x and y is a polynomial $p(x,y)$ which may not be factored as a product of two polynomials, both non-constants. Thus for instance $p(x,y) = x^3 + x^2 - y^2$ is irreducible, as is $r(x,y) = x + y$ and $s(x,y) = x - y$. A special case of an important theorem is the following:

Theorem 2.6 (Unique Factorization of Polynomials) *Any polynomial in x and y with real (respectively complex) coefficients, may be factored as a product of powers of irreducible polynomials with real (respectively complex) coefficients. These irreducible polynomials are unique except for possibly being proportional by constant factors.*

We make the following definition:

Definition 2.5 (The factorization in irreducible polynomials) The irreducible factorization of $f(x,y)$ is defined as an expression

$$f(x,y) = p_1(x,y)^{n_1} \cdots p_r(x,y)^{n_r}$$

where n_i are positive integers and all $p_i(x,y)$ are irreducible and no two are proportional by a constant factor.

This factorization is unique up to constant factors, by the theorem.

Corollary 2.7 *Theorem 2.6 also holds for a polynomial in a number of variables, 1 up to any N. Definition 2.5 is also unchanged in the general case.*

A polynomial may be irreducible as a polynomial with real coefficients, but *reducible* when considered as a polynomial with *complex* coefficients. This is the case for the polynomial

$$g(x,y) = x^2 + y^2,$$

which may not be factored as a polynomial with real coefficients, while

$$x^2 + y^2 = (x + iy)(x - iy).$$

The curve given by this polynomial has another interesting feature: As a curve in $\mathbb{A}^2_{\mathbb{R}}$ it consists only of the origin, while it consists of two (complex) *lines* in $\mathbb{A}^2_{\mathbb{C}}$, with equations $y = \pm ix$. They have only one real point on them, namely their point of intersection which is the origin. We would consider this as a degenerate case, say as a member of a family of circles, where the radius has shrunk to zero.

Definition 2.6 (Real Affine Curve) A real affine plane curve C is the set of points $(a,b) \in \mathbb{A}^2_{\mathbb{R}}$ which are zeroes of a polynomial $f(x,y)$ with real co-efficients. The irreducible polynomials $p_i(x,y)$ occurring in the irreducible factorization of $f(x, y)$ referred to in Definition 2.5 define subsets C_i of C called the irreducible components of C. The exponent n_i of $p_i(x, y)$ in the factorization of $f(x, y)$ is called the multiplicity of the irreducible component.

In other words, C_i occurs with multiplicity n_i in C.

Remark This definition suffices as a first approximation, but it should not be concealed that it does represent a simplification. Indeed, according to the definition the "real affine curve" defined by $x^2 + y^2 = 0$ is the same as the one defined by $x^2 + 2y^2 = 0$. For a variety of reasons this is undesirable. One solution is to simply *define* a curve in $\mathbb{A}_\mathbb{R}^2$ as being an equivalence class of polynomials, two polynomials being regarded as equivalent if one is a non-zero constant multiple of the other. This is mathematically sound, but only applies to a special geometric situation, where one geometric object, here the curve, is contained in another geometric object of one dimension higher, here the plane, and is defined by one "equation". The final clarification of this concept will come when we explain the notion of a *scheme*, which was introduced by *Alexander Grothendieck*.

After a change of variables, which corresponds to a change of coordinate system,

$$\overline{x} = a + \alpha_{1,1}x + \alpha_{1,2}y$$

$$\overline{y} = b + \alpha_{2,1}x + \alpha_{2,2}y$$

the curve given by $f(x, y) = 0$ is expressed by the equation $\overline{f}(\overline{x}, \overline{y}) = 0$, where $\overline{f}(\overline{x}, \overline{y})$ is obtained by substituting the expressions obtained by solving for x and y,

$$x = \overline{a} + \beta_{1,1}\overline{x} + \beta_{1,2}\overline{y}$$

$$y = \overline{b} + \beta_{2,1}\overline{x} + \beta_{2,2}\overline{y}$$

into $f(x, y)$.

There are curves in the affine plane $\mathbb{A}_\mathbb{R}^2$ which are not affine algebraic, but nevertheless form an important subject in geometry. The Archimedean spiral and the quadratrix of Hippias are such curves. They both were invented to solve some of the Classical Problems. They are not defined by a polynomial equation. Some other simple examples are the curves defined by $y = \sin(x)$ or by $y = e^x$. This class of curves is called *the Transcendental Curves*. In this book we will confine the general theory to treating the algebraic curves, that is to say the ones defined by a polynomial equation.

2.8 Singularities and Multiplicities

We now return to some general concepts introduced in Sect. 2.2, where we needed it to understand the degeneracy of conics. Consider an algebraic affine curve K with equation

$$f(x,y) = 0.$$

Furthermore, let (a,b) be a point on the curve, i.e., $f(a,b) = 0$. We note that the following definition relies heavily on the *equation* of the curve, not just the curve as a subset of $\mathbb{A}^2_{\mathbb{R}}$:

Definition 2.7 (a,b) is said to be a smooth, or a non-singular, point on K if

$$\left(\frac{\partial f}{\partial x}(a,b), \frac{\partial f}{\partial y}(a,b) \right) \neq (0,0).$$

Otherwise (a,b) is called a singular point on K. A curve all of whose points are non-singular is referred to as a non-singular curve.

The vector $(\frac{\partial f}{\partial x}, \frac{\partial f}{\partial y})$ is referred to as the Jacobian vector (for short, the *Jacobian*) of the polynomial $f(x,y)$. Thus by definition a singular point is a point on the curve at which the Jacobian evaluates to the zero vector.

In Sect. 2.2 we saw that a non-degenerate conic curve is a non-singular curve. We look at the situation in more detail by the examples below.

It is time to turn to some examples.

(1) We first look at simple *conics*, and start out with a circle of radius $R > 0$, which has the equation

$$x^2 + y^2 = R^2.$$

Here $f(x,y) = x^2 + y^2 - R^2$, and

$$\left(\frac{\partial f}{\partial x}, \frac{\partial f}{\partial y} \right) = (2x, 2y).$$

Evidently no point outside the origin can be a singular point of the circle, and as $R > 0$, every point on the circle is therefore smooth. We note that the same proof shows that an ellipse on standard form,

$$\left(\frac{x}{a} \right)^2 + \left(\frac{y}{b} \right)^2 = 1$$

is smooth everywhere as well.

A (non-degenerate) hyperbola on standard form, which is given as

$$\left(\frac{x}{a} \right)^2 - \left(\frac{y}{b} \right)^2 = 1$$

similarly has Jacobian

$$\left(\frac{2}{a^2} x, -\frac{2}{b^2} y \right)$$

which also does not vanish outside the origin, showing that a hyperbola is smooth.

A degenerate hyperbola is one which has collapsed to the asymptotes, hence a curve with equation

$$\left(\frac{x}{a}\right)^2 - \left(\frac{y}{b}\right)^2 = 0.$$

This curve has the same Jacobian as in the non-degenerate case, but now the origin actually lies on the curve, which therefore has the origin as its only singular point. Of course this degenerate hyperbola consists of two irreducible components which are lines intersecting at the origin, and that point is singular.

Our final conic curve is the parabola with equation

$$ay - x^2 = 0$$

where $a \neq 0$. The Jacobian is $(-2x, a)$, so the only possibility of getting the zero vector at a point would be to have $x = 0$ and $a = 0$. For $a \neq 0$ we therefore have no singular points. If $a = 0$, then the equation yields the y-axis with multiplicity 2, and we see that then *all points on the curve are singular*.

(2) We next turn to the *nodal cubic curve* with equation

$$y^2 - x^3 - x^2 = 0$$

which is plotted in Fig. 2.6. The Jacobian is

$$(-3x^2 - 2x, 2y)$$

and thus (x, y) is a singular point if and only if the two additional equations below are satisfied:

$$-3x^2 - 2x = 0$$

$$2y = 0.$$

Thus $(x, y) = (0, 0)$ or $(x, y) = (-\frac{2}{3}, 0)$, and only the former lies on the curve, so the only singular point is $(0, 0)$.

(3) If $f(x, y)$ is a polynomial, then all points on the curve given by $f(x, y)^n = 0$ for n an integer greater than 1, will have all its points singular. This follows at once, since the Jacobian is

$$\left(nf(x, y)^{n-1}\frac{\partial f}{\partial x}, nf(x, y)^{n-1}\frac{\partial f}{\partial y}\right).$$

It is highly recommended that the reader examines the curves plotted in Sect. 2.4, and determines their singular points.

2.9 Tangency

Let (a, b) be a smooth point on the curve K. Then we may find the equation for the tangent line at that point as follows. We first consider *the parametric form* for a line through (a, b) with direction given by the vector (u, v):

$$L = \left\{ (x, y) \ \middle| \ \begin{array}{l} x = a + ut \\ y = b + vt \end{array} \text{ where } t \in \mathbb{R} \right\}.$$

This line will have the point (a, b) in common with K. We wish to determine other points of intersection. To do so we substitute the expressions for x and y in the parametric form for L into the equation for K, and get

$$f(a + ut, b + vt) = 0.$$

Expanding the left hand side in a Taylor series we obtain

$$f(a,b) + t\left(u\frac{\partial f}{\partial x}(a,b) + v\frac{\partial f}{\partial y}(a,b) \right)$$
$$+ t^2\left(u^2\frac{\partial^2 f}{\partial x^2}(a,b) + 2uv\frac{\partial^2 f}{\partial x \partial y}(a,b) + v^2\frac{\partial^2 f}{\partial y^2}(a,b) \right) + \cdots = 0.$$

which since $f(a, b) = 0$ gives the following equation for t:

$$t\left(u\frac{\partial f}{\partial x}(a,b) + v\frac{\partial f}{\partial y}(a,b) \right)$$
$$+ t^2\left(u^2\frac{\partial^2 f}{\partial x^2}(a,b) + 2uv\frac{\partial^2 f}{\partial x \partial y}(a,b) + v^2\frac{\partial^2 f}{\partial y^2}(a,b) \right) + \cdots = 0. \quad (2.1)$$

The points of intersection between the curve and the line are found by solving this equation for t. Of course we have $t = 0$ as one solution, and we see that this solution will occur with multiplicity 1 if and only if

$$u\frac{\partial f}{\partial x}(a,b) + v\frac{\partial f}{\partial y}(a,b) \neq 0.$$

Such values of u, v exist if and only if (a, b) is a smooth point on the curve. In that case there is exactly one line through $P = (a, b)$ which *does not intersect the curve with multiplicity 1*, namely the line corresponding to u and v such that

$$u\frac{\partial f}{\partial x}(a,b) + v\frac{\partial f}{\partial y}(a,b) = 0.$$

By substituting

$$ut = x - a$$

$$vt = y - b$$

in this equation, we recover the equation for the tangent line to the curve at the point (a, b)

$$(x - a)\frac{\partial f}{\partial x}(a, b) + (y - b)\frac{\partial f}{\partial x}(a, b) = 0.$$

Earlier we used this equation to *define* the tangent line to a curve C at a point $P = (a, b)$, but at that stage without a geometric justification. Now we see the geometric meaning of this definition.

Definition 2.8 We denote the multiplicity of the solution $t = 0$ of (2.1) by $m_{K,P}(L)$. This number is referred to as the multiplicity with which the line L intersects the curve K in the point P.

To sum up what we have so far, the point P is a smooth point on K provided there is exactly one line L intersecting K in P with multiplicity >1, and L is then called the tangent to K in P. The normal situation is that the multiplicity is 2, if it is ≥ 3 then P is referred to as a flex (or an *inflection point*), if $m_{K,P}(L) = 3$ the flex is said to be *an ordinary flex*. The term *inflection point* is also used for curves in $\mathbb{A}^2_{\mathbb{R}}$ as a point where the sign of the curvature changes. A smooth point with this property is an inflection point in our sense, but not conversely. The tangent line at a flex is called *an inflectional tangent*.

We next turn to the question of what happens at *a singular point*. So let $P = (a, b)$ be a singular point on the curve K. Since the situation is more complicated than in the case when P is smooth, we introduce new variables by

$$\overline{x} = x - a, \qquad \overline{y} = y - b.$$

In other words, we shift the variables so that the new origin falls in P, $P = (0, 0)$. We then find a new polynomial g such that

$$f(x, y) = g(\overline{x}, \overline{y})$$

by substituting $x = \overline{x} + a$ and $y = \overline{y} + b$ into $f(x, y)$. The curve is also given by the equation

$$g(\overline{x}, \overline{y}) = 0.$$

Since the origin is a point on the curve given by $g(\overline{x}, \overline{y}) = 0$, it is clear that the polynomial $g(\overline{x}, \overline{y})$ has no constant term. We now collect the terms of $g(\overline{x}, \overline{y})$ which are of lowest total degree, and denote the sum of those terms by $h(\overline{x}, \overline{y})$.

Thus for example, if

$$g(\overline{x}, \overline{y}) = 2\overline{x}\,\overline{y}^2 - 5\overline{x}^2\overline{y} + 10\overline{x}^9\overline{y}^2 + 15\overline{x}^2\overline{y}^{12},$$

then

$$h(\overline{x}, \overline{y}) = 2\overline{x}\,\overline{y}^2 - 5\overline{x}^2\overline{y}.$$

The sum of all terms of lowest total degree of the polynomial g is called the *initial part* of the polynomial, and denoted by $\mathrm{in}(g)$. If the point $P = (a, b)$ is smooth, then the Taylor expansion around the point (a, b) immediately shows that the polynomial $h(\overline{x}, \overline{y})$ is nothing but

$$\frac{\partial g}{\partial x}(0,0)\overline{x} + \frac{\partial g}{\partial y}(0,0)\overline{y} = \frac{\partial f}{\partial x}(a,b)(x-a) + \frac{\partial f}{\partial y}(a,b)(y-b).$$

Thus the concept introduced below generalizes the tangent at a smooth point, to a concept which applies to *singular points as well*.

With notations as above the polynomial $h(\overline{x}, \overline{y})$ defines a curve which is a finite union of lines through the point $(0, 0)$. In terms of x and y, the equation

$$h(x - a, y - b) = 0$$

defines a finite union of lines through $P = (a, b)$, some of them occurring with multiplicity >1. Indeed, we have

$$h(\overline{x}, \overline{y}) = a_0\overline{x}^m + a_1\overline{x}^{m-1}\overline{y} + \cdots + a_i\overline{x}^{m-i}\overline{y}^i + \cdots + a_m\overline{y}^m$$

where not all a_i vanish. If (α_0, β_0) satisfies $h(\alpha_0, \beta_0) = 0$, then we also have $h(s\alpha_0, s\beta_0) = 0$ for all real numbers s, as one immediately verifies since all the monomials of h are of the same total degree m.

These lines are called the *lines of tangency* at the point $P = (a, b)$. If P happens to be smooth, then there is only one line, occurring with multiplicity 1.

Definition 2.9 The curve given by $h(x - a, y - b) = 0$ is referred to as *the (affine) tangent cone* of K at P.

Any line through $P = (a, b)$ may, as we have seen, be written on parametric form as

$$x - a = ut, \qquad y - b = vt$$

and its intersections with the curve is determined by the equation

$$f(a + ut, b + vt) = g(ut, vt) = 0.$$

The multiplicity of the root $t = 0$ in this equation is referred to as *the multiplicity of intersection* between the curve and the line at the point $P = (a, b)$.

All lines through $P = (a, b)$ which do not coincide with one of the lines of tangency, intersect the curve with multiplicity equal to the number m. This number m is of course only dependent upon the polynomial $f(x, y)$ and the point $P = (a, b)$.

In fact, we may assume that $P = (0, 0)$. An arbitrary line through $(0, 0)$ has the parametric form

$$L = \left\{ (x, y) \;\middle|\; \begin{matrix} x = ut \\ y = vt \end{matrix} \text{ where } t \in \mathbb{R} \right\}.$$

To find all points of intersection between this line and the curve K, we substitute the expressions for x and y into $f(x, y)$ and get

$$f(a + ut, b + vt) = 0.$$

This gives

$$h(ut, vt) + R(ut, vt) = 0$$

where $R(x, y)$ denotes $f(x, y) - h(x, y)$. Thus the points of intersection are given by the roots of the equation

$$t^m (h(u, v) + t\varphi(t)) = 0.$$

One of the roots is $t = 0$, and this solution will occur with multiplicity $\geq m$, where equality holds if and only if

$$h(u, v) \neq 0$$

thus if and only if L is not one of the lines of tangency.

We conclude with the

Definition 2.10 (Multiplicity of a point on a curve) The number m referred to above is called the multiplicity of the point P at K.

We thus have the observation

Proposition 2.8 *A point on an affine algebraic curve is smooth if and only if it has multiplicity* 1.

Chapter 3
Higher Geometry in the Projective Plane

We now replace the field of real numbers $\mathbb{R}$ by a general field k. All previous constructions and definitions carry over to general fields with obvious modifications, and we start with the formal definition of an affine or projective (plane) algebraic curve over a field k. Likewise formal definitions of affine restriction and projective closure of such curves are given, and the interplay between these concepts is explored, as well as smooth and singular point on them. The properties of intersection between a line and an affine or projective curve is examined and the *tangent star* of a curve et a point is defined. The concepts of projective equivalence and asymptotes are introduced, and the class of general conchoids is defined, an important example being the *Conchoid of Nicomedes*. The dual curve is defined, this being merely the top of a mighty iceberg, to be explored at a later stage.

3.1 Projective Curves

We define curves in the *projective plane* $\mathbb{P}^2_k$ analogously to curves in the affine plane $\mathbb{A}^2_k$. The difference is that we can not use ordinary polynomials in two variables, but have to work with *homogeneous polynomials in three variables* instead, as we did for conics.

Thus the polynomial

$$X_0 + 5X_0X_1^2$$

is not homogeneous, since one monomial which occurs is X_0, and another is $5X_0X_2^2$. They are of degrees 1 and 3, respectively. On the other hand, the polynomial

$$X_0^3 + 5X_0X_1^2$$

is homogeneous, the two monomials which occur are both of degree 3.

A. Holme, *A Royal Road to Algebraic Geometry*,
DOI 10.1007/978-3-642-19225-8_3, © Springer-Verlag Berlin Heidelberg 2012

Now assume that we have a homogeneous polynomial with real coefficients

$$F(X_0, X_1, X_2) = \sum_{I \in S} c_I X_0^{i_0} X_1^{i_1} X_2^{i_2}$$

where $I = (i_0, i_1, i_2)$, $d = i_0 + i_1 + i_2$, and the symbol $\sum_{I \in S}$ means that we have a sum where I runs through a finite subset S of triples of non-negative integers, when no confusion is possible we usually write just $\sum_I$. c_I is a real number, called the *coefficient* of the *monomial* $X_0^{i_0} X_1^{i_1} X_2^{i_2}$. Let $(a_0, a_1, a_2) \in \mathbb{A}_k^3$. Then we have

$$F(ta_0, ta_1, ta_2) = \sum_I c_I (ta_0)^{i_0} (ta_1)^{i_1} (ta_2)^{i_2}$$

$$= t^d \left(\sum_I c_I a_0^{i_0} a_1^{i_1} a_2^{i_2} \right) = t^d F(a_0, a_1, a_2)$$

since $d = i_0 + i_1 + i_2$. Thus we find that whenever $t \neq 0$, then

$$F(ta_0, ta_1, ta_2) = 0 \quad \text{if and only if} \quad F(a_0, a_1, a_2) = 0.$$

It follows that the zero locus for a homogeneous polynomial in X_0, X_1 and X_2 is well defined in $\mathbb{P}_k^2$. Moreover, we also note the

Theorem 3.1 *In the irreducible factorization of a homogeneous polynomial, given by Corollary 2.7, all the irreducible polynomials occurring are also homogeneous.*

Proof The proof is by induction on $d = \deg(F)$. For $d = 1$ the claim is immediate. Suppose that the claim is true for all homogeneous polynomials of degree $< d$, and let F be homogeneous of degree d. If F is irreducible, there is nothing to prove. Otherwise we may write

$$F = F_1 F_2$$

where F_1 and F_2 are polynomials of degrees $< d$. We may write

$$F_i = H_i + G_i, \quad \text{for } i = 1, 2$$

where H_i is the homogeneous piece of highest degree of F_i. Thus

$$F = H_1 H_2 + G_1 H_2 + G_2 H_1 + G_1 G_2 = H_1 H_2 + G$$

F and $H_1 H_2$ are homogeneous of the same degree. If G were non zero it would be of degree $< d$, which is absurd. Thus we must have

$$F = H_1 H_2,$$

and the claim follows by induction. $\square$

Definition 3.1 (Projective Algebraic Curve) Let k be a field. A plane projective curve $C \subset \mathbb{P}_k^2$ is the zero locus of a homogeneous polynomial in X_0, X_1 and X_2, with coefficients from k. The irreducible components of C, as well as their multiplicities, are defined analogously to the affine case by means of Theorem 3.1.

3.2 Projective Closure and Affine Restriction

Given an affine curve $K \subset \mathbb{A}_k^2$, with equation $f(x,y) = 0$. In the same way as we did for curves of degree 2, we may define the *projective closure* $C \subset \mathbb{P}_k^2$ of K. It is defined by the equation $F(X_0, X_1, X_2) = 0$ where $F(X_0, X_1, X_2)$ is constructed by putting $x = \frac{X_1}{X_0}$ and $y = \frac{X_2}{X_0}$ and substituting this in $f(x,y)$, and writing the result as

$$f\left(\frac{X_1}{X_0}, \frac{X_2}{X_0}\right) = \frac{F(X_0, X_1, X_2)}{X_0{}^m}$$

where X_0 does not divide the numerator. Here $F(X_0, X_1, X_2)$ is a homogeneous polynomial with coefficients from k, uniquely determined by $f(x,y)$ as follows: If

$$f(x,y) = \sum_{I=(i_1,i_2)\in\Phi} a_I x^{i_1} y^{i_2}$$

where Φ denotes a finite set of tuples of non-negative integers (i_1, i_2), then the degree of K is $d = \max\{i_1 + i_2 | (i_1, i_2) \in \Phi\}$, and the projective closure is given by the equation

$$F(X_0, X_1, X_2) = \sum_{I=(i_1,i_2)\in\Phi} a_I X_0^{d-i_1-i_2} X_1^{i_1} X_2^{i_2} = 0.$$

d is the degree of the original affine curve K as well as of its projective closure C.

Definition 3.2 The homogeneous polynomial $F(X_0, X_1, X_2)$ as defined above is denoted by $f^h(X_0, X_1, X_2)$, and referred to as the homogenization of the (non-homogeneous) polynomial $f(x,y)$.

The key to understanding the relation between an affine curve and its projective closure lies in the simple and beautiful relation

$$f(a,b) = f^h(1,a,b)$$

which holds for all a and b.

Thus if K is the affine curve defined by $f(x,y) = 0$, then the projective closure C of K is defined by the equation $f^h(X_0, X_1, X_2) = 0$. Conversely, if we are given a projective curve C by the equation $F(X_0, X_1, X_2) = 0$, then we may define its *affine restriction* to $D_+(X_0)$ as identified with $\mathbb{A}_k^2$ as the curve given by the equation $F(1, x, y) = 0$. But this affine restriction is not always defined: Namely, if $F(X_0, X_1, X_2) = X_0^d$, then C is the line $L_\infty = V_+(X_0)$, the line at infinity, with multiplicity d. Of course the affine restriction of this curve to $D_0(X_0)$ is given by the equation $1 = 0$, so we might say that the *affine restriction of this curve to $D_+(X_0)$ is empty*. On the other hand, if we chose to take the affine restriction to $D_+(X_1)$ instead, and put $x = \frac{X_0}{X_1}$ and $y = \frac{X_2}{X_1}$, then the affine restriction is the curve given by $x^d = 0$, in other words the y-axis counted with multiplicity d.

So the concepts of projective closure and affine restriction are not independent of the coordinate system. The change to another projective coordinate system in $\mathbb{P}_k^2$ has been described in Chap. 1, Sect. 1.3. The equations defining the new coordinate system may also be used to define a bijective mapping of $\mathbb{P}_\mathbb{R}^2$ onto itself, known as *a projective transformation*. This was also explained there, and will not be repeated here.

Even though the concepts of projective closure and affine restriction do depend on the coordinate system, they are very useful in the investigation of properties and concepts which *are* coordinate independent. Normally we perform the projective closure by letting $V_+(X_0)$ contain the added points at infinity, and identify the affine plane $\mathbb{A}_k^2$ with $D_+(X_0)$. When an alternative procedure is used, this will be explicitly stated. Also, if $V_+(aX_0 + bX_1 + cX_2)$ is a projective line in $\mathbb{P}_k^2$, then we may identify $D_+(aX_0 + bX_1 + cX_2)$ with $\mathbb{A}_k^2$ and carry out affine restrictions to $\mathbb{A}_k^2$ by restricting to $D_+(aX_0 + bX_1 + cX_2)$. Again, if this non-standard procedure is used we shall explicitly state so. The most convenient method is to choose a new projective coordinate system by putting

$$\overline{X}_0 = aX_0 + bX_1 + cX_2$$

and choosing linear forms $a_1X_0 + b_1X_1 + c_1X_2$ and $a_2X_0 + b_2X_1 + c_2X_2$ such that the determinant of the coefficients of the three forms is non-zero, so letting

$$\overline{X}_1 = a_1X_0 + b_1X_1 + c_1X_2 \quad \text{and} \quad \overline{X}_2 = a_2X_0 + b_2X_1 + c_2X_2$$

we get a new projective coordinate system. Then the affine restriction is carried out in the standard fashion with respect to it.

Using an affine restriction we are able to study local properties of a curve, like questions of tangency or singularity, with greater precision. Taking projective closure we obtain information on how the curve behaves very far away from the origin, *at infinity*, information crucial to a global understanding of the affine curve itself. An example of this which we shall return to later is the determination of all the *asymptotes* of a curve in $\mathbb{A}_\mathbb{R}^2$. We conclude this section on projective closure and affine restriction with the

Proposition 3.2 1. *Let K be an affine curve in $\mathbb{A}_k^2$, and let C be its projective closure. Then the affine restriction of C is equal to K.*

2. Let C be a projective curve, and let K be its affine restriction. If C is just a multiple of $V_+(X_0)$ then K is empty. Otherwise K is an affine curve, and its projective closure C' consist of all irreducible components of C, with the same multiplicity as before, except possibly for the component $V_+(X_0)$, which is removed when passing from C to C'.

Proof To prove 1., let K be given by

$$f(x,y) = \sum_{I=(i_1,i_2)\in\Phi} a_I x^{i_1} y^{i_2}.$$

Then the projective closure is given by

$$F(X_0,X_1,X_2) = \sum_{I=(i_1,i_2)\in\Phi} a_I X_0^{d-i_1-i_2} X_1^{i_1} X_2^{i_2} = 0.$$

Substituting $X_0 = 1$, $X_1 = x$ and $X_2 = y$ clearly gives us back $f(x,y)$, and 1. is proven.

As for 2., assume that C is given by the homogeneous polynomial

$$F(X_0,X_1,X_2) = X_0^r \left(\sum_{I=(i_1,i_2)\in\Phi} a_I X_0^{d-i_1-i_2} X_1^{i_1} X_2^{i_2} \right)$$

where the polynomial inside the parenthesis is not divisible by X_0. Denoting the latter by $G(X_0,X_1,X_2)$, we find that

$$F(1,x,y) = G(1,x,y)$$

and the affine restriction of C is defined by $G(1,x,y) = 0$. So the projective closure C' of the affine restriction is defined by $G(X_0,X_1,X_2)$ with a possible component $V_+(X_0)$ removed. This completes the proof.   □

3.3 Smooth and Singular Points on Affine and Projective Curves

Let C be given by the equation

$$F(X_0,X_1,X_2) = 0.$$

Moreover, let $P = (a_0 : a_1 : a_2)$ be a point on C.

Definition 3.3 We say that the point P is a smooth point on C if

$$\left(\frac{\partial F}{\partial X_0}(a_0, a_1, a_2), \frac{\partial F}{\partial X_1}(a_0, a_1, a_2), \frac{\partial F}{\partial X_2}(a_0, a_1, a_2)\right) \neq (0, 0, 0).$$

Whenever this condition is not satisfied, the point is referred to as a *singular* point. Correspondingly, a smooth point is also referred to as a *non-singular* point.

Earlier we defined the term *smooth point* for affine curves $K \subset \mathbb{A}^2_k$. Even if this previous definition is similar to the one we have given here, we need to show that they do not contradict one another. Namely, when we form the *projective closure* of the affine curve K, we obtain a *projective curve* $C \subset \mathbb{P}^2_k$. A point $p \in K$ should then be smooth as a point of the affine curve K if and only if it is smooth as a point on the projective curve C.

This problem is disposed of by means of the following proposition:

Proposition 3.3 *With notations as in as in Chap. 1, Sect. 3.2 we have*

$$\left(\frac{\partial f}{\partial x}\right)^h (X_0, X_1, X_2) = \frac{\partial f^h}{\partial X_1}(X_0, X_1, X_2)$$

and

$$\left(\frac{\partial f}{\partial y}\right)^h (X_0, X_1, X_2) = \frac{\partial f^h}{\partial X_2}(X_0, X_1, X_2).$$

Proof We put

$$f(x, y) = \sum_{I = (i_1, i_2) \in \Phi} a_I x^{i_1} y^{i_2}$$

then $F = f^h$ is given by

$$F(X_0, X_1, X_2) = \sum_{I = (i_1, i_2) \in \Phi} a_I X_0^{d - i_1 - i_2} X_1^{i_1} X_2^{i_2} = 0.$$

The verification of the claim is immediate from this.   □

Corollary 3.4 *Let K be an affine curve, and let C be the projective closure of K, where $V_+(X_0)$ is the points at infinity. Then (a, b) is a smooth point on the affine curve K if and only if $(1 : a : b)$ is a smooth point on the projective curve C.*

Proof We apply the relation

$$g(a, b) = g^h(1, a, b)$$

to the partial derivatives.   □

The second important observation concerning smooth or singular points is contained in the

Proposition 3.5 *The concept of smooth point on a projective curve is independent of the projective coordinate system.*

Proof We may write the transition from one coordinate system to another as a matrix multiplication as follows:

$$\begin{bmatrix} \alpha_{0,0} & \alpha_{0,1} & \alpha_{0,2} \\ \alpha_{1,0} & \alpha_{1,1} & \alpha_{1,2} \\ \alpha_{2,0} & \alpha_{2,1} & \alpha_{2,2} \end{bmatrix} \cdot \begin{bmatrix} Y_0 \\ Y_1 \\ Y_2 \end{bmatrix} = \begin{bmatrix} X_0 \\ X_1 \\ X_2 \end{bmatrix}$$

where the matrix has determinant $\neq 0$,

$$\begin{vmatrix} \alpha_{0,0} & \alpha_{0,1} & \alpha_{0,2} \\ \alpha_{1,0} & \alpha_{1,1} & \alpha_{1,2} \\ \alpha_{2,0} & \alpha_{2,1} & \alpha_{2,2} \end{vmatrix} \neq 0.$$

Clearly

$$\frac{\partial X_i}{\partial Y_j} = \alpha_{i,j}.$$

Moreover, if the curve C is given in the original coordinate system as

$$F(X_0, X_1, X_2) = 0$$

then it will be given in the new coordinate system by

$$G(Y_0, Y_1, Y_2) = 0$$

where

$$G(Y_0, Y_1, Y_2) = F(\alpha_{0,0}Y_0 + \alpha_{0,1}Y_1 + \alpha_{0,2}Y_2, \alpha_{1,0}Y_0 + \alpha_{1,1}Y_1 + \alpha_{1,2}Y_2,$$
$$\alpha_{2,0}Y_0 + \alpha_{2,1}Y_1 + \alpha_{2,2}Y_2).$$

Now let the point P be expressed as $(a_0 : a_1 : a_2)$ and $(b_0 : b_1 : b_2)$ in the two coordinate systems. Then we get by the chain rule

$$\frac{\partial G}{\partial Y_0}(b_0, b_1, b_2) = \alpha_{0,0}\frac{\partial F}{\partial X_0}(a_0, a_1, a_2) + \alpha_{1,0}\frac{\partial F}{\partial X_1}(a_0, a_1, a_2)$$
$$+ \alpha_{2,0}\frac{\partial F}{\partial X_2}(a_0, a_1, a_2)$$

$$\frac{\partial G}{\partial Y_1}(b_0, b_1, b_2) = \alpha_{0,1}\frac{\partial F}{\partial X_0}(a_0, a_1, a_2) + \alpha_{1,1}\frac{\partial F}{\partial X_1}(a_0, a_1, a_2)$$
$$+ \alpha_{2,1}\frac{\partial F}{\partial X_2}(a_0, a_1, a_2)$$

$$\frac{\partial G}{\partial Y_2}(b_0, b_1, b_2) = \alpha_{0,2}\frac{\partial F}{\partial X_0}(a_0, a_1, a_2) + \alpha_{1,2}\frac{\partial F}{\partial X_1}(a_0, a_1, a_2)$$

$$+ \alpha_{2,2}\frac{\partial F}{\partial X_2}(a_0, a_1, a_2).$$

Since the determinant of the matrix of the α's is non-zero, it follows that the vector of the evaluated partials to the left will not all vanish if and only if the vector of the evaluated partials to the right do not all vanish. Thus smoothness or singularity for a point on C is independent of the coordinate system in which the corresponding condition is expressed. $\square$

3.4 The Tangent to a Projective Curve

Before we deduce the equation for *the tangent line to a projective curve,* we need to make some comments on *lines and other curves on parametric form* in $\mathbb{P}_k^2$. We first consider the case of lines. A line $L \subset \mathbb{P}_k^2$ which passes through the points $(a_0 : a_1 : a_2)$ and $(b_0 : b_1 : b_2)$ may be expressed as follows, on parametric form:

$$L = \left\{ (X_0 : X_1 : X_2) \left| \begin{array}{l} X_0 = ua_0 + vb_0 \\ X_1 = ua_1 + vb_1 \\ X_2 = ua_2 + vb_2 \end{array} \right. \right\}.$$

Here u and $v \in k$ are two parameters which yield all the points on the line L, but as we see, it is only the ratio $(u : v)$ which distinguish between the points. In particular we have that $(u : v) = (1 : 0)$ yields the point $(a_0 : a_1 : a_2)$, while $(u : v) = (0 : 1)$ yields $(b_0 : b_1 : b_2)$.

More generally we may consider a curve in $\mathbb{P}_k^2$ given on parametric form:

$$C = \left\{ (X_0 : X_1 : X_2) \left| \begin{array}{l} X_0 = \xi_0(u, v) \\ X_1 = \xi_1(u, v) \\ X_2 = \xi_2(u, v) \end{array} \right. \right\}.$$

Here we assume that the polynomials $\xi_0(u, v), \xi_1(u, v)$ and $\xi_2(u, v)$ are *homogeneous of the same degree* in the variables u and v. The class of curves which may be so described do not contain all projective curves in $\mathbb{P}_k^2$, there are curves which are *not* parameterizable by polynomials. But it does include all lines in $\mathbb{P}_k^2$.

If we choose

$$\xi_0(u, v) = u^2, \quad \xi_1(u, v) = uv \quad \text{and} \quad \xi_2(u, v) = v^2$$

we get a curve of degree 2, in other words *a projective conic curve:* It has the equation

$$X_1^2 - X_0 X_2 = 0.$$

If we choose $\xi_0(u,v), \xi_1(u,v)$ and $\xi_2(u,v)$ as general homogeneous polynomials of degree 2, then we get general projective curves of degree 2 in $\mathbb{P}^2_k$: The class of projective curves of degree 2 consists of parameterizable ones. But even if we let $\xi_0(u,v), \xi_1(u,v)$ and $\xi_2(u,v)$ be general homogeneous polynomials of degree 3, we only obtain a special class of degree 3 projective curves in $\mathbb{P}^2_k$, namely the *rational cubics* in $\mathbb{P}^2_k$.

In general, the curves are parameterizable as described above by homogeneous polynomials of the same degree d are referred to as *the rational degree d-curves in* $\mathbb{P}^2_k$. Here an explanation should be interjected: This number d, the common degree of the polynomials $\xi_0(u,v)$, $\xi_1(u,v)$ and $\xi_2(u,v)$, turns out to be the degree of the equation

$$F(X_0, X_1, X_2) = 0$$

which expresses the relation between the polynomials $\xi_0(u,v)$, $\xi_1(u,v)$ and $\xi_2(u,v)$.

We now come to the concept of *tangent line of a general projective curve* in $\mathbb{P}^2_k$. We consider a point $P = (a_0 : a_1 : a_2)$ on the curve C defined by the homogeneous polynomial $F(X_0, X_1, X_2)$,

$$F(X_0, X_1, X_2) = 0.$$

As we did in the affine case, we consider the collection of all lines passing through P, as we saw above these lines are all given on parametric form as

$$L = \left\{ (X_0 : X_1 : X_2) \;\middle|\; \begin{array}{l} X_0 = ua_0 + vb_0 \\ X_1 = ua_1 + vb_1 \\ X_2 = ua_2 + vb_2 \end{array} \right\}$$

where $P = (a_0 : a_1 : a_2)$ is the fixed point on C, and $Q = (b_0 : b_1 : b_2)$ is another point $\neq P$ in $\mathbb{P}^2_k$ and u and v are the parameters describing the line L passing through P and Q, the points on L corresponding to the ratio $u : v$. The point P corresponds to $u : v = 1 : 0$, while Q corresponds to $u : v = 0 : 1$. We wish to examine the points of intersection of the line L with C, as well as the multiplicities with which they occur. We then have to find all u and v which satisfy the equation

$$F(ua_0 + vb_0, ua_1 + vb_1, ua_2 + vb_2) = 0.$$

But our objective now is *not* to find all the other points of intersection between L and C. Instead, we are interested in examining *how the line intersects the curve in the point P*, in other words we wish to study the solution $(u,v) = (1,0)$ of the equation, and since only the ratios count, this amounts to studying the solution $t = 0$ of the equation

$$\varphi(t) = F(a_0 + tb_0, a_1 + tb_1, a_2 + tb_2) = 0.$$

Since $P \in C$, $t = 0$ certainly is a solution. As in the affine case the multiplicity of the solution $t = 0$ is referred to as the *multiplicity with which the line L intersects C at P*.

If $k = \mathbb{R}$, then expanding $\varphi(t)$ in a Taylor series around $t = 0$ we actually get a polynomial of degree d, the degree of the curve C. We get

$$\varphi(t) = \varphi(0) + \varphi'(0)t + \frac{1}{2}\varphi''(0)t^2 + \cdots + \frac{1}{i!}\varphi^{(i)}(0)t^i + \cdots + \frac{1}{d!}\varphi^{(d)}(0)t^d.$$

Here $\varphi(0) = 0$, and using the general Chain Rule we obtain

$$\varphi'(t) = b_0 \frac{\partial}{\partial X_0} F(a_0 + tb_0, a_1 + tb_1, a_2 + tb_2)$$

$$+ b_1 \frac{\partial}{\partial X_1} F(a_0 + tb_0, a_1 + tb_1, a_2 + tb_2)$$

$$+ b_2 \frac{\partial}{\partial X_2} F(a_0 + tb_0, a_1 + tb_1, a_2 + tb_2)$$

$$= \left(\left(b_0 \frac{\partial}{\partial X_1} + b_1 \frac{\partial}{\partial X_I} + b_2 \frac{\partial}{\partial X_2} \right) F \right) (a_0 + tb_0, a_1 + tb_1, a_2 + tb_2)$$

and hence

$$\varphi'(0) = \left(\left(b_0 \frac{\partial}{\partial X_1} + b_1 \frac{\partial}{\partial X_1} + b_2 \frac{\partial}{\partial X_2} \right) F \right) (a_0, a_1, a_2).$$

Taking the derivative of $\varphi'(t)$ and using the Chain Rule again, we similarly get the expression

$$\varphi''(0) = \left(\left(b_0 \frac{\partial}{\partial X_0} + b_1 \frac{\partial}{\partial X_1} + b_2 \frac{\partial}{\partial X_2} \right)^2 F \right) (a_0, a_1, a_2).$$

The expression

$$\left(b_0 \frac{\partial}{\partial X_0} + b_1 \frac{\partial}{\partial X_1} + b_2 \frac{\partial}{\partial X_2} \right)^2 F$$

is short term for the more elaborate

$$b_0^2 \frac{\partial^2 F}{\partial X_0^2} + b_1^2 \frac{\partial^2 F}{\partial X_1^2} + b_2^2 \frac{\partial^2 F}{\partial X_2^2}$$

$$+ 2b_0 b_1 \frac{\partial^2 F}{\partial X_0 \partial X_1} + 2b_0 b_2 \frac{\partial^2 F}{\partial X_0 \partial X_2} + 2b_1 b_2 \frac{\partial^2 F}{\partial X_1 \partial X_2}.$$

The point is that we have the general formula

$$\varphi^{(m)}(0) = \left(\left(b_0 \frac{\partial}{\partial X_0} + b_1 \frac{\partial}{\partial X_1} + b_2 \frac{\partial}{\partial X_2} \right)^i F \right) (a_0, a_1, a_2)$$

where the expression

$$\left(b_0\frac{\partial}{\partial X_0} + b_1\frac{\partial}{\partial X_1} + b_2\frac{\partial}{\partial X_2}\right)^m F$$

has a similar meaning as in the case $m = 2$: We multiply out the polynomial in D_0, D_1 and D_2,

$$\Delta_m(D_0, D_1, D_2) = (b_0 D_0 + b_1 D_1 + b_2 D_2)^m$$

and then replace the monomials

$$D_0^{j_0} D_1^{j_1} D_2^{j_2}$$

by

$$\frac{\partial^{j_0 + j_1 + j_2} F}{\partial X_0^{j_0}\partial X_1^{j_1}\partial X_2^{j_2}}.$$

It is a reasonably straightforward exercise to prove this by induction on the exponent m. We thus have the following formula:

$$\varphi(t) = F(a_0, a_1, a_2) + (D_{(b_0,b_1,b_2)}F)(a_0, a_1, a_2)t$$

$$+ \frac{1}{2}(D_{(b_0,b_1,b_2)}^2 F)(a_0, a_1, a_2)t^2 + \cdots + \frac{1}{i!}(D_{(b_0,b_1,b_2)}^i F)(a_0, a_1, a_2)t^i$$

$$+ \cdots + \frac{1}{d!}(D_{(b_0,b_1,b_2)}^d F)(a_0, a_1, a_2)t^d$$

where

$$D_{(b_0,b_1,b_2)} = b_0\frac{\partial}{\partial X_0} + b_1\frac{\partial}{\partial X_1} + b_2\frac{\partial}{\partial X_2}.$$

Actually we can give a precise formula for $(D_{(b_0,b_1,b_2)}^m F)(a_0, a_1, a_2)$. In fact, there is a generalization of the familiar *binomial formula*

$$(D_0 + D_1)^m = \sum \frac{m!}{i_0!i_1!}D_0^{i_0}D_1^{i_1}$$

where the sum runs over all non-negative i_0, i_1 such that $i_0 + i_1 = m$, to the case of any number of indeterminates $D_0, \ldots, D_r$. Indeed, we have the formula

$$(D_0 + D_1 + \cdots + D_r)^m = \sum \frac{m!}{i_0!i_1!\cdots i_r!}D_0^{i_0}D_1^{i_1}\cdots D_r^{i_r}$$

where the sum runs over all non-negative $i_0, i_1, \ldots, i_r$ such that $i_0 + \cdots + i_r = m$. We may prove this formula by induction by first noting that it holds for

$m = 0$ or 1. Then assuming it for $m - 1$ we need only verify the multiplication

$$\left(\sum_{i_1 + \cdots + i_r = m-1} \frac{(m-1)!}{i_0! i_1! \cdots i_r!} D_0^{i_0} D_1^{i_1} \cdots D_r^{i_r} \right) (D_0 + D_1 + \cdots + D_r)$$

$$= \sum_{i_1 + \cdots + i_r = m} \frac{m!}{i_0! i_1! \cdots i_r!} D_0^{i_0} D_1^{i_1} \cdots D_r^{i_r}$$

which we leave to the reader.

Using this *Multinomial Formula* we obtain the important identity, valid for any number of variables but stated here only for three:

$$(D_{(b_0, b_1, b_2)}^m F)(a_0, a_1, a_2)$$

$$= \sum b_0^{i_0} b_1^{i_1} b_2^{i_2} \frac{m!}{i_0! i_1! i_2!} \left(\frac{\partial^m}{\partial X_0^{i_0} \partial X_1^{i_1} \partial X_2^{i_2}} F \right)(a_0, a_1, a_2)$$

where the sum runs over all non-negative i_0, i_1, i_2 such that $i_0 + i_1 + i_2 = m$.

We have $F(a_0, a_1, a_2) = 0$, so the constant term of $\varphi(y)$ is zero for all choices of (b_0, b_1, b_2). It may happen that the coefficient of t vanishes as well, for all choices of (b_0, b_1, b_2), and so on, up to a certain t^m. We make the following definition:

Definition 3.4 (Multiplicity of Points on Projective Curves) The point $P = (a_0 : a_1 : a_2)$ on the projective curve C given by $F(X_0, X_1, X_2) = 0$ is said to be of multiplicity m if for all $n < m$ and all i_0, i_1, i_2

$$\left(\frac{\partial^n F}{\partial X_0^{i_0} \partial X_1^{i_1} \partial X_2^{i_2}} \right)(a_0, a_1, a_2) = 0$$

while for at least one choice of i_0, i_1, i_2

$$\left(\frac{\partial^m F}{\partial X_0^{i_0} \partial X_1^{i_1} \partial X_2^{i_2}} \right)(a_0, a_1, a_2) \neq 0.$$

This definition is independent of the projective coordinate system, the proof is straightforward but a little complicated. We omit it here. Clearly we have the following result:

Proposition 3.6 *The point $P = (a_0 : a_1 : a_2)$ on the projective curve C is of multiplicity 1 if and only if it is smooth.*

We also note the following:

Proposition 3.7 *The point $P = (a_0 : a_1 : a_2)$ on the projective curve C given by $F(X_0, X_1, X_2) = 0$ is of multiplicity m if and only if for all $n < m$ and all*

$(b_0, b_1, b_2) \notin C$

$$\sum_{i_0+i_1+i_2=n} b_0^{i_0} b_1^{i_1} b_2^{i_2} \frac{n!}{i_0! i_1! i_2!} \left(\frac{\partial^n F}{\partial X_0^{i_0} \partial X_1^{i_1} \partial X_2^{i_2}} \right) (a_0, a_1, a_2) = 0$$

while for at least one tuple $(b_0, b_1, b_2) \notin C$

$$\sum_{i_0+i_1+i_2=m} b_0^{i_0} b_1^{i_1} b_2^{i_2} \frac{m!}{i_0! i_1! i_2!} \left(\frac{\partial^m F}{\partial X_0^{i_0} \partial X_1^{i_1} \partial X_2^{i_2}} \right) (a_0, a_1, a_2) \neq 0.$$

This last proposition shows that a line L through the point P will intersect C at P with multiplicity at least equal to m, the multiplicity of the point P on C. And there exists at least one line through P which intersects C at P with multiplicity m.

This analysis of multiplicity goes through unchanged for any field of characteristic zero.

We now return to a general ground field k. We need the following result:

Proposition 3.8 *Let* $F(X_0, X_1, X_2)$ *be a homogeneous polynomial of degree* d. *Then the following identity holds:*

$$X_0 \frac{\partial F}{\partial X_0} + X_1 \frac{\partial F}{\partial X_1} + X_2 \frac{\partial F}{\partial X_2} = dF(X_0, X_1, X_2).$$

Proof We consider the following identity which holds for the variables $X_0, \ldots, X_n$ and t

$$F(tX_0, tX_1, \ldots, tX_n) = t^d F(X_0, X_1, \ldots, X_n).$$

This identity is verified by substituting tX_i for X_i in the polynomial, and observing that by definition all the monomials which occur are of the same degree d.

We now compute the derivative with respect to t. The right hand side yields

$$dt^{d-1} F(X_0, X_1, \ldots, X_n)$$

while the chain rule applied to the left hand side yields

$$\frac{\partial F}{\partial X_0}(tX_0, tX_1, tX_2)X_0 + \frac{\partial F}{\partial X_1}(tX_0, tX_1, tX_2)X_1 + \frac{\partial F}{\partial X_2}(tX_0, tX_1, tX_2)X_2.$$

These two polynomials are equal, and putting $t = 1$ we get the claimed formula.   □

Now assume that $P = (a_0 : a_1 : a_2) \in C$ has multiplicity m on C. We define the curve $T_{C,P}$, referred to as the *tangent cone* to C at the point P, by the equation

$$H_{C,P}(X_0, X_1, X_2)$$
$$= \sum_{i_0+i_1+i_2=m} X_0^{i_0} X_1^{i_1} X_2^{i_2} \frac{m!}{i_0! i_1! i_2!} \left(\frac{\partial^m F}{\partial X_0^{i_0} \partial X_1^{i_1} \partial X_2^{i_2}} \right) (a_0, a_1, a_2) = 0.$$

This equation actually does define a curve of degree m in $\mathbb{P}_k^2$, as it is non-zero and homogeneous of degree m. We have the following important result:

Theorem 3.9 $T_{C,P}$ *is the union of a finite number of lines through* P. *The irreducible components of* $T_{C,P}$ *are exactly the lines through* P *which intersect* C *with multiplicity* $> m$.

Proof We start out by noticing that $P \in T_{C,P}$. Indeed, we have the identity

$$d(d-1)\ldots(d-m+1)F(X_0, X_1, X_2)$$
$$= \sum_{i_0+i_1+i_2=m} X_0^{i_0} X_1^{i_1} X_2^{i_2} \frac{m!}{i_0! i_1! i_2!} \left(\frac{\partial^m F}{\partial X_0^{i_0} \partial X_1^{i_1} \partial X_2^{i_2}} \right) \overset{!}{=} 0$$

as is easily seen by repeated application of Proposition 3.8, first to F, then to the first order partial derivatives, so to the second order ones and so on, up to the partials of order m.

Since by definition of the multiplicity m there exists a point $(b_0 : b_1 : b_2)$ such that $H_{C,P}(b_0, b_1, b_2) \neq 0$, $T_{C,P}$ is a curve. Moreover, if $H_{C,P}(b_0, b_1, b_2) = 0$ then the line through P and $Q = (b_0, b_1, b_2)$ intersects C at P with multiplicity $> m$, since the corresponding $\varphi(t)$ has $t = 0$ as a root occurring with multiplicity $> m$. Thus any point $Q' = (b_0', b_1', b_2')$ on that line will also satisfy $H_{C,P}(b_0', b_1', b_2') = 0$. Thus the curve $T_{C,P}$ consists of lines passing through P. $\qquad\square$

Definition 3.5 The curve $T_{C,P}$ is referred to as the *projective tangent cone* to C at P.

We have earlier defined the concept of multiplicity and tangent cone in the affine case. These concepts are completely compatible under affine restriction and under projective closure:

Proposition 3.10 *Let* K *be an affine curve, and* $p = (a, b) \in K$. *Let* C *be the projective closure, and put* $P = (1 : a : b) \in C$. *Then the point* p, *as a point on an affine curve, is of the same multiplicity as the point* P *on the projective curve* C. *Moreover the affine restriction of* $T_{C,P}$ *is equal to the affine tangent cone of* K *at* p, *as given in Definition 2.9.*

Proof We may change the projective coordinate system to one in which $P = (1:0:0)$, this corresponds to a change of affine coordinate system to one in which $(a,b) = (0,0)$. In this case the claim is easily checked. □

The irreducible components of the curve $T_{C,P}$ are referred to as the *lines of tangency* to C at P. If the point $P \in C$ is smooth, then $m = 1$ and there is only one line of tangency, which we refer to as the *tangent line to C at P*, and denote as before by $T_{C,P}$. The equation is

$$X_0 \frac{\partial F}{\partial X_0}(a_0, a_1, a_2) + X_1 \frac{\partial F}{\partial X_1}(a_0, a_1, a_2) + X_2 \frac{\partial F}{\partial X_2}(a_0, a_1, a_2) = 0.$$

We finally compute an example. Consider the projective curve given by

$$F(X_0, X_1, X_2) = X_0 X_2^2 - X_1^3 - X_0 X_1^2 = 0$$

which is the projective closure of the affine curve defined by

$$y^2 - x^3 - x^2 = 0$$

in other words, the nodal cubic curve. To find the projective tangent cone at the point $(1:0:0)$, it is most convenient to pass to the affine restriction. Then we immediately see that the affine tangent cone is defined by

$$y^2 - x^2 = 0$$

which has the projective closure given by

$$X_2^2 - X_1^2 = 0.$$

This is the fastest way to proceed. But we could also use the definition directly. Then we compute

$$\frac{\partial F}{\partial X_0} = X_2^2 - X_1^2, \qquad \frac{\partial F}{\partial X_1} = 3X_1^2 - 2X_0 X_1, \qquad \frac{\partial F}{\partial X_2} = 2X_0 X_2.$$

They all evaluate to zero at $(1:0:0)$, so this point is singular (as we know from the affine restriction). We differentiate again, and obtain

$$\frac{\partial^2 F}{\partial X_0^2} = 0, \qquad \frac{\partial^2 F}{\partial X_1^2} = 6X_1 - 2X_0, \qquad \frac{\partial^2 F}{\partial X_2^2} = 2X_0$$

$$\frac{\partial^2 F}{\partial X_0 \partial X_1} = -2X_1, \qquad \frac{\partial^2 F}{\partial X_0 \partial X_2} = 2X_2, \qquad \frac{\partial^2 F}{\partial X_1 \partial X_2} = 0.$$

Evaluating at P, we get

$$\frac{\partial^2 F}{\partial X_0^2} = 0, \qquad \frac{\partial^2 F}{\partial X_1^2} = -2, \qquad \frac{\partial^2 F}{\partial X_2^2} = 2$$

$$\frac{\partial^2 F}{\partial X_0 \partial X_1} = 0, \qquad \frac{\partial^2 F}{\partial X_0 \partial X_2} = 0, \qquad \frac{\partial^2 F}{\partial X_1 \partial X_2} = 0$$

and thus according to our formula the equation for $T_{C,P}$ is

$$\frac{2!}{0!2!0!}(-2)X_1^2 + \frac{2!}{0!0!2!}2X_2^2 = 0$$

confirming what we found above.

3.5 Projective Equivalence

Definition 3.6 Two irreducible projective curves C and C' are called projectively equivalent if C is mapped to C' by a projective transformation. Two irreducible affine curves K and K' are said to be projectively equivalent if their projective closures are projectively equivalent, and an affine curve K is projectively equivalent to its projective closure C.

Remark Frequently we replace the sentence C *is mapped to C' by a projective transformation* by C *becomes C' by a projective change of coordinate system.* This way of expressing the equivalence is perfectly legitimate when we think of the curves as given by explicit equations.

Thus for example the affine version of the nodal cubic

$$y^2 - x^3 - x^2 = 0$$

is projectively equivalent to the projective version defined by

$$X_0 X_2^2 - X_1^3 - X_0 X_1^2 = 0.$$

However, note that *we do not assert* that its affine tangent cone at the origin

$$y^2 - x^2 = 0$$

is projectively equivalent to the projective tangent cone at $(1:0:0)$,

$$X_2^2 - X_1^2 = 0$$

as we use this terminology for irreducible curves only.

The first case to study in light of the above definition would be the non-degenerate conics. This has been completely clearified by Theorem 2.5 in Sect. 2.3.

There we found that up to projective equivalence there is only one non-degenerate conic curve in $\mathbb{P}_k^2$, provided that k is algebraically closed. For $k = \mathbb{R}$, where the hypothesis of being algebraically closed does not hold, we still have the same result for non-degenerate conics with more than one real point, the details may be found in [27], Sect. 14.5 or [28], Sect. 15.5.

When $k = \mathbb{R}$ there is a very close connection between *projective transformations* on one hand and actual *projections* on the other. In fact, it is fair to say that these two concepts are practically equivalent. This is also given a detailed explanation in [27] and [28].

We conclude this section with an examination of two classes of curves studied in Sect. 2.4. A semi-cubic parabola may, by a change of affine coordinate system, be brought on the form

$$x^3 - y^2 = 0.$$

The usual projective closure of this curve is given by

$$X_0 X_2^2 - X_1^3 = 0$$

in $\mathbb{P}^2_{\mathbb{R}}$. Taking the affine restriction to $D_+(x_2)$ and letting $\overline{y} = \frac{X_0}{X_2}$, $\overline{x} = \frac{X_1}{X_2}$, we get the equation

$$\overline{y} - \overline{x}^3 = 0$$

which is a cubic parabola. Thus cubic and semi-cubic parabolas are projectively equivalent.

3.6 Asymptotes

We may now give a simple treatment of a subject which often appears rather mysterious. We take $k = \mathbb{R}$, and recall that an *asymptote* to a given curve is defined as a line such that the distance from a point on the curve to the line tends to zero as the point on the curve moves further and further away from the origin.

This definition renders it quite mysterious how to actually compute all asymptotes to a given curve. Another drawback is that it defines the concept in terms of *distance*, thus the concept defined in this way is not an algebraic one.

The following definition is equivalent to the one given above for algebraic affine curves in $\mathbb{A}^2_{\mathbb{R}}$:

Definition 3.7 Let K be the affine curve defined by

$$f(x, y) = 0.$$

Let C be the projective closure in $\mathbb{P}^2_{\mathbb{R}}$ obtained by letting $x = \frac{X_1}{X_0}$, $y = \frac{X_2}{X_0}$ as usual. Let $P_1, \ldots, P_m$ be the points at infinity of C, and let $L_1, \ldots, L_r$ be all lines in $\mathbb{P}^2_{\mathbb{R}}$ different from $V_+(X_0)$ and appearing as a line of tangency to C at one of the points $P_1, \ldots, P_m$. Let $\ell_1, \ldots, \ell_r$ be the affine restrictions of $L_1, \ldots, L_r$. Then $\ell_1, \ldots, \ell_r$ are all the asymptotes of K in $\mathbb{A}^2_{\mathbb{R}}$.

The Trisectrix of Maclaurin has equation $x^3 + xy^2 + y^2 - 3x^2 = 0$. It was treated in Sect. 2.6, and its appearance makes one wonder if it might have a vertical asymptote, crossing the x-axis somewhere to the left of the origin. We shall now check this.

The projective closure of the trisectrix is given by the equation $X_1^3 + X_1X_2^2 + X_0X_2^2 - 3X_0X_1^2 = 0$. We find the points at infinity by substituting $X_0 = 0$ into this equation, we get $X_1^3 + X_1X_2^2 = 0$. This yields one real point, given by $X_1 = 0$, and two complex points determined by $X_1^2 + X_2^2 = 0$, which do not concern us as we are dealing with the real points only. Thus the one (real) point at infinity is $(0 : 0 : 1)$. We now take the affine restriction to $D_+(X_2)$ by putting $x' = \frac{X_0}{X_2}$ and $y' = \frac{X_1}{X_2}$. This affine restriction is given by $y'^3 + y' + x' - 3x'y'^2 = 0$. Hence the origin is a smooth point, the tangent there is given by $x' + y' = 0$. Going back to the projective plane, this line has the equation $X_0 + X_1 = 0$, and taking the affine restriction to the original affine xy-plane, we get the equation $x = -1$: This, then, is the asymptote of the curve, affirming our suspicion that such a line might exist.

An even simpler example, but an important one, is to verify the asymptotes of a general hyperbola. Assume it is given on standard form, as

$$\left(\frac{x}{a}\right)^2 - \left(\frac{y}{b}\right)^2 = 1.$$

To show is that the asymptotes are given by

$$\left(\frac{x}{a}\right)^2 - \left(\frac{y}{b}\right)^2 = 0.$$

We leave this verification as an exercise.

3.7 General Conchoids

The *Conchoid of Nicomedes* (280–210 B.C.) is the curve generated by taking a fixed point at a given distance a from a given line, and and fixing a certain distance b. A variable line through P intersects the fixed line in a variable point Q as it rotates, and from this point Q the distance b is marked off on the rotating line, yielding the point R. R then generates a curve, called the Conchoid of Nicomedes, as shown in the figure.

We choose the coordinate system with origin at the fixed point P, y-axis parallel to ℓ and x-axis normal to ℓ as shown in Fig. 3.1. A line through P has the equation $y = tx$, and a point $R = (x, y)$ on it, at distance b from its intersection with ℓ must satisfy

$$y = tx$$
$$(x - a)^2 + (y - ta)^2 = b^2$$

Fig. 3.1 The Conchoid of
Nicomedes with
$b = 3 > a = 1$

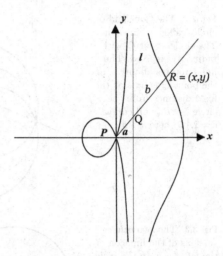

which when $t = \frac{y}{x}$ from the former is substituted in the latter, yields

$$(x - a)^2 x^2 + y^2(x - a)^2 = b^2 x^2$$

or

$$(x - a)^2(x^2 + y^2) = b^2 x^2.$$

The line ℓ is an asymptote for the conchoid. This discovery is attributed
to Nicomedes himself. We check the result with our method for finding all
asymptotes. The projective closure of the conchoid is given by

$$(X_1 - aX_0)^2(X_1^2 + X_2^2) - b^2 X_0^2 X_1^2 = 0.$$

The points at infinity are determined by

$$X_1^2(X_1^2 + X_2^2) = 0$$

thus as we only consider real points, the only point at infinity is $(0 : 0 : 1)$. We
now take the affine restriction to $D_+(X_2)$ by letting $x' = \frac{X_0}{X_2}$ and $y' = \frac{X_1}{X_2}$.
The equation in the $x'y'$-plane becomes

$$(y' - ax')^2(y'^2 + 1) - b^2 x'^2 y'^2 = 0.$$

The homogeneous part of lowest degree of this polynomial is $H(x', y') = (y' - ax')^2$, so the tangent cone at the point $(0, 0)$ is the line $y' = ax'$, with
multiplicity 2. Taking the projective closure again we get the projective line
$X_1 - aX_0 = 0$, and its affine restriction to our original affine plane $\mathbb{A}^2_{\mathbb{R}} = D_+(X_0)$ is $x = a$. Thus we have proved that the line $x = a$ is an asymptote
of the conchoid.

The Conchoid of Nicomedes is a special case of a general class of curves.
We make the following

Fig. 3.2 The Conchoid of a circle, called a Limaçon. The circle about C with radius AC is fixed. The line ℓ rotates about A, and the point Q is on ℓ at the fixed distance from the circle $b = PQ$. Of course $d = 2AC$, here $d > b$

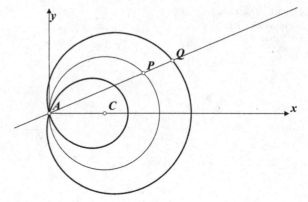

Fig. 3.3 The two other versions of the limaçon. To the *left* $d = b$, to the *right* $d < b$

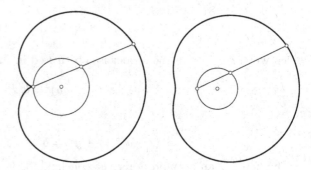

Definition 3.8 Given an affine curve K and a fixed point P. Consider the collection of all lines through P. The conchoid of K for the pole P and constant b is then the locus of all points Q such that Q lies on one of these lines at a distance b from its intersection with K.

A conchoid of a circle for a fixed point on it is called a *Limaçon of Pascal*, the first part of the name picked by Étienne Pascal, the name meaning *snail* in French. When b is equal to the diameter d of the circle, the curve is called the *cardioid*, in other words the *heart curve*, and if the constant b is equal to the radius of the circle, we get a curve which may be used to trisect an angle in equal parts, often referred to as a trisectrix. We shall now analyze the different cases. Depending on the relation between b and d we get three versions of the Limaçon, one being shown in Fig. 3.2, the two others in Fig. 3.3.

These curves are simple to describe in polar coordinates as $r = b + d\cos(\varphi)$, where d is the diameter of the circle and b is the constant, see Fig. 3.2.

A simple computation yields

$$x^2 + y^2 - dx = db\cos(\varphi) + b^2$$

and hence

$$(x^2 + y^2 - dx)^2 = d^2 b^2 \cos^2(\varphi) + 2db^3\cos(\varphi) + b^4 = b^2(x^2 + y^2).$$

Thus we obtain the somewhat less transparent form in usual xy-coordinates

$$(x^2 + y^2 - dx)^2 - b^2(x^2 + y^2) = 0.$$

3.8 The Dual Curve

In projective algebraic geometry the *principle of duality* acquires a very precise meaning. For $\mathbb{P}_k^2$ we have the following:

Every line in $\mathbb{A}_k^2$ is given by an equation

$$a_0 X_0 + a_1 X_1 + a_2 X_2 = 0.$$

If we multiply each coefficient by a common non-zero constant, then we get the same equation. Thus the line may be associated with a uniquely determined point of another copy of the projective plane,

$$L^\vee = (a_0 : a_1 : a_2) \in \mathbb{P}_k^2.$$

Conversely, to a point $P \in \mathbb{P}_k^2$ we may associate a *line* $P^\vee \subset \mathbb{P}_k^2$. The correspondence $(\ )^\vee$ preserves *incidence*, as already explained in Sect. 1.5.

We now extend this to *projective, algebraic curves*. We get the following concept of duality: For any projective curve $C \subset \mathbb{P}_k^2$, consider the subset

$$C^\vee = \{(L)^\vee \,|\, L \text{ is a line of tangency to } C\}.$$

We denote this set by $C^\vee$, and refer to it as the *dual curve of* C. Indeed, it turns out that this subset of $\mathbb{P}_k^2$ is actually a projective curve, in $\mathbb{P}_k^2$, except for the case when C is a projective line, in which case $C^\vee$ consists of just one point.

Assume that C has the equation $F(X_0, X_1, X_2) = 0$. The equation for the dual curve is then expressed in terms of the indeterminates Y_0, Y_1 and Y_2 when we eliminate X_0, X_1, X_2 in the system

$$\frac{\partial F}{\partial X_0}(X_0, X_1, X_2) = Y_0$$

$$\frac{\partial F}{\partial X_1}(X_0, X_1, X_2) = Y_1$$

$$\frac{\partial F}{\partial X_2}(X_0, X_1, X_2) = Y_2$$

$$F(X_0, X_1, X_2) = 0.$$

Here we have six variables X_0, X_1, X_2 and Y_0, Y_1, Y_2, and four relations among them. In general we may then *eliminate* any three of them, and obtain *one relation* between the remaining variables. We now eliminate X_0, X_1 and X_2. This will give as result a single equation

$$G(Y_0, Y_1, Y_2) = 0$$

which defines the dual curve $C^\vee$.

We may think of what we are doing here in the following way: We write

$$\frac{\partial F}{\partial X_0}(X_0, X_1, X_2) = Y_0(X_0, X_1, X_2)$$

$$\frac{\partial F}{\partial X_1}(X_0, X_1, X_2) = Y_1(X_0, X_1, X_2)$$

$$\frac{\partial F}{\partial X_2}(X_0, X_1, X_2) = Y_2(X_0, X_1, X_2).$$

We then *solve this system of equations* for X_0, X_1 and X_2:

$$X_0 = X_0(Y_0, Y_1, Y_2)$$

$$X_1 = X_1(Y_0, Y_1, Y_2)$$

$$X_2 = X_2(Y_0, Y_1, Y_2)$$

and then get

$$G(Y_0, Y_1, Y_2) = F(X_0(Y_0, Y_1, Y_2), X_1(Y_0, Y_1, Y_2), X_2(Y_0, Y_1, Y_2)).$$

Of course, usually we are not able to find X_0, X_1 and X_2 as homogeneous polynomials in the Y's, not even as single valued functions. To put these considerations on a mathematically sound basis, it was necessary to develop the machinery of *elimination theory*. But we shall bypass this, and work for a while with such fictitious entities as the X_0, X_1 and X_2 as functions of Y_0, Y_1 and Y_2. In the end they are gone, and only the $G(Y_0, Y_1, Y_2)$, which does exist thanks to elimination theory, remains. But in some happy cases the X_0, X_1 and X_2 *do exist* as homogeneous polynomials in Y_0, Y_1 and Y_2, and then they simplify the situation considerably.

Since questions of tangency are independent of projective coordinate system, the same is true for questions of duality. We shall use this important observation in proving the following result:

Theorem 3.11 *The dual curve of a non-degenerate conic curve in $\mathbb{P}^2_{\mathbb{R}}$ is again a non-degenerate conic curve.*

Proof We showed in Theorem 2.5 that any non-degenerate conic curve is projectively equivalent to the one given by

$$X_1^2 + X_2^2 - X_0^2 = 0$$

thus we may assume that $F(X_0, X_1, X_2) = X_1^2 + X_2^2 - X_0^2$. Then

$$\frac{\partial F}{\partial X_0} = -2X_0$$

$$\frac{\partial F}{\partial X_0} = -2X_0$$

$$\frac{\partial F}{\partial X_1} = 2X_1$$

$$\frac{\partial F}{\partial X_2} = 2X_2.$$

So we have to eliminate X_0, X_1, X_2 in

$$-2X_0 = Y_0$$
$$-2X_0 = Y_0$$
$$2X_1 = Y_1$$
$$2X_2 = Y_2$$
$$X_1^2 + X_2^2 - X_0^2 = 0.$$

In this case we may solve for X_0, X_1 and X_2, which yields

$$Y_1^2 + Y_2^2 - Y_0^2 = 0$$

which is a non-degenerate conic curve. $\square$

Clearly the proof shows more than stated, in fact the theorem holds over any field k of characteristic $\neq 2$.

As we see, the equation is essentially the same as the one we started with. Just looking at this one example, one might be tempted to draw the conclusion that $C = C^\vee$. But this is far from true, even for general conics. The point is that the property of *having a non-degenerate conic curve as dual* is independent of the coordinate system, while the property of *having a dual which is defined by a fixed homogeneous polynomial* certainly very much depends on the coordinate system. However, we have the following important theorem, which is true in much greater generality than the version we give here:

Theorem 3.12 *Let k be a field of characteristic 0, and let C be a curve given by an irreducible homogeneous polynomial in $\mathbb{P}_k^2$. Then the dual of the dual of C equals C.*

Proof We use the simplified form given above, avoiding elimination theory. We also carry out the proof for $k = \mathbb{R}$, the reader may then easily see that the argument applies to the general case stated in the theorem.

G is really nothing but the original F, but expressed in terms of the new, dual, variables Y_0, Y_1, Y_2 instead of the original X_0, X_1, X_2. We therefore only have to prove that if we put

$$\overline{X}_0 = \frac{\partial F}{\partial Y_0}$$

$$\overline{X}_1 = \frac{\partial F}{\partial Y_1}$$

$$\overline{X}_2 = \frac{\partial F}{\partial Y_2}$$

then

$$\overline{X}_0 = \alpha X_0, \quad \overline{X}_1 = \alpha X_1 \quad \text{and} \quad \overline{X}_2 = \alpha X_2$$

for some real number $\alpha \neq 0$. This is done as follows: By Proposition 3.8, we have that

$$X_0 Y_0 + X_1 Y_1 + X_2 Y_2 = dF(X_0, X_1, X_2)$$

where d is the degree of F in X_0, X_1, X_2. Hence

$$\overline{X}_i = \frac{1}{d} \frac{\partial}{\partial Y_i} (X_0 Y_0 + X_1 Y_1 + X_2 Y_2) = X_i$$

for $i = 0$, 1 and 2. Thus the claim follows. $\qquad\qquad\square$

Chapter 4
Plane Curves and Algebra

The affine and homogeneous coordinate rings are introduced, as well as multiplicity of points and intersection multiplicity. A complete treatment of intersection multiplicity for curves in the projective plane is given here. This treatment goes over with minor adaptations to an intersection theory for curves on a smooth projective surface. Also treated in this chapter are a rudimentary start on linear systems of curves, Bézout's theorem, simple elimination theory with application to the twisted cubic curve, points of inflexion and the Hessian.

4.1 Affine and Homogeneous Coordinate Rings

Let K be an affine curve in $\mathbb{A}^2_k$, given by the polynomial $f(X,Y)$ with coefficients from k. Then the quotient ring of $k[X,Y]$ modulo the ideal generated by $f(X,Y)$, $k[X,Y]/(f)k[X,Y] = k[x,y]$, is referred to as the *affine coordinate ring* of the curve K, and denoted by $\Gamma(K)$. Similarly, if C is a projective curve in $\mathbb{P}^2_k$ given by the homogeneous polynomial $F(X_0,X_1,X_2)$ then the ring $k[X_0,X_1,X_2]/(F)k[X_0,X_1,X_2] = k[x_0,x_1,x_2]$ is denoted by $\Gamma_+(C)$ and referred to as *the homogeneous coordinate ring* of C.

The geometry of the affine or projective curves and the algebra of these corresponding coordinate rings are intimately connected. The fundament of modern algebraic geometry, namely *Alexander Grothendieck's* theory of schemes, is an immense generalization of this simple starting point for plane curves. We shall now explain some of the algebra of plane curves in more detail.

If A is a commutative ring with 1, we let $\text{Max}(A)$ denote the set of all maximal ideals in A. We start with the following important fact, which ties the points on K to the maximal ideals in the coordinate ring:

Theorem 4.1 1. *If $P = (a,b)$ is a point on the affine curve K, then it corresponds to a maximal ideal in the coordinate ring $\Gamma(K)$ of K in the following*

A. Holme, *A Royal Road to Algebraic Geometry*,
DOI 10.1007/978-3-642-19225-8_4, © Springer-Verlag Berlin Heidelberg 2012

manner:

$$K \longrightarrow \operatorname{Max}(\Gamma(K)) \quad by \ P = (a,b) \mapsto \mathfrak{m}(a,b) = (x - a, y - b)\Gamma(K).$$

This map is injective.

2. *If k is algebraically closed, then the map is bijective.*

Proof 1. The k-algebra homomorphism

$$\varphi_{(a,b)} : k[X,Y] \longrightarrow k \quad \text{given by } X \mapsto a, Y \mapsto b$$

sends a polynomial $p(X,Y)$ with coefficients from k to $p(a,b)$, thus $f(X,Y)$ is mapped to 0 and hence the ideal generated by $f(X,Y)$ is contained in $\ker(\varphi_{(a,b)})$. Therefore $\varphi_{(a,b)}$ induces a surjective homomorphism $\overline{\varphi}_{(a,b)} : \Gamma(K) \longrightarrow k$ whose kernel is $\mathfrak{m}(a,b)$, it induces an isomorphism $\Gamma(K)/\mathfrak{m}(a,b) \cong k$, and since k is a field $\mathfrak{m}(a,b)$ is a maximal ideal.

2. If $\mathfrak{m}$ is a maximal ideal in $\Gamma(K)$, then $L = \Gamma(K)/\mathfrak{m}$ is a field containing k. L is also finitely generated as a *ring* over k, which is a strong condition. In fact, it implies that L is finite as a *k-vector space*, as follows from the theorem which we state below. But then for all $\alpha \in L$ there exists m such that the elements $1, \alpha, \ldots, \alpha^m$ are linearly dependent over k, thus there exists a polynomial Q with coefficients from k such that $Q(\alpha) = 0$. Thus L is an algebraic extension of k, but as k is algebraically closed by assumption, $k = L$. Now let a be the image of x in $L = k$, and b be the image of y. Since $x - a$ and $y - b$ both lie in $\mathfrak{m}$ and generate a maximal ideal by 1. above, they must generate $\mathfrak{m}$. This completes the proof, modulo our next theorem. $\square$

Theorem 4.2 *Let L be a field which is finitely generated as a ring over the subfield k. Then L is a finite dimensional vector space over k.*

Proof Assume that L is generated by n elements as a ring over k. We claim that L is then a finite vector space over k. For $n = 1$, $L = k[x_1]$ and since this is a field, $\frac{1}{x_1} \in k[x_1]$ and thus

$$\frac{1}{x_1} = a_0 + a_1 x_1 + a_2 x_1^2 + \cdots + a_m x_1^m,$$

hence

$$-1 + a_0 x_1 + a_1 x_1^2 + a_2 x_1^3 + \cdots + a_m x_1^{m+1} = 0,$$

so x_1^{m+1} is a linear combination in $1, x_1, \ldots, x_1^m$ with coefficients from k. Similarly x_1^{m+2} may be expressed as such a linear combination, and so on. Thus $k[x_1]$ is generated as a k-vector space by $1, x_1, \ldots, x_1^m$.

Assume that $n \geq 2$ and that the claim is true for all fields generated as a ring over a base field k by $n - 1$ or fewer elements. Let $L = k[x_1, x_2, \ldots, x_n]$, we wish to show that the induction assumption implies that L is a finite vector space over k.

For this it suffices to show that the subring $k[x_1]$ of L is actually a *subfield*. Indeed, suppose we have shown this, and put $k' = k[x_1]$. Then $L = k'[x_2, \ldots, x_n]$, and thus by the induction assumption L is a finite vector space over the field k'. But k' is a finite dimensional vector space over k, by the case $n = 1$ which has already been proven above.

To show that $k[x_1]$ is a field, let $k[X]$ be the polynomial ring in one variable over k, and consider the homomorphism

$$f : k[X] \longrightarrow L \quad \text{where } X \mapsto x_1.$$

The ideal $\mathfrak{p} = \ker(f)$ in $k[X]$ is then prime, and $k[X]/\mathfrak{p} \cong k' = k[x_1]$. To show that k' is a field, we must show that $\mathfrak{p}$ is a maximal ideal. But if it is not maximal, then it must be the zero ideal, hence $x_1 \in L$ is transcendental over k. The proof will be completed by showing that this leads to a contradiction.

By the induction assumption the field $k(x_1)$ still has the property that L is a finite vector space over it. Thus for all $i = 2, \ldots, n$ we have relations

$$x_i^{m_i} + a_{i,1} x_i^{m_i - 1} + \cdots + a_{i,m_i} = 0$$

where $a_{i,j} \in k(x_1)$. We may assume that $a_{i,j} = \frac{b_{i,j}}{b}$, where b and all $b_{i,j}$ are in $k[x_1]$. Thus we obtain

$$b x_i^{m_i} + b_{i,1} x_i^{m_i - 1} + \cdots + b_{i,m_i} = 0,$$

and multiplying with the appropriate power of b we get $c_{i,j} \in k[x_1]$ such that

$$(b x_i)^{m_i} + c_{i,1} (b x_i)^{m_i - 1} + \cdots + c_{i,m_i} = 0,$$

in other words $b x_i$ is integral over $k[x_1]$ for all i. Now recall that the set of all elements in L which are integral over $k[x_1]$ is a subring of L containing $k[x_1]$, referred to as the *integral closure* of $k[x_1]$ in L. Thus a polynomial in $b x_2, \ldots, b x_m$ with coefficients from $k[x_1]$ is also integral over $k[x_1]$. Now let $z \in L$, it may be written as $z = p(x_1, \ldots, x_n)$ where $p(X_1, \ldots, X_n)$ is a polynomial with coefficients from k. Then $b^N z$ may be written as a polynomial in $b x_1, \ldots, b x_n$ for a sufficiently large integer N. Thus we have shown that $b^N z$ is integral over $k[x_1]$ for some integer N. In particular this is true for all $z \in k(x_1)$.

At this point we use the assumption that x_1 is transcendental over k. So $k[x_1]$ is integrally closed in its field of fractions. Thus if $b^N z$ is integral over $k[x_1]$, then it lies in $k[x_1]$. Hence the proof is reduced to showing that the following assertion is *false*:

There exist a polynomial $\zeta(X)$ with coefficients from k such that for all rational expressions $R(X) = \frac{P(X)}{Q(X)}$ there exist a natural number N such that $\zeta(X)^N R(X) \in k[X]$.

But this assertion is obviously false since $k[X]$ is a unique factorization domain. Indeed, we factor our polynomial $\zeta(X)$ in irreducible polynomials

$$\zeta(X) = p_1(X)^{r_1} \cdots p_s(X)^{r_s}$$

and let $q(X)$ be an irreducible polynomial not a constant multiple of any of these factors. Then the rational expression

$$\frac{\zeta(X)^N}{q(X)} = \frac{p_1(X)^{Nr_1} \cdots p_s(X)^{Nr_s}}{q(X)}.$$

cannot be simplified to a polynomial for any N.   □

We now have arrived at the stage where it is necessary to sharpen some of the concepts we have been working with, in a rather naive way so far. We have seen that a non-constant polynomial $P(X,Y)$ defines an *affine plane curve*. Two polynomials P and Q define the same curve if there is a non-zero element $a \in k$ such that $P = aQ$. (a,b) is a point on the curve K given by the polynomial $p(X,Y)$ if $p(a,b) = 0$. Two different curves may have the same points, as we have seen for the curves given by $P(X,Y) = Y$ and $Q(X,Y) = Y^2$, the former defining the x-axis and the latter the x-axis with multiplicity 2. Moreover, the need to consider complex points on a real curve strains even further the naive definition of a curve as the graph of an equation $P(X,Y) = 0$. Thus the curves given in $\mathbb{A}^2_{\mathbb{R}}$ by $Q(X,Y) = X^2 + Y^2 + 1$ and $Q(X,Y) = X^4 + Y^4 + 1$ should be viewed as two different *real curves*, even though they have no (real) points at all. We now make the following formal definition:

Definition 4.1 A curve in $\mathbb{A}^2_k$ is an equivalence class of polynomials with coefficients from k under the equivalence relation that

$$P(X,Y) \sim Q(X,Y) \quad \Longleftrightarrow \quad \exists a \in k^* \quad \text{such that} \quad P(X,Y) = aQ(X,Y).$$

The curve $K = [P(X,Y)]$, the equivalence class of the polynomial $P(X,Y)$, is said to "have the equation $P(X,Y) = 0$", and if k' is a field containing k, and $a, b \in k'$, then (a,b) is called a k'-point on the curve K if $P(a,b) = 0$. The set of all k'-points on K is denoted by $K(k')$. When

$$P(X,Y) = P_1(X,Y)^{r_1} \cdots P_s(X,Y)^{r_s}$$

is the factorization as a product of irreducible polynomials, the curves K_i given by $P_i(X,Y)$ for $i = 1, \ldots, s$ are called the irreducible components of K, and the exponents are referred to as the multiplicities of the components. A curve with only one component occurring with multiplicity one is said to be irreducible. If the curve is still irreducible when considered as a curve in $\mathbb{A}^2_k(\overline{k})$, where $\overline{k}$ denotes the algebraic closure of k, then it is said to be geometrically irreducible.

We note that

$$K(k') = K_1(k') \cup \cdots \cup K_s(k')$$

for all fields k' containing k. Moreover, we note that Theorem 4.1 immediately implies the following:

Proposition-Definition 4.3 *Let $\overline{k}$ be the algebraic closure of k, and let K be an affine curve over k. Then there is a bijective correspondence between* $\mathrm{Max}(\Gamma(K))$ *and* $K(\overline{k})$. $K(\overline{k})$ *is referred to as the set of geometric points of K.*

Assume that K is irreducible, and define the field $k(K)$ as the field of quotients of the affine coordinate ring, so $k(K) = k(x,y)$. Then (x,y) is a $k(K)$-point of K, this point is called *a generic point* on the curve K.

Now let $P = (a,b)$ be a k-point on the affine curve K, given by the polynomial $F(X,Y) \in k[X,Y]$. The *local ring* of K at P is the ring

$$\mathcal{O}_{K,P} = \Gamma(K)_{\mathfrak{m}(a,b)}.$$

More generally, if P is a $\overline{k}$-point, then by Theorem 4.1 it corresponds to a maximal ideal $\mathfrak{m}$ in $\Gamma(K)$, and the local ring is defined as above with this ideal instead of $\mathfrak{m}(a,b)$.

We now turn to a projective curve $C \subset \mathbb{P}^2_k$ and its homogeneous coordinate ring $\Gamma_+(C)$. We shall explain how to derive the affine coordinate ring of any affine restriction of C, and start out by considering the standard restriction to $\mathbb{A}^2_k = D_+(X_0)$.

The homogeneous coordinate ring is defined as

$$\Gamma_+(C) = k[X_0, X_1, X_2]/(F(X_0, X_1, X_2))k[X_0, X_1, X_2]$$

where $F(X_0, X_1, X_2)$ is the homogeneous polynomial defining C. Since this homogeneous polynomial generates a homogeneous ideal, the quotient is itself a graded k-algebra:

$$\Gamma_+(C) = \bigoplus_{n=0}^{\infty} \Gamma_+(C)_n$$

where $\Gamma_+(C)_n$ denotes the elements of degree n. We now localize the ring $\Gamma_+(C)$ in the multiplicative subset

$$S = \{x_0^n \mid n = 0, 1, 2, \dots\}$$

and obtain

$$\Gamma_+(C)_{x_0} = \left\{ \frac{P(x_0, x_1, x_2)}{x_0^n} \,\middle|\, \begin{matrix} n = \\ 0, 1, 2, \dots \end{matrix} \right\} = \bigoplus_{n=-\infty}^{\infty} (\Gamma_+(C)_{x_0})_n.$$

In particular we put

$$\Gamma_+(C)_{(x_0)} = (\Gamma_+(C)_{x_0})_0,$$

and note that we have

$$(\Gamma_+(C)_{x_0})_0 = k\left[\frac{x_1}{x_0}, \frac{x_2}{x_0}\right].$$

Replacing $\frac{x_1}{x_0}, \frac{x_2}{x_0}$ with x, y, we thus obtain the affine coordinate ring of the affine restriction.

More generally, let $L = a_0X_0 + a_1X_1 + a_2X_2$, we wish to find the affine coordinate ring of the affine restriction to $D_+(L)$. We proceed as before, and get

$$\Gamma_+(C)_L = \left\{\frac{P(x_0, x_1, x_2)}{L^n} \middle| \begin{array}{l} n = \\ 0, 1, 2, \ldots \end{array}\right\} = \bigoplus_{n=-\infty}^{\infty} (\Gamma_+(C)_L)_n.$$

In particular we put

$$\Gamma_+(C)_{(L)} = (\Gamma_+(C)_L)_0,$$

and note that if $a_0 \neq 0$, then we have

$$(\Gamma_+(C)_L)_0 = k\left[\frac{x_1}{L}, \frac{x_2}{L}\right].$$

Replacing $\frac{x_1}{L}, \frac{x_2}{L}$ with x, y, we have the affine coordinate ring of the affine restriction to $D_+(L)$.

The relation between points and ideals in the affine case is simpler than what we get in the projective case: The points in $\mathbb{P}^2_k$ correspond to *homogeneous prime ideals* of height 1 in $k[X_0, X_1, X_2]$, so these ideals are not maximal. Indeed, the only homogeneous maximal ideal in this ring is $(X_0, X_1, X_2)k[X_0, X_1, X_2]$, which is often referred to as the *irrelevant ideal*. The situation is similar for $\Gamma_+(C)$, the only homogeneous maximal ideal in this ring is the irrelevant ideal generated by x_0, x_2, x_2. The point $P = (a_0 : a_1 : a_2) \in \mathbb{P}^2_k$ corresponds to the homogeneous prime ideal

$$\mathfrak{p}(P) = (\{a_ix_j - a_jx_i \mid 0 \leq i, j \leq 2\})\Gamma(C).$$

If k is algebraically closed, then the set of homogeneous prime ideals of height 1 in $k[X_0, X_1, X_2]$, respectively in $\Gamma(K)$, may be identified with the set of points in $\mathbb{P}^2_k$, respectively on C.

If $\mathfrak{p}$ is a homogeneous prime of height 1 in $\Gamma_+(C)$, then the localization $\Gamma_+(C)_\mathfrak{p}$ is graded by $\mathbb{Z}$,

$$\Gamma_+(C)_\mathfrak{p} = \sum_{i \in \mathbb{Z}} (\Gamma_+(C)_\mathfrak{p})_i$$

and we find the local ring of C by taking the homogeneous part of degree 0:

Proposition 4.4 *The local ring of the projective curve C at the point $P = (a_0 : a_1 : a_2)$ is given by*

$$\mathcal{O}_{C,P} = (\Gamma_+(C)_{\mathfrak{p}(P)})_0.$$

Proof After a change of projective coordinate system we may assume that $P = (1 : a : b)$, i.e. that $a_0 = 1, a_1 = a$ and $a_2 = b$. Then $\mathfrak{p}(P) = (x_1 - ax_0, x_2 - bx_0)k[x_0, x_1, x_2]$. Since $x_0 \notin \mathfrak{p}(P)$, we may perform the localization $\Gamma_+(C)_{\mathfrak{p}(P)}$ by first localizing in x_0 and then in the ideal of $\Gamma_+(C)_{x_0}$ corresponding to $\mathfrak{p}(P)$. But

$$(\Gamma_+(C)_{x_0})_0 = k\left[\frac{x_1}{x_0}, \frac{x_2}{x_0}\right],$$

as is easily seen. The rest of the verification is left to the reader.   □

4.2 Multiplicity and the Local Rings

Assume for simplicity that k is an algebraically closed field, and let K be an affine curve in $\mathbb{A}_k^2$ and P be a point on it. In the local ring $\mathcal{O}_{K,P}$ we denote the maximal ideal by $\mathfrak{m}_{K,P}$. Then we have the following result:

Theorem 4.5 *For all n the $\mathcal{O}_{K,P}$-module $\mathfrak{m}_{K,P}^n/\mathfrak{m}_{K,P}^{n+1}$ is a finite dimensional k-vector space. It is of dimension equal to the multiplicity $m_P(K)$ of the point P on the curve K, for all $n \geq m_P(K)$.*

Proof By a change of affine coordinate system, which corresponds to a change of variables in the coordinate ring $\Gamma(K)$, we may assume that $P = (0,0)$. For all n the ideal $\mathfrak{m}_{K,P}^n$ in $\mathcal{O}_{K,P}$ is an $\mathcal{O}_{K,P}$-module, hence so is $\mathfrak{m}_{K,P}^n/\mathfrak{m}_{K,P}^{n+1}$. Let α and $\beta \in k = \mathcal{O}_{K,P}/\mathfrak{m}_{K,P}$ be the images of the two elements a and $b \in \mathcal{O}_{K,P}$, and assume that f and $g \in \mathfrak{m}_{K,P}^n$ are such that $f - g \in \mathfrak{m}_{K,P}^{n+1}$, so $\overline{f} = \overline{g}$ in $\mathfrak{m}_{K,P}^n/\mathfrak{m}_{K,P}^{n+1}$. Then $af - bg = af - bf + bf - bg = (a - b)f + b(f - g) \in \mathfrak{m}_{K,P}^{n+1}$, so $\alpha\overline{f}$ is well defined. Thus $\mathfrak{m}_{K,P}^n/\mathfrak{m}_{K,P}^{n+1}$ becomes a k-vector space, and since the images $\overline{x}$ and $\overline{y}$ generate the ideal $\mathfrak{m}_{K,P}$, this vector space is generated by the images modulo $\mathfrak{m}_{K,P}^{n+1}$ of all monomials in x and y of degree n. Hence $\mathfrak{m}_{K,P}^n/\mathfrak{m}_{K,P}^{n+1}$ is finite dimensional.

The proof will be complete once we have shown the following

Lemma 4.6 *There is an integer c such that for all sufficiently large values of n (in fact for all $n \geq m_P(K)$),*

$$\dim_k(\mathcal{O}_{K,P}/\mathfrak{m}_{K,P}^n) = m_P(K)n + c.$$

Indeed, since

$$(\mathcal{O}_{K,P}/\mathfrak{m}_{K,P}^{n+1})/(\mathfrak{m}_{K,P}^n/\mathfrak{m}_{K,P}^{n+1}) \cong \mathcal{O}_{K,P}/\mathfrak{m}_{K,P}^n$$

it follows from this lemma that

$$\dim_k(\mathfrak{m}_{K,P}^n/\mathfrak{m}_{K,P}^{n+1}) = m_P(K)(n+1) + c - (m_P(K)n - c) = m_P(K).$$

To prove the lemma, we note the canonical isomorphisms

$$\mathcal{O}_{K,P}/\mathfrak{m}_{K,P}^n \cong k[X,Y]_{(X,Y)}/(X^n, X^{n-1}Y, \ldots, Y^n, F(X,Y))k[X,Y]_{(X,Y)}$$

$$\cong (k[X,Y]_{(X,Y)}/(X^n, X^{n-1}Y, \ldots, Y^n)k[X,Y]_{(X,Y)})/(F(\overline{X},\overline{Y}))$$

where $\overline{X}, \overline{Y}$ denote X and Y modulo the ideal

$$(X^n, X^{n-1}Y, \ldots, Y^n)k[X,Y]_{(X,Y)}.$$

We now consider the ring

$$k[X,Y]_{(X,Y)}/(X^n, X^{n-1}Y, \ldots, Y^n)k[X,Y]_{(X,Y)}$$

and note that instead of first localizing and then dividing out by the ideal $(X,Y)^n k[X,Y]_{(X,Y)}$, we may first form $k[X,Y]/(X,Y)^n$ and then localize in the ideal $(\overline{X},\overline{Y})$. However, the ring $k[X,Y]/(X,Y)^n$ is an Artinian local ring with maximal ideal $(\overline{X},\overline{Y})$, thus it is unchanged by this localization. Hence we conclude that

$$\mathcal{O}_{K,P}/\mathfrak{m}_{K,P}^n \cong k[X,Y]/(X^n, X^{n-1}Y, \ldots, Y^n, F(X,Y))k[X,Y]$$

$$= k[X,Y]/((X,Y)^n, F(X,Y))k[X,Y].$$

We next construct an exact sequence of vector spaces

$$0 \to k[X,Y]/(X,Y)^{n-m} \xrightarrow{\alpha} k[X,Y]/(X,Y)^n$$

$$\xrightarrow{\beta} k[X,Y]/((X,Y)^n, F(X,Y))k[X,Y] \to 0$$

as follows: The mapping β to the right is the canonical homomorphism, which is surjective and has kernel equal to the ideal generated by the image of $F(X,Y)$. The mapping α is the k-linear mapping defined by

$$\alpha(\overline{A(X,Y)}) = \overline{A(X,Y)F(X,Y)}.$$

Thus the image of α is equal to the kernel of β. Finally α is injective whenever $n \geq m$, since

$$F(X,Y) = h(X,Y) + g(X,Y),$$

where $h(X,Y)$ is homogeneous of degree $m = m_P(K)$ and $g(X,Y)$ is of degree greater than m. In fact, if $AF \in (X,Y)^n$, then $Ah \in (X,Y)^n$, so that $A \in (X,Y)^{n-m}$ since h is homogeneous of degree m.

Now $k[X,Y]/(X,Y)^n$ has a base as k-vector space consisting of the elements

$$1, \overline{X}, \overline{Y}, \ldots, \overline{X^{n-1}}, \overline{X^{n-2}Y}, \ldots, \overline{Y^{n-1}}$$

thus the dimension is $1 + 2 + \cdots + n = \frac{n(n+1)}{2}$. Hence

$$\dim_k (k[X,Y]/((X,Y)^n, F(X,Y))k[X,Y]) = \frac{n(n+1)}{2} - \frac{(n-m)(n-m+1)}{2}$$

$$= mn - \frac{m(m-1)}{2},$$

and the proof of the lemma is complete.   □

Corollary 4.7 *Assume that k is algebraically closed. Then a point P on K is non-singular if and only if $\mathcal{O}_{K,P}$ is a discrete valuation ring.*

The concept of valuation rings is important in algebra, and played a very central role in *André Weil's* development of the subject in his fundamental work [41]. Even though Weil's *Foundations* have been superseded by the modern treatment in Grothendieck's [15] and [17], the basic theory of valuation rings is a vital prerequisite for understanding even the elementary theory of algebraic curves. We shall therefore explain some of the basic facts about valuation rings in Sect. 4.5.

For the time being, we make a preliminary definition based on the following proposition:

Proposition-Definition 4.8 *Let $\mathcal{O}$ be a Noetherian integral domain which is not a field. The following are equivalent:*

(i) *$\mathcal{O}$ is a local ring, and the maximal ideal is generated by one element,*

$$\mathfrak{m}_{\mathcal{O}} = (\pi)\mathcal{O}.$$

(ii) *$\mathcal{O}$ contains an irreducible element π such that all $a \in \mathcal{O}$ can be written uniquely as*

$$a = u\pi^n$$

where $u \in \mathcal{O}^*$, in other words, u in is invertible in $\mathcal{O}$, or as we also say, u is a unit in $\mathcal{O}$.

(iii) *The $k = k_{\mathcal{O}} = \mathcal{O}/\mathfrak{m}_{\mathcal{O}}$-vector space $\mathfrak{m}_{\mathcal{O}}/\mathfrak{m}_{\mathcal{O}}^2$ is of dimension 1.*

Under these conditions, $\mathcal{O}$ is called a discrete valuation ring of rank 1.

Proof of the proposition (i) $\Longrightarrow$ (ii): If $a \notin \mathfrak{m}_{\mathcal{O}}$, then it is a unit since in this case $(a)\mathcal{O} = \mathcal{O}$, thus for some $b \in \mathcal{O}$ $ab = 1$. Thus (ii) is true with $n = 0$. If $a \in \mathfrak{m}_{\mathcal{O}}$, then there is an element $a_1 \in \mathcal{O}$ such that $a = a_1\pi$. a_1 is unique since $\mathcal{O}$ is an integral domain. If $a_1 \notin \mathfrak{m}$, then as above (ii) holds with $n = 1$. If $a_1 \in \mathfrak{m}_{\mathcal{O}}$ then $a_1 = a_2\pi$. When $a_2 \notin \mathfrak{m}_{\mathcal{O}}$ then (ii) holds with $n = 2$, otherwise we continue. We claim that the process will have to stop, in other words, sooner or later we arrive at some a_N which is a unit. Assume the converse. Then we get a chain of ideals

$$(a) \subseteq (a_1) \subseteq (a_2) \subseteq \cdots .$$

Since $\mathcal{O}$ is Noetherian, this chain becomes stationary, in particular we find m such that $(a_m) = (a_{m+1})$. We then have

$$a_m = a_{m+1}\pi \quad \text{and there exists} \quad t \in \mathcal{O} \quad \text{such that} \quad a_{m+1} = ta_m.$$

But this yields $a_m = (ta_m)\pi$, thus $1 = t\pi$ since $\mathcal{O}$ is an integral domain. But then π is a unit, which is a contradiction.

(ii) $\Longrightarrow$ (iii): By (ii) it follows that $\mathfrak{m}_{\mathcal{O}} = (\pi)\mathfrak{m}_{\mathcal{O}}$, and thus $\mathfrak{m}_{\mathcal{O}} \neq \mathfrak{m}_{\mathcal{O}}^2$. Hence the vector space $\mathfrak{m}_{\mathcal{O}}/\mathfrak{m}_{\mathcal{O}}^2$ is non-zero and generated by one element, hence it is of dimension 1.

(iii) $\Longrightarrow$ (i): Follows immediately by the more general

Lemma 4.9 *Let R be a local ring with maximal ideal $\mathfrak{m}$, and let M be an R-module. Let $x_1, \ldots, x_n$ be elements of M such that $\overline{x_1}, \ldots, \overline{x_n}$ is a base for the $R/\mathfrak{m}$-vector space $M/\mathfrak{m}M$. Then $x_1, \ldots, x_n$ generate the R-module M.*

To prove this lemma, we let N be the sub-R-module of M generated by $x_1, \ldots, x_n$, and wish to show that $M/N = 0$. The composition of R-linear maps

$$N \longrightarrow M \longrightarrow M/\mathfrak{m}M$$

sends N onto $M/\mathfrak{m}M$. Thus $\mathfrak{m}M + N = M$, and since $\mathfrak{m}(M/N) = (\mathfrak{m}M + N)/N$, we conclude that $\mathfrak{m}(M/N) = M/N$. Now we apply a useful special case of a general result known as *Nakayama's lemma*:

Proposition 4.10 *If $\overline{M}$ is a finitely generated R-module for the local ring R, such that $\mathfrak{m}\overline{M} = \overline{M}$, then $\overline{M} = 0$.*

This is easily seen as follows: Let $\overline{x}_1, \ldots, \overline{x}_n$ be a set of generators for $\overline{M}$. By the assumption there exist $m_{i,j} \in \mathfrak{m}$ such that

$$\overline{x}_1 = m_{1,1}\overline{x}_1 + \cdots + m_{1,n}\overline{x}_n$$

$$\vdots$$

$$\overline{x}_n = m_{n,1}\overline{x}_1 + \cdots + m_{n,n}\overline{x}_n$$

or in other words

$$(m_{1,1} - 1)\bar{x}_1 + \cdots + m_{1,n}\bar{x}_n = 0$$
$$\vdots$$
$$m_{n,1}\bar{x}_1 + \cdots + (m_{n,n} - 1)\bar{x}_n = 0,$$

which may be written as

$$\begin{bmatrix} m_{1,1} - 1 & \cdots & m_{1,n} \\ \vdots & \ddots & \vdots \\ m_{n,1} & \cdots & m_{n,n} - 1 \end{bmatrix} \begin{bmatrix} \bar{x}_1 \\ \vdots \\ \bar{x}_n \end{bmatrix} = \begin{bmatrix} 0 \\ \vdots \\ 0 \end{bmatrix}.$$

The determinant of this matrix $\mathcal{M}$ does not lie in $\mathfrak{m}$, and is therefore a unit. Thus the inverse matrix $\mathcal{M}^{-1} = \mathcal{N}$ also has entries from R, and hence we get

$$\begin{bmatrix} \bar{x}_1 \\ \vdots \\ \bar{x}_n \end{bmatrix} = \mathcal{N}\mathcal{M} \begin{bmatrix} \bar{x}_1 \\ \vdots \\ \bar{x}_n \end{bmatrix} = \mathcal{N} \begin{bmatrix} 0 \\ \vdots \\ 0 \end{bmatrix} = \begin{bmatrix} 0 \\ \vdots \\ 0 \end{bmatrix}$$

which completes the proof of Nakayama's lemma and hence of our Lemma 4.9 and Proposition 4.8. $\qquad\qquad\qquad\qquad\qquad\qquad\qquad\qquad\qquad\qquad\square$

Proof of Corollary 4.7 If the point is non-singular, then it has multiplicity 1, so $\mathfrak{m}_{K,P}/\mathfrak{m}_{K,P}^2$ is of dimension 1, which is one of the equivalent characterizations of a discrete valuation ring. Conversely, if the ring is a discrete valuation ring, then $\mathfrak{m}_{K,P}/\mathfrak{m}_{K,P}^2$ is of dimension 1, thus not both partial derivatives of the equation $F(X, Y)$ can vanish at P, thus P is non-singular. $\qquad\square$

For a general algebraic curve $K \subset \mathbb{A}_k^n$ we defined the coordinate ring $\Gamma(K)$ and local rings $\mathcal{O}_{K,P}$ at the points in Sect. 4.1. We now define the Zariski tangent space $ZT_{K,P}$ at the point $P = (a, b)$ as the dual k-vector space of $\mathfrak{m}_{\mathcal{O}_{K,P}}/\mathfrak{m}_{\mathcal{O}_{K,P}}^2 \colon ZT_{K,P} = (\mathfrak{m}_{\mathcal{O}_{K,P}}/\mathfrak{m}_{\mathcal{O}_{K,P}}^2)^*$. The former space is referred to as the *Zariski cotangent space* of K at P. Similarly the Zariski tangent space of $\mathbb{A}_k^2$ at P is the dual vector space of $\mathfrak{m}_{\mathcal{O}_{\mathbb{A}_k^2,P}}/\mathfrak{m}_{\mathcal{O}_{\mathbb{A}_k^2,P}}^2$, where $\mathcal{O}_{\mathbb{A}_k^2,P}$ is the local ring of $\mathbb{A}_k^2$ at P, so

$$\mathcal{O}_{\mathbb{A}_k^2,P} = k[X, Y]_{(X-a, Y-b)}.$$

Evidently $\mathfrak{m}_{\mathcal{O}_{\mathbb{A}_k^2,P}}/\mathfrak{m}_{\mathcal{O}_{\mathbb{A}_k^2,P}}^2$ is a k-vector space with base ξ, η, the images of $X - a$ and $Y - b$. Denoting the images of $x - a$ and $y - b$ in $\mathfrak{m}_{\mathcal{O}_{K,P}}/\mathfrak{m}_{\mathcal{O}_{K,P}}^2$ by $\bar{\xi}, \bar{\eta}$, we have a surjective k-linear map of Zariski cotangent spaces

$$\mathfrak{m}_{\mathcal{O}_{\mathbb{A}_k^2,P}}/\mathfrak{m}_{\mathcal{O}_{\mathbb{A}_k^2,P}}^2 \longrightarrow\!\!\!\!\rightarrow \mathfrak{m}_{\mathcal{O}_{K,P}}/\mathfrak{m}_{\mathcal{O}_{K,P}}^2$$

which sends ξ, η to $\overline{\xi}, \overline{\eta}$. Dualizing, we get the Zariski tangent space of K at P as a linear subspace of the Zariski tangent space of $\mathbb{A}_k^2$ at P. The latter being of dimension 2, the Zariski tangent space for a plane curve is either of dimension 1, in the smooth case, or of dimension 2 in the non-smooth case.

4.3 Intersection Multiplicities for Affine Plane Curves

Let K be a curve and let $P = (a, b)$ be a point on it. We may assume, after a change of coordinate system, that $P = (0, 0)$, and thus that the curve is given by a polynomial of the type

$$F(X, Y) = a_{1,0}X + a_{0,1}Y + a_{2,0}X^2 + a_{1,1}XY + a_{0,1}Y^2 + \cdots .$$

We have defined the multiplicity with which a line L intersects K in $P = (0, 0)$ as the multiplicity of the solution $t = 0$ in the equation

$$\varphi(t) = F(ut, vt) = 0,$$

where the line has the parametric form

$$L = \{(x, y) \mid x = ut, y = vt\}.$$

If the line has equation

$$aX + bY = 0,$$

then we may take $u = b, v = -a$, and it is easily seen that the multiplicity is nothing but the dimension as k-vector space of the k-algebra

$$R_P(K \cap L) = k[X, Y]_{(X,Y)}/(F(X, Y), aX + bY)k[X, Y]_{(X,Y)}.$$

Indeed, by changing the coordinate system we may assume $a = 0, b = 1$, then L gets the equation $Y = 0$, and

$$\varphi(t) = F(t, 0),$$

while

$$R_P(K \cap L) = k[X, Y]_{(X,Y)}/(F(X, Y), Y)k[X, Y]_{(X,Y)} \cong k[t]_{(t)}/(F(t, 0))k[t]_{(t)}.$$

We also note that this is an Artinian local ring, hence we have

$$R_P(K \cap L) \cong \mathcal{O}_{\mathbb{A}_k^2, P}/(F(X, Y), aX + bY)\mathcal{O}_{\mathbb{A}_k^2, P}.$$

This definition applies to a more general situation. In fact, if $P \notin K \cap L$, then the intersection multiplicity is zero, and conversely.

We now have the definition of intersection multiplicity on a form which suggests a very natural generalization:

Definition 4.2 Let the curve K_1 and the curve K_2 be general affine curves over k, given by the polynomials $F_1(X,Y)$ and $F_2(X,Y)$, respectively. Then their order of intersection in the point $P = (a,b) \in \mathbb{A}_k^2$ is defined as the dimension of the k-vector space

$$R_P(K_1 \cap K_2) = \mathcal{O}_{\mathbb{A}_k^2,P}/(F_1(X,Y),F_2(X,Y))\mathcal{O}_{\mathbb{A}_k^2,P}.$$

This ring is a k-algebra in the natural way, and we denote its dimension as a vector space over k by

$$I(P,K_1 \cap K_2) = \dim_k(R_P(K_1 \cap K_2)).$$

This number is referred to as the intersection number of K_1 and K_2 at the point P.

We first note the

Proposition 4.11 *The intersection number $I(P,K_1 \cap K_2)$ is independent of coordinate system.*

Proof Explicitly, we have the following expression for the k-algebra introduced above: Let $P = (a,b)$. Then

$$R_P(K_1 \cap K_2)$$
$$= (k[X,Y]_{(X-a,Y-b)})/(F_1(X,Y),F_2(X,Y))(k[X,Y]_{(X-a,Y-b)})$$
$$= k[\xi,\zeta]_{(\xi,\zeta)}$$

where ξ,ζ are the images of $X - a, Y - b$. A new coordinate system just yields new generators for the k-algebra $R_P(K_1 \cap K_2)$. □

Proposition 4.12 1. $I(P,K_1 \cap K_2) = 0$ if and only if $P \notin K_1 \cap K_2$.
    2. $I(P,K_1 \cap K_2) = \infty$ if and only if P lies in a common component of K_1 and K_2.

Proof 1. If $P = (a,b) \notin K_1 \cap K_2$, then the maximal ideal $(X-a,Y-b)k[X,Y]$ in $k[X,Y]$ does not contain both $F_1(X,Y)$ and $F_2(X,Y)$. Thus at least one of these polynomials is a unit in $\mathcal{O}_{\mathbb{A}_k^2,P}$, hence $R_P(K_1 \cap K_2)$ is no k-algebra, but the "zero-ring" in this case.[1] Conversely, if $R_P(K_1 \cap K_2)$ is the zero-ring, then at least one of the two polynomials is a unit in $\mathcal{O}_{\mathbb{A}_k^2,P}$, so (a,b) can not lie in $K_1 \cap K_2$.

---

[1] Usually a commutative ring A with 1, and in particular a k-algebra over a field k, is required to have $0_A \neq 1_A$. But in some cases it is convenient to admit the *zero-ring* consisting of just one element as a "ring". This will be clear in the second part of this book, when we get to the notion of *an affine scheme* and its relation to the *fiber of a morphism*.

2. If P lies in a common component of K_1 and K_2, then there is a polynomial $G(X,Y)$ such that $F_i(X,Y) = G(X,Y)H_i(X,Y)$ for $i = 1,2$, and $G(a,b) = 0$. Thus

$$R_P(K_1 \cap K_2) = \mathcal{O}_{\mathbb{A}_k^2, P}/(G(X,Y)H_1(X,Y), G(X,Y)H_2(X,Y))\mathcal{O}_{\mathbb{A}_k^2, P}$$

and this ring contains the prime ideal $\mathfrak{p}$ generated by the image of $G(X,Y)$. Thus the local ring $R_P(K_1 \cap K_2)$ is of Krull dimension > 0, hence infinite dimensional as a vector space over k. Conversely, if P does not lie in a common component, then $F_1(X,Y), F_2(X,Y) \in \mathcal{O}_{\mathbb{A}_k^2, P}$ have no common factor. Now a prime ideal $\mathfrak{p}$ in $\mathcal{O}_{\mathbb{A}_k^2, P}$ containing $F_1(X,Y), F_2(X,Y)$ must contain an irreducible factor of each of them, say $p_1(X,Y)$ and $p_2(X,Y)$. But $p_1(X,Y)$ generates a non-zero prime ideal $\mathfrak{q}$ contained in $\mathfrak{p}$, and since $p_2(X,Y)$ lies outside it $\mathfrak{p}$ is strictly bigger than $\mathfrak{q}$. As $\mathcal{O}_{\mathbb{A}_k^2, P}$ is of Krull-dimension 2, we therefore have that $\mathfrak{p}$ is the maximal ideal. Hence the local ring $\mathcal{O}_{\mathbb{A}_k^2, P}/(F_1(X,Y), F_2(X,Y))\mathcal{O}_{\mathbb{A}_k^2, P}$ has only one prime ideal, namely the maximal ideal, thus it is Artinian and therefore a finite dimensional vector space over k. $\square$

4.4 Intersection Theory for Curves in $\mathbb{P}_k^2$

Let C_1 and $C_2 \subset \mathbb{P}_k^2$ be two projective plane curves, and let $P \in \mathbb{P}_k^2$. We say that C_1 and C_2 *intersect properly* at P if they have no common component through P. If C_1 and C_2 intersect properly at all points, then we say that they *intersect properly*.

Proposition 4.13 *Let k be algebraically closed. Then two plane projective curves C_1 and C_2 intersect properly if and only if the set $C_1 \cap C_2$ is a finite subset of $\mathbb{P}_k^2$.*

Proof We may assume that C_1 and C_2 are irreducible, i.e. that they are given by irreducible polynomials.

In that case, if they have a common component then $C_1 = C_2$, thus $C_1 \cap C_2 = C_1$. We show that C_1 has an infinite number of points. Indeed, for any $\alpha \in k$ let $F_\alpha(Y) = F(\alpha, Y)$. $F_\alpha(Y)$ is a constant polynomial for at most a finite number of $\alpha \in k$. Thus $F_\alpha(Y)$ has at least one zero in k, say β_α. Then $F(\alpha, \beta_\alpha) = 0$, and as there is an infinite number of such points $(\alpha, \beta_\alpha) \in \mathbb{P}_k^2$, C_1 is infinite as claimed.

Conversely, assume that C_1 and C_2 do not have a common component, i.e. under the assumption of irreducibility, that $C_1 \neq C_2$. To show that $C_1 \cap C_2$ is a finite subset, it suffices to show that this is so for all affine restrictions. Thus we may assume that C_1 and C_2 are affine curves in $\mathbb{A}_k^2$. If these two curves are given by the polynomials $F(X,Y)$ and $G(X,Y)$, respectively, then

$C_1 \cap C_2$ consists of all points $(a, b) \in \mathbb{A}_k^2$ such that the corresponding maximal ideal in $k[X, Y]$ contains F and G:

$$\mathfrak{m}_{(a,b)} = (X - a, Y - b)k[X, Y] \ni F(X, Y), G(X, Y).$$

These maximal ideals correspond to the set of all maximal ideals in the Noetherian ring $k[X, Y]/(F, G)k[X, Y]$, which is of Krull dimension zero, hence they are also minimal primes, thus their number is finite. $\qquad \square$

The intersection cycle of the two curves is defined as a formal sum

$$\sum_{P \in C_1 \cap C_2} I(P, C_1 \cap C_2)P$$

provided the two curves intersect properly.

Remark 4.14 If k is not algebraically closed, then a projective irreducible curve may not have an infinite number of points in $\mathbb{P}_k^2$, even for an infinite field, such as $k = \mathbb{R}$.

In fact, a very simple example is provided by *"a circle with an imaginary radius"* such as the curve in $\mathbb{P}_\mathbb{R}^2$ given by

$$X_1^2 + X_2^2 + X_0^2$$

which have no points in $\mathbb{P}_\mathbb{R}^2$, or the union of two imaginary lines through the point $(0 : 0 : 1)$, given by

$$X_1^2 + X_2^2.$$

In Definition 4.2 we defined the intersection multiplicity of two affine curves. We now define the intersection multiplicity, also denoted by $I(P, C_1 \cap C_2)$, for the two projective curves at P by passing to any affine restriction such that P is not at infinity.

We are now going to established seven basic and very important properties of intersection numbers. Remarkably, these quite simple properties determine the intersection theory *uniquely*, and moreover they imply other properties, appearing to be of a considerably less trivial nature. We first state and prove the properties.

By Proposition 4.11 the definition we have given above is independent of the choice of affine restriction. Moreover, we showed in Proposition 4.12 that the intersection number at a point is finite if and only if the two curves intersect properly in the point, and that it is zero if and only if the point lies outside their intersection. We thus have established the first two of our properties:

(I 1) $I(P, C_1 \cap C_2)$ is a non negative integer if the intersection is proper at P, otherwise $I(P, C_1 \cap C_2) = \infty$.

(I 2) $I(P, C_1 \cap C_2) = 0$ if and only if $P \notin C_1 \cap C_2$.

By Proposition 4.11 we immediately establish the third property:

(I 3) $I(P, C_1 \cap C_2)$ is independent of projective coordinate system.

The fourth property is immediate by the definition:

(I 4) $I(P, C_1 \cap C_2) = I(P, C_2 \cap C_1)$.

The next property constitutes the starting point for computations of intersection numbers:

(I 5) If L_1 and L_2 are two distinct lines through P, then

$$I(P, L_1 \cap L_2) = 1.$$

To be able to compute from this starting point, we need the two final properties:

(I 6) $I(P, (\ ), (\ ))$ is bilinear in both variables, i.e. we have

$$I\left(P, \sum m_i C_i, \sum n_j C_j'\right) = \sum_{i,j} m_i n_j I(P, C_i, C_j').$$

(I 7) If the curves C_1, C_2 and C_3 are given by the homogeneous polynomials $P_1(X_0, X_1, X_2), P_2(X_0, X_1, X_2)$ and $P_3 = P_2 + F P_1$, where F is some homogeneous polynomial, then

$$I(P, C_1 \cap C_2) = I(P, C_1 \cap C_3).$$

Proof of the final three properties: (I 7) is immediate, as the k-algebras whose lengths give the intersection numbers are equal.

(I 5) is also immediate, we chose an affine restriction in which $P = (0,0)$ and L_1 corresponds to $X = 0$ and L_2 corresponds to $Y = 0$. In this case we clearly obtain

$$R_P(L_1 \cap L_2) = k.$$

(I 6) is easy but not immediate. Indeed, we may assume that C_1 and C_2 have no common component through P, and it suffices to show that

$$I(P, C_1 \cap C_2) = I(P, C_1 \cap C_2') + I(P, C_1 \cap C_2''),$$

if C_2' and C_2'' are given by the factors G and H in $F_2 = G \cdot H$. By a change of projective coordinate system and taking affine restriction we may assume that the curves are affine and that $P = (0,0)$.

In the following we let $\mathcal{O} = k[X, Y]_{(X,Y)}$. We have to show that

$$\dim_k(\mathcal{O}/(f, gh)) = \dim_k(\mathcal{O}/(f, g)) + \dim_k(\mathcal{O}/(f, h))$$

where f, g and h are the polynomials in X and Y which correspond to the original homogeneous ones. Since C_1 and C_2 have no common component through $P = (0,0)$, we may assume that the polynomial f has no common factor with g or with h. Indeed, such factors would have non zero constant terms, thus be units in the local ring $\mathcal{O}$, and may therefore be disregarded without changing the k-algebras we are working with here.

The claim now follows by the exact sequence of k-vector spaces

$$0 \longrightarrow \mathcal{O}/(f,h) \overset{\alpha}{\to} \mathcal{O}/(f,gh) \overset{\beta}{\to} \mathcal{O}/(f,g) \longrightarrow 0$$

where the surjective map β to the right is the canonical k-algebra homomorphism, and α to the left the mapping defined by

$$\alpha(a(X,Y) \bmod (f,h)) = a(X,Y)g(X,Y) \bmod (f,gh).$$

This is well defined, since if $a - b \in (f,h)$ means that $a - b = Af + Bh$, so that $(a - b)g = Afg + Bhg \in (f,gh)$. Moreover, α is k-linear, and clearly the image of α is equal to the kernel of β, $\mathrm{im}(\alpha) = \ker(\beta)$. It remains to show that α is injective. For this, assume that $\alpha(a(X,Y) \bmod (f,h)) = 0$, i.e., that $a(X,Y)g(X,Y) \in (f,gh)$, so

$$ag = uf + vgh \quad \text{where } u,v \in \mathcal{O}.$$

This is a rational identity in X and Y with coefficients from k, where f,g and h are polynomials in X and Y, while a, u and v are rational expressions, the denominators being polynomials in X,Y with non-zero constant terms, since we assume that $P = (0,0)$:

$$a = \frac{A}{S}, \quad u = \frac{B}{S} \quad \text{and} \quad v = \frac{C}{S}$$

where S is a polynomial with non zero constant term. Thus

$$Ag = Bf + Cgh, \quad \text{thus } g(A - Ch) = Bf.$$

Now f and g are relatively prime, so f will have to divide $A - CH$, thus $A - Ch = Df$. But then

$$a = \frac{A}{S} = \frac{C}{S}h + \frac{D}{S}f$$

thus $a(X,Y) \bmod (f,h) = 0$, as claimed.   □

It is convenient to write $I(P, C_1 \cap C_2)$ as $I(P, F \cap G)$, when F and G are the polynomials in X and Y defining the affine restriction of the curves, P not being at infinity.

We now prove uniqueness of the intersection theory. This will be done by describing an algorithm which makes it possible to compute the intersection number of two arbitrary curves at any point, using only the properties

(I 1)–(I 7) listed above. Before we present the formal proof, we illustrate the procedure by a very simple example.

Example We wish to compute $I(P,(Y^3+X)\cap(X^3+Y^2))$ for $P=(0,0)$. We have the polynomials

$$F(X,Y)=Y^3+X$$

and

$$G(X,Y)=X^3+Y^2.$$

Write

$$F(X,Y)=Y^3+X, \quad \text{the "pure X-part" is of degree } N=1$$

and

$$G(X,Y)=Y^2+X^3, \quad \text{where the pure X-part is of degree } M=3.$$

We want $N \geq M$. Here this fails, therefore we switch the roles of F and G, using (I 4):

$$F(X,Y)=Y^2+X^3 \text{ so } N=3,$$

and

$$G(X,Y)=Y^3+X \text{ so } M=1.$$

Now we reduce the degrees of the pure X-parts, with the intention of arriving at degree zero eventually:

$$\overline{F}(X,Y)=F(X,Y)-X^{N-M}G(X,Y)=-X^2Y^3+Y^2.$$

Thus we already have degree zero. Moreover

$$\overline{F}(X,Y)=Y\overline{F}_1(X,Y)=Y^2(-X^2Y+1).$$

Then by (I 7),

$$I(P,F\cap G)=I(P,\overline{F}\cap G)=I(P,Y\overline{F}_1(X,Y)\cap G)$$

and by (I 6),

$$I(P,Y\overline{F}_1(X,Y)\cap G)=I(P,Y\cap G)+I(P,\overline{F}_1(X,Y)\cap G).$$

Using (I 6) a second time we have,

$$I(P,\overline{F}_1(X,Y)\cap G)=I(P,Y\cap G)+I(P,(-X^2Y+1)\cap G).$$

By (I 2) $I(P,(-X^2Y+1)\cap G)=0$ and by (I 7) followed by (I 5),

$$I(P,Y\cap G)=I(P,Y\cap(Y^3+X))=I(P,Y\cap X)=1.$$

Thus

$$I(P,(Y^3 + X) \cap (X^3 + Y^2)) = 2.$$

We are now ready to prove the uniqueness-result:

Proposition 4.15 *The properties* (I 1)–(I 7) *determine the intersection numbers uniquely. In fact, these properties make it possible to compute intersection numbers in a finite number of steps.*

Proof We may assume that no common component passes through P, since the intersection number is ∞ otherwise.

Let $n \geq 0$ be an integer and consider the following statement:

$P(n)$ Let C_1 and C_2 be two curves in $\mathbb{P}^2_k$ and P be any point $\in \mathbb{P}^2_k$. Assume that $\deg(C_1) \leq n$. Then there exists a constructive procedure for computing $I(P, C_1 \cap C_2)$ using only (I 1)–(I 7).

The proof will be complete once we show $P(n)$ for all n. We do so by induction. For $n = 0$ we must have $P \notin C_1$, hence $P \notin C_1 \cap C_2$ and $I(P, C_1 \cap C_2) = 0$ by (I 2).

So assume $P(n-1)$ and let C_1 and C_2 be such that $\deg(C_1) \leq n$. We take an affine restriction such that $P = (0,0)$, and C_1 and C_2 correspond to the polynomials $F(X,Y)$ and $G(X,Y)$, respectively. By (I 4) the proof will be complete once we succeed in expressing $I(P, C_1 \cap C_2)$ in terms of $I(P, \hat{C}_1 \cap \hat{C}_2)$ for new curves where $\hat{C}_1$ is of smaller degree than n.

We do so using the procedure in the example above. Accordingly, we put

$$F(X,Y) = YF_1(X,Y) + \varphi_1(X)$$
$$G(X,Y) = YG_1(X,Y) + \psi_1(X)$$

where

$$\varphi_1(X) = a_N X^N + a_{N-1} X^{N-1} + \cdots + a_1 X$$
$$\psi_1(X) = b_M X^M + b_{M-1} X^{M-1} + \cdots + b_1 X.$$

We may assume that $N \geq M$, if necessary after interchanging C_1 and C_2, justified by (I 4). Now observe that

$$\overline{\varphi}_1(X) = \varphi_1(X) - \frac{a_N}{b_M} X^{N-M} \psi_1(X) = a_{N'} X^{N'} + \cdots$$

where $N' < N$. Thus if we let

$$\overline{F}(X,Y) = F(X,Y) - \frac{a_N}{b_M} X^{N-M} G(X,Y)$$

then by (I 4) and (I 7)

$$I(P, F \cap G) = I(P, \overline{F} \cap G)$$

where

$$\overline{F}(X,Y) = Y\overline{F}_1(X,Y) + \overline{\varphi}_1(X)$$
$$G(X,Y) = YG_1(X,Y) + \psi_1(X).$$

If still $N' \geq M$ we repeat this, otherwise we interchange the roles of F and G by (I 4), and repeat.

Clearly, after a finite number of steps with repetitions and interchanging roles as needed, we obtain polynomials $\hat{F}(X,Y)$ and $\hat{G}(X,Y)$ such that

$$I(P, F \cap G) = I(P, \hat{F} \cap \hat{G})$$

where

$$\hat{F}(X,Y) = Y\hat{F}_1(X,Y) + \hat{\varphi}(X)$$
$$\hat{G}(X,Y) = Y\hat{G}_1(X,Y).$$

Thus by (I 6)

$$I(P, F \cap G) = I(P, \hat{F} \cap \hat{G}_1) + I(P, \hat{F} \cap Y).$$

Now the curve given by $\hat{G}_1$ is of degree $\leq n - 1$, and using (I 7) we find $I(P, \hat{F} \cap Y) = I(P, \hat{\varphi}(X) \cap Y)$. Writing $\hat{\varphi}(X) = X^m(a_0 + a_1 X + \cdots + a_r X^r)$ where $a_0 \neq 0$ we find

$$I(P, \hat{\varphi}(X) \cap Y) = I(P, X^m \cap Y) = mI(P, X \cap Y) = m$$

by (I 7), (I 1), (I 6) and (I 5). Hence we are done. $\square$

The case of coinciding tangent lines is regulated by the following

Proposition 4.16 *We have*

$$I(P, C_1 \cap C_1) \geq m_P(C_1)m_P(C_2)$$

with equality holding if and only if the tangent lines of C_1 are distinct from the tangent lines of C_2 at P.

Proof Let $m = m_P(C_1)$ and $n = m_P(C_2)$. We take an affine restriction, and may assume that $P = (0,0)$, let C_1 and C_2 be given in the affine plane by $F(X,Y) = 0$ and $G(X,Y) = 0$, respectively. Put $I = (X,Y)k[X,Y]$. We first note the following exact sequence:

$$k[X,Y]/I^n \times k[X,Y]/I^m \xrightarrow{\psi} k[X,Y]/I^{m+n} \xrightarrow{\varphi} k[X,Y]/(I^{m+n}, F, G) \to 0.$$

In fact, φ is the canonical homomorphism, which is surjective. ψ is defined by

$$\psi(A \bmod I^n, B \bmod I^m) = AF + BG \bmod I^{m+n},$$

which makes sense since if $A - A' \in I^n$ and $B - B' \in I^m$ then $F(A - A') + G(B - B') \in I^{m+n}$. Note that here we use that $m = m_P(C_1)$ and $n = m_P(C_2)$.

So we have the exact sequence, from which follows that $\mathrm{im}(\psi) = \ker(\varphi)$. Thus we have

$$\dim_k(k[X,Y]/I^n \times k[X,Y]/I^m) = \dim_k(\ker(\psi)) + \dim_k(\ker(\varphi)),$$

and

$$\dim_k(k[X,Y]/(I^{m+n}, F, G)) = \dim_k(k[X,Y]/I^{m+n}) - \dim_k(\ker(\varphi)).$$

Now the k-algebra $k[X,Y]/(I^{m+n}, F, G)$ is an Artinian local ring. Hence it is a quotient of the k-algebra $\mathcal{O}/(F,G)\mathcal{O}$, so we have the exact sequence

$$0 \to \ker(\pi) \to \mathcal{O}/(F,G)\mathcal{O} \xrightarrow{\pi} k[X,Y]/(I^{m+n}, F, G) \to 0$$

and thus

$$I(P, C_1 \cap C_2) = \dim_k(k[X,Y]/(I^{m+n}, F, G)) + \dim(\ker(\pi)).$$

Still following Fulton [10] (p. 78) we now conclude as follows:
Putting the above together, we get the following string of inequalities:

$$\begin{aligned} I(P, C_1 \cap C_2) &\geq \dim_k(k[X,Y]/(I^{m+n}, F, G)) \\ &\geq \dim_k(k[X,Y]/I^{m+n}) - \dim_k(k[X,Y]/I^m) \\ &\quad - \dim_k(k[X,Y]/I^n) \\ &= mn, \end{aligned}$$

the latter inequality being a simple exercise with binomial coefficients, and the first part of the proposition is proven.

The last part is proven by the lemma below, again following [10]:

Lemma 4.17 (a) *ψ is injective if and only if the tangent lines of C_1 and C_2 at $P = (0,0)$ are different.*

(b) *In this case*

$$I^t \subset (F,G)\mathcal{O}$$

provided that $t \geq m + n - 1$.

Proof of the lemma (a): Using subscripts to indicate the homogeneous parts, we have

$$F = F_m + F_{m+1} + \cdots \quad \text{and} \quad G = G_n + G_{n+1} + \cdots,$$

in other words F_m and G_n are the initial forms. Assume that there is a common tangential line through $(0,0)$, we may assume it to be the Y-axis. Then X is a common factor of F_m and G_n, we have $F_m = XB, G_n = XA$, and we have $\psi(A \mod I^n, -B \mod I^m) = AF - BG \mod I^{m+n}$. But

$$AF - BG = \frac{G_n}{X}F - \frac{F_m}{X}G \in I^{m+n}$$

and therefore ψ is not injective.

Conversely, assume that C_1 and C_2 have no common tangential line, and that A and B are polynomials such that $\psi(A \mod I^n, B \mod I^m) = 0$, i.e. $AF + BG \in I^{m+n}$. If A and B have the initial forms A_r and B_s, respectively, then $A_r F_m + B_s G_n$ is zero or of initial degree $\geq m + n$. In the latter case $r \geq n$ and $s \geq m$, thus $(A \mod I^n, B \mod I^m) = (0,0)$. In the former case $A_r F_m = -B_s G_n$. Thus $B_s = CF_m$ and $A_r = DG_n$, since F_m and G_n have no common factors. Thus $s \geq m$ and $r \geq n$. Again we find $(A \mod I^n, B \mod I^m) = (0,0)$. Thus (a) is proven.

To show (b), let $L_1, L_2, \ldots, L_m$ be the linear forms corresponding to the tangential lines of C_1 at $P = (0,0)$, repeated as many times as the multiplicity indicates, so $F_m = L_1 \cdot L_2 \cdots L_m$. Similarly $M_1, M_1, \ldots, M_n$ correspond to the tangential lines of C_2 at P. Since the tangential lines are all different, it follows that $L_i \neq \lambda M_j$ for all $\lambda \in k$ and all i and j. For all $i > m$ and $j > n$ we let $L_i = L_m$ and $M_j = M_n$. We then put

$$A_{i,j} = L_1 \cdots L_i M_1 \cdots M_j,$$

and claim that the $A_{i,j}$ with $i + j = d$ generate the k-vector space $k[X,Y]_d$. We now use the assumption that

$$L_i = \alpha_i X + \beta_i Y \quad \text{and} \quad M_j = \gamma_j X + \delta_j Y$$

are not proportional, thus

$$\begin{vmatrix} \alpha_i & \beta_i \\ \gamma_j & \delta_j \end{vmatrix} \neq 0.$$

Therefore by Cramer's Rule we have $\alpha'_{i,j}, \ldots, \delta'_{i,j} \in k$ such that

$$X = \alpha'_{i,j} L_i + \beta'_{i,j} M_j \quad \text{and} \quad Y = \gamma'_{i,j} L_i + \delta'_{i,j} M_j.$$

Using this, we easily see by induction on d that all $X^i Y^{d-i}$ may be expressed by the $A_{i,j}$'s.

To prove the claim, it therefore suffices to show that if $i + j \geq N \geq m+n-1$, then $A_{i,j} \in (F,G)\mathcal{O}$. Now we have either $i \geq m$ or $j \geq n$, suppose that $i \geq m$. Moreover $A_{i,j} = A_{m,0} B$, where B is homogeneous of degree $t = i+j-m$ and $A_{m,0} = F_m$. Thus $F = A_{m,0} + F'$, where F' has initial degree $\geq m+1$. Then $A_{i,j} = BF - BF'$, and the term BF' have initial degree $\geq N+1$. This term can be expressed by $A_{i,j}$'s with $i + j \geq N+1$. Thus it suffices to show that if $i + j \geq N+1$, then $A_{i,j} \in (F,G)\mathcal{O}$.

Repeating this we find that it suffices to prove that all $A_{i,j} \in (F,G)\mathcal{O}$ for $i+j \geq N+2$, and so on.

Thus we have to show that $I^\ell \subset (F,G)\mathcal{O}$ for $\ell \gg 0$, in other words that in the Artinian local ring $\mathcal{A} = \mathcal{O}/(F,G)\mathcal{O}$ we have $\mathfrak{m}_{\mathcal{A}}^\ell = (0)$ for $\ell \gg 0$. But this is a well known property of Artinian local rings, easily proved by Nakayama's lemma, in the form of Proposition 4.10. Indeed, we have the descending chain of ideals in $\mathcal{A}$

$$\mathfrak{m}_{\mathcal{A}} \supseteq \mathfrak{m}_{\mathcal{A}}^2 \supseteq \mathfrak{m}_{\mathcal{A}}^3 \supseteq \cdots$$

thus by the descending chain condition of Artinian rings there is $\ell \gg 0$ such that $\mathfrak{m}_{\mathcal{A}}^\ell = \mathfrak{m}_{\mathcal{A}}^{\ell+1}$. With $\overline{M} = \mathfrak{m}_{\mathcal{A}}^\ell$ we then have $\overline{M} = \mathfrak{m}\overline{M}$, thus by Proposition 4.10 $\overline{M} = \mathfrak{m}_{\mathcal{A}}^\ell = 0$. $\qquad\square$

The following observation is important and useful:

Proposition 4.18 *If P is a simple point on the curve C_1, then*

$$I(P, C_1 \cap C_2) = \operatorname{ord}_P(C_2).$$

Proof We may assume that F is irreducible. Let g be the image of G in $\mathcal{O}_P(C_1)$, the local ring of C_1 at P. Then $\operatorname{ord}_P(C_2) = v_1(g) = \dim_k(\mathcal{O}_P(C_1)/(g))$, by the definition of the valuation v_1 corresponding to $P \in C_1$. $\qquad\square$

Another important property is given in the

Proposition 4.19 *Let the affine curves K and L be defined by the polynomials $F(X,Y)$ and $G(X,Y)$. Assume that the curves have no common component. Then*

$$\sum_P I(P, K \cap L) = \dim_k(k[X,Y]/(F,G)k[X,Y]).$$

Proof Since K and L have no common components and k is algebraically closed, $K \cap L$ consists of a finite number of points, $P_1, \ldots, P_r$, corresponding to the maximal ideals $\mathfrak{m}_1, \ldots, \mathfrak{m}_r$ of the semi-Artinian ring $k[X,Y]/(F,G)k[X,Y]$. $\qquad\square$

4.5 Valuations and Valuation Rings

We now return to the basic theory of *valuation rings*.

Let V be an integral domain which is not a field, and let K be its field of fractions. We make the following

Definition 4.3 V is called a valuation ring if for all non-zero $x \in K$ we have $x \in V$ or $\frac{1}{x} \in V$.

Evidently a *discrete valuation ring* as defined in Proposition-Definition 4.8, is a valuation ring. But this concept is much more general. We first note the following:

Proposition 4.20 *Let V be a valuation ring. Then*

(i) V *is a local ring.*
(ii) *If $V' \subset K$ is a ring containing V, then V' is a valuation ring.*
(iii) V *is integrally closed in K.*

Proof (i) follows since the set $\mathfrak{m}$ of all elements in V which are not invertible, is an ideal in V (necessarily maximal). In fact, $a \in \mathfrak{m}$ if and only if either $a = 0$ or $\frac{1}{a} \notin V$. If $a = 0$ then $ra = 0$ for all $r \in V$, and if $\frac{1}{a} \notin V$, then $\frac{1}{ra} \notin V$ for all $r \in V$. Thus the multiplicative property of an ideal follows. To show that $\mathfrak{m}$ is an additive subgroup of $V+$ it suffices to show that if a and b are in $\mathfrak{m}$, then so is $a - b$. We may assume that a and b are non zero elements of V, thus either $\frac{a}{b}$ or $\frac{b}{a} \in V$. Assume $\frac{a}{b} \in V$, then by the multiplicative property which we have already proven, $a - b = (\frac{a}{b} - 1)b \in \mathfrak{m}$.

(ii) is obvious. To show (iii), let $x \in K$ be integral over V, so there are elements $a_1, \ldots, a_n \in V$ such that $x^n + a_1 x + \cdots + a_{n-1}x + a_n = 0$. If now $x \notin V$, then we have $\frac{1}{x} \in V$, so that

$$x = -\left(a_1 + a_2 \frac{1}{x} + \cdots + a_n \left(\frac{1}{x} \right)^{n-1} \right) \in V,$$

which is a contradiction. Thus $x \in V$, and (iii) is proven.   □

We note that $V = K$ satisfies the definition of a valuation ring. This case is referred to as *the trivial valuation ring of K*.

Valuation rings are characterized by a remarkable property:

Proposition 4.21 *The following are equivalent for an integral domain R and its quotient field K.*

(i) R *is a valuation ring of K.*
(ii) *If $\mathfrak{a}$ and $\mathfrak{b}$ are two ideals in R, then either $\mathfrak{a} \subseteq \mathfrak{b}$ or $\mathfrak{b} \subseteq \mathfrak{a}$.*

Proof (i) $\Longrightarrow$ (ii): Assume the converse of (ii), and let $a \in \mathfrak{a}$ be such that $a \notin \mathfrak{b}$, and $b \notin \mathfrak{a}$ be such that $b \in \mathfrak{b}$. We have either $\frac{a}{b} \in R$ or $\frac{b}{a} \in R$ by definition of valuation rings. But if $\frac{a}{b} \in R$, then $a = b\frac{a}{b} \in \mathfrak{b}$, while if $\frac{b}{a} \in R$ we find that $b = a\frac{b}{a} \in \mathfrak{a}$. The contradiction proves (ii).

(ii) $\Longrightarrow$ (i): Let $\mathfrak{a} = (a)R$, and $\mathfrak{b} = (b)R$, by (ii) we have $(a)R \subseteq (b)R$ or $(b)R \subseteq (a)R$. In the latter case $b = ra$, so $\frac{b}{a} \in R$, while the former gives $a = sb$, so $\frac{a}{b} \in R$.   □

A partial ordering on a set Σ is a relation $\geq$ such that

(PO 1) $a \geq a$
(PO 2) If $a \geq b$ and $b \geq c$ then $a \geq c$
(PO 3) If $a \geq b$ and $b \geq a$, then $a = b$.

If Σ is *totally ordered* if in addition

(TO 1) For all a and b we have either $a \geq b$ or $b \geq a$.

Thus by Proposition 4.21 the set of ideals in a valuation ring V is a totally ordered set, and in particular this is so for *the set of prime ideals*. We make the following definition:

Definition 4.4 The ordinal type of the set of proper prime ideals in a valuation ring V is referred to as the rank of V. In particular, if V is of Krull dimension n, then it is said to have rank n.

An *ordered group* is an Abelian group Γ where there is defined a partial ordering such that

(POG) If $a \geq b$ then for all c $a + c \geq b + c$.

A totally ordered group is an ordered group under a total ordering.
If $a \geq b$ and $a \neq b$ we write $a > b$. Immediate examples of totally ordered Abelian groups are $\mathbb{Z}+, \mathbb{Q}+$ and $\mathbb{R}+$, the integers, rational numbers or the real numbers under the usual addition.

Definition 4.5 Let K be a field and Γ be a *totally ordered Abelian group*. A valuation of K with values in Γ is a surjective mapping

$$v : K \longrightarrow \Gamma \cup \{\infty\}$$

where ∞ is an additional symbol greater than all the element in Γ, such that for all a and b in K

(V 1) $v(ab) = v(a) + v(b)$
(V 2) $v(a + b) \geq \min\{v(a), v(b)\}$
(V 3) $v(a) = \infty \Longleftrightarrow a = 0$.

According to this definition the mapping $v_0 : K \cup \{\infty\} \longrightarrow \{0, \infty\}$ given by $v_0(a) = 0$ for all non-zero elements a and $v_0(0) = \infty$ is a valuation, a rather uninteresting one, as a matter of fact. v_0 is called the *trivial* valuation of K. Two valuations v and v' with value groups Γ and Γ', of the same field K, are said to be equivalent if there is an isomorphism as ordered groups $\varphi : \Gamma \longrightarrow \Gamma'$ such that $v' = \varphi \circ v$. The relationship between valuations and valuation rings is given by the following

Proposition 4.22 *If v is a valuation of the field K, then the set $V_v = \{x \in K | v(x) \geq 0\}$ is a valuation ring with quotient field K. Conversely, if V is a valuation ring with quotient field K, then there is a valuation v of K such that $V_v = V$. The valuation v is unique up to equivalence. The maximal ideal of non-units in V_v is $\mathfrak{m}_v = \{x \in K | v(x) > 0\}$.*

Proof We start out by collecting some basic facts we need in the

Lemma 4.23 *Let v be a valuation of the field K. Then*

(i) $v(\frac{1}{a}) = -v(a)$
(ii) $v(\frac{a}{b}) = v(a) - v(b)$
(iii) $v(a - b) \geq \min\{v(a), v(b)\}$
(iv) $v(a) < v(b) \Longrightarrow v(a + b) = v(a)$.

Proof of lemma Since $1 = 1 \cdot 1$ we have $v(1) = v(1) + v(1)$, hence $v(1) = 0$, the zero element in the value group Γ. Thus $0 = v(1) = v(a \cdot (\frac{1}{a})) = v(a) + v(\frac{1}{a})$, so $v(\frac{1}{a}) = -v(a)$ and (i) follows. (i) immediately implies (ii). Applying (V 2) to a and $-b$, we get (iii). To prove (iv), note first that by (V 2) $v(a+b) \geq v(a)$, while on the other hand $v(a) = v(a + b - b) \geq \min\{v(a + b), v(b)\}$ by (iii). Since $v(a) < v(b)$ we therefore must have $v(a) \geq v(a + b)$. By (V 2) it then follows that $v(a) = v(a + b)$, as claimed. □

Returning to the proof of the proposition, we prove first that V_v is a ring. In fact, by (i) V_v is closed under the operation of taking additive differences, thus it is a subgroup of $K+$. V_v also contains 1, and is closed under multiplication by (V 1). So V_v is a subring of K. It is a valuation ring since for all non zero element $a \in K$ we either have $v(a) > 0$ or $v(a) < 0$. In the former case $a \in V_v$, in the latter case $\frac{1}{a} \in V_v$ by (iii) in the lemma.

Now let V be a valuation ring for K, we shall construct a valuation v such that $V = V_v$. We consider the *relation of association* in $K - \{0\}$:

$$a \sim b \iff \exists u \in V^* \text{ such that } a = ub.$$

This is an equivalence relation in $K - \{0\}$, and we define $\Gamma = K - \{0\}/\sim$. Now $K - \{0\}$ is an Abelian group under multiplication, and $\sim$ is a congruence relation for this multiplication, thus Γ becomes an Abelian group under the operation $[a] + [b] = [ab]$. Moreover, it is easily checked that the relation

$$[a] \geq [b] \iff \exists r \in V \text{ such that } a = rb$$

makes Γ into a totally ordered Abelian group. This gives a valuation of K with value group Γ, and we get $V_v = V$. These routine verifications are left to the reader. □

Example For the rings which we gave the preliminary definition as discrete valuation rings of rank 1, we had that the maximal ideal $\mathfrak{m} \subset \mathcal{O}$ is generated by one element, $\mathfrak{m} = (\pi)\mathcal{O}$. Then a non-zero element $a \in \mathcal{O}$ may be written uniquely as $a = u\pi^n$, where $u \in \mathcal{O}^*$, with the integer $n \geq 0$, while any non-zero element from K may be written in this way with a possibly negative integer. We then define v by letting

$$v(u\pi^n) = n, \quad \text{and} \quad v(0) = \infty.$$

It is easily seen that this is a valuation, with value group $\mathbb{Z}$, and since $\mathfrak{m}$ is the only proper prime ideal, the rank is 1.

We next come to the general relation between the value group of a valuation and the set of ideals in the corresponding valuation ring. We start off by making some definitions:

Definition 4.6 Let v be a valuation of the field K with value group Γ and valuation ring V_v.

(i) A non-empty subset $\Delta \subset \Gamma$ is called a *segment* if whenever $\alpha \geq 0$ and $\alpha \in \Delta$, we have

$$\{\beta \mid -\alpha \leq \beta \leq \alpha\} \subseteq \Delta.$$

A segment which is a proper subgroup of Γ is called an isolated subgroup of Γ.

(ii) If A is a subset of the valuation ring V_v, we put $Av = v(A) \subset \Gamma$, and put $\Gamma_A = \Gamma - (Av \cup -Av)$, the complement in Γ of $Av \cup -Av$.

We have the following result:

Theorem 4.24 Let $\mathfrak{A}$ be a proper ideal in V_v. Then $\Gamma_{\mathfrak{A}}$ is a segment in Γ. If $\mathfrak{P}$ is a prime ideal, then $\Gamma_{\mathfrak{P}}$ is an isolated subgroup, and conversely. The correspondence between segments and ideals is inclusion reversing.

Example If V_v is a discrete valuation ring as preliminary defined earlier, then all ideals may be written as $\mathfrak{A} = (\pi^n)V_v$, thus $\mathfrak{A}v = \{n, n+1, n+2, \dots\}$, so $\Gamma_{\mathfrak{A}} = \langle -(n-1), n-1 \rangle$, all integers between $-(n-1)$ and $n-1$. The only possibility for an isolated subgroup is when $n - 1 = 0$, that is $n = 1$, so we get the maximal ideal as expected since there are no other proper prime ideals in V_v.

Proof If $\mathfrak{A}$ is a proper ideal in V_v, then $\mathfrak{A}v$ is non-empty and contains only positive elements. Thus $\Gamma_{\mathfrak{A}}$ at least contains the zero element, so it is non-empty. $\Gamma_{\mathfrak{A}}$ is also a proper subset of Γ.

Since $\mathfrak{A}V_v \subset \mathfrak{A}$, we get $\mathfrak{A}v + \Gamma_+ \subset \mathfrak{A}v$, thus $\alpha \in \mathfrak{A}v$ and $\beta \geq \alpha$ implies $\beta \in \mathfrak{A}v$. Hence $\Gamma_{\mathfrak{A}}$ is a segment. Conversely one shows easily that if Γ_A is a segment, then A is an ideal. Being a prime ideal is equivalent to the complement in the ring being closed under multiplication, this implies that the corresponding segment is an isolated subgroup.   □

Our final result to be given here ties up with the preliminary definition:

Theorem 4.25 *The valuation ring R is Noetherian if and only if the corresponding valuation v has $\mathbb{Z}+$ as valuation group.*

Proof An ascending chain of ideals in a valuation ring R corresponds to a descending chain of segments in the value group Γ. If $\Gamma = \mathbb{Z}$, the segments are of the form $\langle -n, n \rangle$, thus any descending chain of segments in $\mathbb{Z}$ becomes stationary. So R is Noetherian.

Assume conversely that R is Noetherian. We show first that then R is of rank 1. Assume the converse. Since the zero subgroup of Γ is always an isolated subgroup, this means that there is a non-zero isolated proper subgroup Δ of Γ. Let $\alpha > 0$ be a positive element of Δ. Then

$$\alpha < 2\alpha < 3\alpha < \cdots < m\alpha \cdots .$$

Let $\beta > 0$ be an element outside Δ. If there existed n such that $\beta < n\alpha$, then β would be an element of Δ since Δ is a segment. Thus $\beta > m\alpha$ for all m. Hence Γ contains an infinite descending sequence of positive elements

$$\beta > \beta - \alpha > \beta - 2\alpha > \cdots > \beta - m\alpha > \cdots .$$

But a descending sequence of positive elements determine a descending chain of segments, which yields an ascending chain of ideals in R. Since R is Noetherian by assumption, this chain becomes stationary, thus so does the chain of segments in Γ, contradicting that the chain of positive elements is strictly descending. This proves that the rank is 1.

Since R is Noetherian, Γ has a minimal positive element α: Otherwise we could construct an infinite strictly descending chain of segments, thus an infinite strictly ascending chain of ideals in R. Now consider the set

$$\Delta = \{ m\alpha \mid m \in \mathbb{Z} \}.$$

No element of Γ can lie between $m\alpha$ and $(m + 1)\alpha$ for any m, since such an element would yield a positive element strictly smaller than α. Thus the subgroup Δ is a segment, so since the rank is 1 we must have $\Delta = \Gamma$, and the claim follows.   □

A valuation of rank 1 is said to be *discrete* if the valuation group is $\mathbb{Z}$. Thus the last theorem states that a valuation ring is Noetherian if and only if the corresponding valuation is discrete of rank 1.

4.6 Linear Systems of Plane Curves

Let k be a field, and let $N \geq 0$ be an integer. In Chap. 1 we defined affine and projective space over k of dimension N, $\mathbb{A}_k^n$ and $\mathbb{P}_k^n$, respectively. We have now explored some of the basic properties of affine and projective plane curves. We now study properties of such curves which lead to subvarieties of higher dimensional spaces, and start with some linear systems of plane projective curves. The main source for this section is Chap. 5 of [10].

So let $d \geq 1$ be an integer, and let C_d denote the set of all projective plane curves in $\mathbb{P}_k^2$ of degree d. Each such curve can be viewed as an equivalence class of a homogeneous polynomial in X_0, X_1, X_2 with coefficients from k, $F(X_0, X_1, X_2)$, under the equivalence relation given by multiplication with non zero elements form k,

$$F(X_0, X_1, X_2) \sim G(X_0, X_1, X_2) \iff \exists \alpha \in k^* \text{ such that } F = \alpha G.$$

The set of all homogeneous polynomials in X_0, X_1, X_2 constitutes a vector space V_d over k of dimension $N = \binom{d+2}{2}$, we fix an ordering of the of the monomials of degree d, for instance the lexicographical ordering

$$X_0^d, X_0^{d-1}X_1, \ldots, X_0^{i_0}X_1^{i_1}X_2^{i_2}, \ldots, X_3^d.$$

In this manner the set C_d is parametrized by the projective space $\mathbb{P}_k^{\frac{d(d+3)}{2}}$.

Definition 4.7 Let $\mathcal{S}$ be a subset of C_d which corresponds to an algebraic subset of $\mathbb{P}_k^{\frac{d(d+3)}{2}}$. Then $\mathcal{S}$ is referred to as an algebraic system of plane curves of degree d. If the algebraic subset is actually linear, so it is defined by linear equations in $\mathbb{P}_k^{\frac{d(d+3)}{2}}$, then $\mathcal{S}$ is called a linear system of plane curves of degree d. The dimension of the linear system is the dimension of the corresponding linear subspace.

We have the

Proposition 4.26 *Let there be given* $m \leq \frac{d(d+3)}{2}$ *points in* $\mathbb{P}_k^2$, $Q_1, \ldots, Q_m$. *The set of projective curves i* $\mathbb{P}_k^2$ *which contain* $Q_1, Q_2, \ldots, Q_m$ *is a non empty linear system.*
If

$$m \leq d + 1$$

then this linear system is of dimension

$$\frac{d(d+3)}{2} - m.$$

Proof Let $C_d(Q)$ denote the set of curves in C_d which contain the point $Q = (q_0 : q_1 : q_2) \in \mathbb{P}_k^2$. For the curve given by the polynomial

$$F(X_0, X_1, X_2) = \sum_{i_0 + i_1 + i_2 = d} a_{i_0, i_1, i_2} X_0^{i_0} X_1^{i_1} X_2^{i_2}$$

the corresponding point in $\mathbb{P}_k^{\frac{d(d+3)}{2}}$ has coordinates $(a_{i_0, i_1, i_2} | i_0 + i_1 + i_2 = d)$, and the condition of being in $C_d(Q)$ is expressed as

$$\sum_{i_0 + i_1 + i_2 = d} a_{i_0, i_1, i_2} q_0^{i_0} q_1^{i_1} q_2^{i_2} = 0.$$

Thus $C_d(Q)$ constitutes a hyperplane in $\mathbb{P}_k^{\frac{d(d+3)}{2}}$, and the first part of the claim follows.

Then assume that $m \le d + 1$. We may assume that

$$Q_i = (1 : \alpha_i : \beta_i)$$

where all $\alpha_i's$ are different. To show that the m hyperplanes in $\mathbb{P}_k^{\frac{d(d+3)}{2}}$ cut out a linear subspace of codimension m, we must verify that

$$\mathrm{rk} \begin{bmatrix} 1, & \alpha_1, & \beta_1, & ,\alpha_1^2, & \alpha_1\beta_1, & \beta_1^2, & \ldots, & \beta_1^d \\ 1, & \alpha_2, & \beta_2, & ,\alpha_2^2, & \alpha_2\beta_2, & \beta_2^2, & \ldots, & \beta_2^d \\ & & \ldots & & \ldots & & & \\ 1, & \alpha_m, & \beta_m, & ,\alpha_m^2, & \alpha_m\beta_m, & \beta_m^2, & \ldots, & \beta_m^d \end{bmatrix} = m.$$

For $m - 1 \le d$ this is clear, since

$$\begin{vmatrix} 1, & \alpha_1, & \alpha_1^2, & \ldots, & \alpha_1^{m-1} \\ 1, & \alpha_2, & \alpha_2^2, & \ldots, & \alpha_2^{m-1} \\ & & \ldots & \ldots & \\ 1, & \alpha_m, & \alpha_m^2, & \ldots, & \alpha_m^{m-1} \end{vmatrix} \neq 0.$$

$\square$

Proposition 4.27 *Let $P \in \mathbb{P}_k^2$ and $r \le d + 1$. The set of projective curves i $C \subset \mathbb{P}_k^2$ which contain P with multiplicity $m_P(C) \ge r$ is a linear system of dimension $\frac{d(d+3)}{2} - \frac{r(r+1)}{2}$.*

Proof Obviously a change of coordinate system in $\mathbb{P}_k^2$ just corresponds to a change of coordinate system in $\mathbb{P}_k^{\frac{d(d+3)}{2}}$. Thus we may assume that $P = (1 : 0 : 0)$. Write

$$F(X_0, X_1, X_2) = \sum F_i(X_1, X_2) X_0^{d-i}$$

where $F_i(X_1, X_2)$ is homogeneous in X_1 and X_2 of degree i. Then the affine restriction to $D_+(X_0) = \mathbb{A}_k^2$ is given by the polynomial

$$f(X, Y) = \sum F_i(X, Y)$$

and thus

$$m_P(C) \geq r \iff F_0 = F_1 = \cdots = F_{r-1} = 0$$
$$\iff a_{i_0, i_1, i_2} = 0 \quad \text{provided } i_1 + i_2 < r.$$

These conditions may be arranged as follows:

Conditions: Vanishing coordinates		
$F_0 = 0$	$a_{d,0,0} = 0$	1
$F_1 = 0$	$a_{d-1,1,0} = a_{d-1,0,1} = 0$	2
$F_2 = 0$	$a_{d-2,2,0} = a_{d-2,1,1} = a_{d-2,0,2} = 0$	3
$\vdots$	$\vdots$	$\vdots$
$F_{r-1} = 0$	$a_{d-r+1,r-1,0} = a_{d-r+1,r-2,1} = \cdots = a_{d-r+1,0,r-1} = 0$	r

As the total number of vanishing coordinates is $\frac{r(r+1)}{2}$, the proof is complete.   □

We are driving at the important concept of the linear system of degree d-curves with prescribed multiplicities:

Definition 4.8 Let $P_1, P_2, \ldots, P_n \in \mathbb{P}_k^2$, and let $r_1, r_2, \ldots, r_n \geq 0$ be integers. Denote by $L(d; r_1 P_1, \ldots, r_n P_n)$ the set of all degree d curves in $\mathbb{P}_k^2$ such that

$$m_{P_i}(C) \geq r_i$$

for all $i = 1, \ldots, n$.

We then have the following result, see [10], p. 110:

Theorem 4.28 (i) $L(d; r_1 P_1, \ldots, r_n P_n)$ *is a linear system of dimension* $\geq$ $\frac{d(d+3)}{2} - \sum_{i=1}^n \frac{r_i(r_i+1)}{2}$.
(ii) *If* $d \geq (\sum_{i=1}^n r_i) - 1$ *then equality holds in* (i).

Proof (i) follows from Proposition 4.27: For two linear subspaces L_1 and L_2 of some $\mathbb{P}_k^N$ we have, with $\operatorname{codim}(L_i) = N - \dim(L_i)$,

$$\operatorname{codim}(L_1 \cap L_2) \leq \operatorname{codim}(L_1) + \operatorname{codim}(L_2),$$

and thus in general

$$\text{codim}\left(\bigcap_{i_1}^m L_i\right) \le \sum_{i=1}^m \text{codim}(L_i).$$

In the present situation $N = \frac{d(d+3)}{2}$ and $\text{codim}(L(d; r_i P_i)) = \frac{r_i(r_i+1)}{2}$, thus the claim follows.

For (ii) we use induction on $m = \sum_{i=1}^n r_i$. We may assume that $d > 1$. Indeed, if $d = 1$ then the curves C are lines. In this case the curves are parametrized by $\mathbb{P}_k^2$.[2] The condition

$$d = 1 \ge \left(\sum r_i\right) - 1$$

yields

$$\sum r_i = \begin{cases} 0 & \text{then } L(1; 0P_1, \dots, 0P_n) = \mathbb{P}_k^2 \\ 1 & \text{then } \dim(L(d; 1P_1, 0P_2, \dots, 0P_n)) = 1 \\ 2 & \text{then } L(1; 2P_1, 0P_2, \dots, 0P_n) = \emptyset, \dim = -1 \\ 2 & \text{then } \dim(L(1; 1P_1, 1P_2, 0P_3, \dots, 0P_n)) = 0. \end{cases}$$

In all these cases the claim in (ii) holds.

We may also assume that $\sum_{i=1}^n r_i > 1$, since $\sum_{i=1}^n r_i = 1$ imposes no condition on d and yields $r_1 = 1$ and all other $r_i = 0$. The claim follows then by Proposition 4.27.

So assume the claim for $\sum_{i=1}^n r_i = 0, 1, \dots, m-1$. We may assume that all $r_i \ge 1$. We distinguish between two cases.

1. All $r_i = 1$. Then $\sum_{i=1}^n \frac{r_i(r_i+1)}{2} = n$, and $d \ge (\sum r_i) - 1 = n - 1$ so the claim follows in this case from Proposition 4.26.
2. Next assume that one of the r_i, say r_1, is greater than 1. We may assume that $P_1 = (1 : 0 : 0)$. Put

$$L_0 = L(d, (r_1 - 1)P_1, r_2 P_2, \dots, r_n P_n).$$

By the induction assumption we have

$$\dim(L_0) = \frac{d(d+3)}{2} - \left(\frac{1}{2}(r_1 - 1)r_1 + \sum_{i=2}^n \frac{r_i(r_i+1)}{2}\right).$$

Now take a curve $C \in L_0$, corresponding to the homogeneous degree d polynomial $F_C(X_0, X_1, X_2)$, and consider the polynomial

$$F_C(1, X, Y) = \sum_{i=0}^{r_1-1} a(C)_i X^i Y^{r_1-1-i} + \text{terms of higher degrees}.$$

---

[2] As we encountered before in Chap. 1, Sect. 1.5.

Then put

$$L_i = [C \in L_0 \,|\, a(C)_j = 0 \text{ for all } j < i].$$

We then have

$$L_0 \supset L_1 \supset \cdots \supset L_{r_1} = L(d, r_1 P_1, \ldots, r_n P_n),$$

and clearly

$$\dim(L_i) - \dim(L_{i+1}) = \begin{cases} 0 \\ \text{or} \\ 1. \end{cases}$$

It now suffices to show that

$$\dim(L_i) - \dim(L_{i+1}) \neq 0:$$

Indeed, if so we have

$$\dim(L_{r_1}) + r_1 = \dim(L_0)$$

thus

$$\begin{aligned}
\dim(L_{r_1}) &= \dim(L_0) - r_1 \\
&= \frac{d(d+3)}{2} - \left(\frac{1}{2} r_1(r_1 - 1) + \sum_{r \geq 2} r_i(r_i + 1) \right) - r_1 \\
&= \frac{d(d+3)}{2} - \frac{1}{2} \left(r_1(r_1 - 1) + 2r_1 + \sum_{r \geq 2} r_i(r_i + 1) \right) \\
&= \frac{d(d+3)}{2} - \frac{1}{2} \sum_{r \geq 1} r_i(r_i + 1))
\end{aligned}$$

and the claim follows.

So it remains to show that for all $i = 0, \ldots, r-1$ we have $L_i \neq L_{i+1}$. Let

$$M_0 = L(d - 1, (r_1 - 2)P_1, r_2 P_2, \ldots, r_n P_n).$$

In an analogous manner to what we did above, take a $C \in M_0$, denote the corresponding homogeneous polynomial by $F_C(X_0, X_1, X_2)$ and put

$$F_C(1, X, Y) = \sum_{i=0}^{r_1 - 1} a(C)_i X^i Y^{r_1 - 2 - i} + \text{terms of higher degrees}$$

and

$$M_i = \{C \in M_0 \,|\, a(C)_j = 0 \text{ for all } j < i\}.$$

As above the induction assumption yields strict inclusions

$$M_0 \supset M_1 \supset \cdots \supset M_{r_1-1} = L(d-1, (r_1-1)P_1, r_2 P_2, \ldots, r_n P_n)$$

and we let F_i be a homogeneous polynomial corresponding to a cure in the complement $M_i - M_{i+1}$ of M_{i+1} in M_i, for $i = 0, \ldots, r_1 - 2$. Then the homogeneous polynomial $X_2 F_i$ corresponds to a curve in $L_i - L_{i+1}$ for $i = 0, \ldots, r_1 - 2$ and $X_1 F_{r_1-2}$ corresponds to a curve in $L_{r_1-1} - L_{r_1}$.

This completes the proof of the theorem.   □

4.7 Affine Restriction and Projective Closure

At this point it is convenient to introduce a useful algebraic notation, which will be employed extensively in the following. For more details, see Zariski and Samuel [42], vol. II, Chap. VII. Fulton [10] uses a different notation. Let $F(X_0, X_1, \ldots, X_n)$ be a homogeneous polynomial over any field (or more generally commutative ring) k. We then put

$$^aF(X_1, X_2, \ldots, X_n) = F(1, X_1, X_2, \ldots, X_n).$$

This makes correspond a (non-homogeneous) polynomial in the n indeterminates $X_1, \ldots, X_n$ to any homogeneous polynomial in the $n+1$ indeterminates $X_0, X_1, \ldots, X_n$.

Conversely, given a polynomial in $X_1, X_2, \ldots, X_n$, we define a homogeneous polynomial $^hf(X_0, X_1, \ldots, X_n)$ in the indeterminates $X_0, X_1, \ldots, X_n$ as follows: Let $\partial(f)$ denote the total degree of f, then

$$^hf(X_0, X_1, \ldots, X_n) = X_0^{\partial(f)} f\left(\frac{X_1}{X_0}, \frac{X_2}{X_0}, \ldots, \frac{X_n}{X_0}\right)$$

where evidently the right hand side is a polynomial and not merely a rational expression. ^hf is referred to as the *homogenization* of f, while ^aF will be referred to as the *affine restriction* of F. We note the following simple facts, easily verified:

Proposition 4.29

(i) $^h(fg) = {^hf} \cdot {^hg}$
(ii) $X_0^{\partial(f)+\partial(g)} \cdot {^h(f+g)} = X_0^{\partial(f+g)}(X_0^{\partial(g)} \cdot {^hf} + X_0^{\partial(f)} \cdot {^hg})$
(iii) $^a(FG) = {^aF} \cdot {^aG}$
(iv) $^a(F+G) = {^aF} + {^aG}$
(v) $^a({^hf}) = f$
(vi) $X_0^m \cdot {^h}({^a}(F)) = F$ where X_0^m is the highest power of X_0 which divides F.

We shall return to this later, and extend these concepts to analyze the relation between ideals and homogeneous ideals. For now, we shall see how we may now simplify the interplay between projective curves and their affine restrictions.

Let the projective curve $C \subset \mathbb{P}_k^2$ be given by the homogeneous polynomial $F(X_0, X_1, X_2)$. Then the affine restriction to $\mathbb{A}_k^2$ identified with $D_+(X_0) \subset \mathbb{P}_k^2$ is given by the polynomial ${}^a F(X, Y)$, and is denoted by ${}^a C$. An affine curve $K \subset \mathbb{A}_k^2$ given by a polynomial $f(X, Y)$ has a *projective closure* denoted by ${}^h K$ in $\mathbb{P}_k^2$, under the same identification $\mathbb{A}_k^2 = D_+(X_0)$. It is given as the projective curve defined by the homogeneous polynomial ${}^h f(X_0, X_1, X_2)$. In accordance with Proposition 4.29 we then have

$$ {}^a({}^h K) = K $$

and

$$ {}^h({}^a C) = C' $$

where C' is obtained by removing multiples of $V_+(X_0)$ which might be present in C.

4.8 Bézout's Theorem

We have now arrived at the following important result, actually found by *Colin Maclaurin*, 1698–1746, but first given a satisfactory proof by *Etienne Bézout*, 1730–1783.

Theorem 4.30 *Let C_1 and C_2 be two curves in $\mathbb{P}_k^2$, without common components. Then*

$$ \sum_{P \in \mathbb{P}_k^2} I(P, C_1 \cap C_2) = \deg(C_1) \cdot \deg(C_2). $$

Proof By Proposition 4.13 the set $C_1 \cap C_2$ is finite, hence we may assume that it is disjoint from $V_+(X_0)$. Indeed, given a finite set $F \subset \mathbb{P}_k^2$ for an infinite field k we may always choose a new projective coordinate system

$$ \overline{X}_0 = \alpha_{0,0} X_0 + \alpha_{0,1} X_1 + \alpha_{0,2} X_2 $$
$$ \overline{X}_1 = \alpha_{1,0} X_0 + \alpha_{1,1} X_1 + \alpha_{1,2} X_2 $$
$$ \overline{X}_2 = \alpha_{2,0} X_0 + \alpha_{2,1} X_1 + \alpha_{2,2} X_2 $$

such that $F \cap V_+(\overline{X}_0) = \emptyset$. To see this, suppose that

$$ F = \{ B_i = (\beta_{0,i} : \beta_{1,i} : \beta_{2,i}) \mid i = 1, \ldots, s \}. $$

We first wish to find

$$\alpha_{0,0}, \alpha_{0,1}, \alpha_{0,2}$$

such that

$$\alpha_{0,0}\beta_{0,i} + \alpha_{0,1}\beta_{1,i} + \alpha_{0,2}\beta_{2,i} \neq 0 \quad \text{for all } i = 1, \ldots, s.$$

For this, note that the subset of k^3

$$\{(x, y, z) \in k^3 \mid x\beta_0 + y\beta_1 + z\beta_2 = 0\}$$

is a linear sub-vector space over k of dimension 2, provided that $(\beta_0, \beta_1, \beta_2) \neq (0, 0, 0)$. A finite union of such sub-vector spaces can never fill up the entire vector space, as guaranteed by the lemma we prove below. So we find a linear form

$$L = \alpha_{0,0}X_0 + \alpha_{0,1}X_1 + \alpha_{0,2}X_2$$

such that $F \cap V_+(L) = \emptyset$, we then let $\overline{X}_0 = L$, and proceed to find the remaining $\alpha_{i,j}$ with $i = 1, 2$ and $j = 0, 1, 2$ such that the matrix

$$\begin{bmatrix} \alpha_{0,0} & \alpha_{0,1} & \alpha_{0,2} \\ \alpha_{1,0} & \alpha_{1,1} & \alpha_{1,2} \\ \alpha_{2,0} & \alpha_{2,1} & \alpha_{2,2} \end{bmatrix}$$

has determinant $\neq 0$. We now state and prove the lemma.

Lemma 4.31 *Let V be an n-dimensional vector space over a field with infinitely many elements. Let $V_i, i = 1, \ldots, s$ be a set of proper linear subspaces. Then*

$$\bigcup_{i=1}^{s} V_i \neq V.$$

Proof of Lemma 4.31 We may assume that all V_i are of dimension $n - 1$, enlarging some of them if necessary. We also may identify V with k^n. Then there are elements $\beta_{i,j} \in k$ such that for all $i = 1, 2, \ldots, s$

$$V_i = \{(a_1, a_2, \ldots, a_n) \in k^n \mid a_1\beta_{1,i} + a_2\beta_{2,i} + \cdots + a_n\beta_{n,i} = 0\}.$$

Writing

$$L_i(Z_1, Z_2, \ldots, Z_n)) = Z_1\beta_{1,i} + Z_2\beta_{2,i} + \cdots + Z_n\beta_{n,i}$$

we then have

$$V_i = \{(a_1, a_2, \ldots, a_n) \in k^n \mid L_i(a_1, a_2, \ldots, a_n) = 0\}.$$

Now let

$$F(Z_1, Z_2, \ldots, Z_n) = \prod_{i=1}^{s} L_i(Z_1, Z_2, \ldots, Z_n).$$

The proof of Lemma 4.31 is now completed by the

Lemma 4.32 *Let k be an infinite field, and let $F(Z_1, Z_2, \ldots, Z_n)$ be a non zero polynomial. Then there exists a point $(a_1, a_1, \ldots, a_n) \in \mathbb{A}_k^n$ such that $F(a_1, a_2, \ldots, a_n) \neq 0$.*

Proof of Lemma 4.32 We use induction on n. For $n = 1$ the claim follows since a polynomial equation of degree d has at most d roots. Assume the claim for $n - 1$, and write

$$F(Z_1, Z_2, \ldots, Z_n)$$
$$= A_0(Z_1, Z_2, \ldots, Z_{n-1}) Z_n^N + A_1(Z_1, Z_2, \ldots, Z_{n-1}) Z_n^{N-1}$$
$$+ \cdots + A_{N-1}(Z_1, Z_2, \ldots, Z_{n-1}) Z_n + A_N(Z_1, Z_2, \ldots, Z_{n-1}).$$

By the induction assumption we find

$$(a_1, \ldots, a_{n-1}) \in \mathbb{A}_k^{n-1}$$

such that not all

$$A_0(a_1, \ldots, a_{n-1}), \quad A_1(a_1, \ldots, a_{n-1}), \quad \ldots, \quad A_N(a_1, \ldots, a_{n-1})$$

vanish, and then by the case $n = 1$ we find a_n such that

$$A_0(a_1, a_2, \ldots, a_{n-1}) a_n^N + A_1(a_1, a_2, \ldots, a_{n-1}) a_n^{N-1}$$
$$+ \cdots + A_{N-1}(a_1, a_2, \ldots, a_{n-1}) a_n + A_N(a_1, a_2, \ldots, a_{n-1}) \neq 0.$$

This completes the proof of Lemma 4.32 and thus of Lemma 4.31.   □

Since we may assume that $V_+(X_0)$ and $C_1 \cap C_2$ are disjoint, we get $C_1 \cap C_2 = K_1 \cap K_2$ where the affine restrictions K_1, K_2 are given by $f(X, Y) = {}^a F(X, Y)$ and similarly $g(X, Y) = {}^a G(X, Y)$.

By Proposition 4.19 we then find that

$$\sum_{P \in \mathbb{P}_k^2} I(P, C_1 \cap C_2) = \sum_{P \in \mathbb{A}_k^2} I(P, K_1 \cap K_2)$$

$$= \dim_k(k[X, Y]/(f, g)k[X, Y]).$$

We put $m = \deg(F), n = \deg(G)$. We introduce the following notations:

$$S = k[X_0, X_1, X_2]$$
$$\overline{S} = S/(F, G)S.$$

We let $S_d, \overline{S}_d$ denote the vector spaces of homogeneous elements of degree d. To prove Bézont's theorem, it clearly suffices to show the following:

Lemma 4.33 *For all $d \gg 0$ we have*

1. $\dim_k(\overline{S})_d = mn$.
2. $\dim_k(\overline{S}_d) = \dim_k(k[X,Y]/(f,g)k[X,Y])$.

Proof of Lemma 4.33 1. To prove that

$$d \geq m + n \quad \Longrightarrow \quad \dim_k(\overline{S}_d) = mn$$

we first claim that there is an exact sequence

$$0 \to S \overset{\psi}{\to} S \times S \overset{\varphi}{\to} S \overset{\pi}{\to} \overline{S} \to 0$$

where π is the canonical surjection, $\varphi(A,B) = AF + BG$ and $\psi(C) = (GC, -FC)$. Indeed, π is surjective and ψ is injective by the implications:

$$\psi(C) = 0 \quad \Longrightarrow \quad (GC, -FC) = (0,0) \quad \Longrightarrow \quad GC = FC = 0 \quad \Longrightarrow \quad C = 0,$$

the last one since S is an integral domain. Clearly $\mathrm{im}(\varphi) = (F,G)S = \ker(\pi)$, and it remains to show that

$$\mathrm{im}(\psi) = \ker(\varphi):$$

If $(D,E) \in \mathrm{im}(\psi)$ then $D = GC, E = -FC$ for some polynomial C. Then $\varphi(D,E) = DF + BG = GCF - FCG = 0$, thus

$$\mathrm{im}(\psi) \subseteq \ker(\varphi).$$

Conversely, if $(D,E) \in \ker(\varphi)$ then $\varphi(D,E) = DF + EG = 0$ so $DF = -EG$. As F and G have no common factor, this implies

$$F|E \quad \text{so } E = H_1 F$$

$$G|D \quad \text{so } D = H_2 G$$

thus

$$H_2 GF = -H_1 FG$$

and since S is an integral domain $H_2 = -H_1$. Hence

$$(D,E) = \psi(H_2)$$

thus

$$\mathrm{im}(\psi) \supseteq \ker(\varphi).$$

Thus exactness of the sequence is proven.

Since F and G are homogeneous of degree m and n, respectively, the mappings ψ, φ and π induce mappings for all $d \geq m + n$, denoted by the same letters, which make the following diagram of k-vector spaces exact:

$$0 \to S_{d-m-n} \xrightarrow{\psi} S_{d-m} \times S_{d-n} \xrightarrow{\varphi} S_d \xrightarrow{\pi} \overline{S}_d \to 0.$$

This yields

$$\dim_k(S_{d-(m+n)}) - (\dim_k(S_{d-m}) + \dim_k(S_{d-n})) + \dim_k(S_d) - \dim_K(\overline{S}_d) = 0$$

and since $\dim_k(S_r) = \frac{(r+1)(r+2)}{2}$, we get

$$\dim_k(\overline{S}_d) = \dim(S_{d-(m+n)}) - \dim_k(S_{d-m}) - \dim_k(S_{d-n}) + \dim_k(S_d)$$

$$= \frac{1}{2}\{(d - (m+n) + 1)(d - (m+n) + 2) - (d - m + 1)(d - m + 2)$$

$$- (d - n + 1)(d - n + 2) + (d + 1)(d + 2)\} = mn.$$

This completes the proof of part 1. of Lemma 4.33.

To prove 2., define

$$\alpha : \overline{S} \longrightarrow \overline{S}$$

$$\overline{H} = H \bmod (F, G) \mapsto \overline{X_0 H} = X_0 H \bmod (F, G)$$

i.e., α is multiplication by $\overline{X_0}$. This is a k linear mapping. We claim that α is injective. Indeed, assume that $\alpha(\overline{H}) = 0$, i.e. that $X_0 H \in (F, G)S$, so

$$X_0 H(X_0, X_1, X_2)$$

$$= A(X_0, X_1, X_2)F(X_0, X_1, X_2) + B(X_0, X_1, X_2)G(X_0, X_1, X_2)$$

for some homogeneous polynomials A and B. This yields

$$A(0, X_1, X_2)F(0, X_1, X_2) = -B(0, X_1, X_2)G(0, X_1, X_2).$$

The polynomials $F(0, X_1, X_2)$ and $G(0, X_1, X_2)$ are still homogeneous of degrees m and n, and they have no common zeroes since $C_1 \cap C_2$ lies outside $V_+(X_0)$. Thus they have no common factors, and we conclude that

$$B(0, X_1, X_2) = F(0, X_1, X_2)C(X_1, X_2)$$

$$A(0, X_1, X_2) = -G(0, X_1, X_2)C(X_1, X_2).$$

Now put

$$A_1(X_0, X_1, X_2) = A(X_0, X_1, X_2) + C(X_1, X_2)G(X_0, X_1, X_2)$$

and

$$B_1(X_0, X_1, X_2) = B(X_0, X_1, X_2) - C(X_1, X_2)F(X_0, X_1, X_2).$$

Then evidently

$$A_1(0, X_1, X_2) = B_1(0, X_1, X_2) = 0$$

and thus

$$A_1(X_0, X_1, X_2) = X_0 A'(X_0, X_1, X_2) \quad \text{and}$$
$$B_1(X_0, X_1, X_2) = X_0 B'(X_0, X_1, X_2).$$

Now

$$A_1 F + B_1 G = (A + CG)F + (B - CF)G$$
$$= AF + BG = X_0 H.$$

This implies

$$X_0 A' F + X_0 B' G = X_0 H$$

thus

$$A'F + B'G = H$$

and we have shown that α is injective.

Thus α induces for all d an injection of k-vector spaces

$$\alpha_d : \overline{S}_d \hookrightarrow \overline{S}_{d+1}.$$

Let $d \geq m + n$. In this case we have already shown that $\dim_k(\overline{S}_d) = mn$, thus α_d is a k-isomorphism.

Now choose $A_1, \ldots, A_{mn} \in S_d$ such that $\overline{A}_1, \ldots, \overline{A}_{mn} \in \overline{S}_d$ is a basis for $\overline{S}_d$. Moreover, put

$$a_i = {}^a A_i(X, Y) \bmod (f, g)k[X, Y].$$

We finally claim that these elements constitute a basis for the k-vector space $k[x, y]/(f, g)k[X, Y]$. This will complete the proof of Lemma 4.33.

As α_d is an isomorphism, $\overline{X_0 A}_1, \ldots, \overline{X_0 A}_{mn} \in \overline{S}_{d+1}$ is a k-basis for $\overline{S}_{d+1}$, as α_{d+1} is an isomorphism, $\overline{X_0^2 A}_1, \ldots, \overline{X_0^2 A}_{mn} \in \overline{S}_{d+2}$ is a k-basis for $\overline{S}_{d+2}$, and so on, until

$$\overline{X_0^r A}_1, \ldots, \overline{X_0^r A}_{mn} \in \overline{S}_{d+r} \quad \text{is a } k\text{-basis for } \overline{S}_{d+r} \text{ for all } r \geq 0.$$

We may now show that $a_1, \ldots, a_{mn}$ generate $k[x, y]/(f, g)k[X, Y]$. Namely, let $u = \overline{U} = U \bmod (f, g)k[X, Y]$ where $U \in k[X, Y]$. We consider the homogeneous polynomial ${}^h U(X_0, X_1, X_2)$. For $N \gg 0$ we have

$$(X_0^N)({}^h U(X_0, X_1, X_2)) \in \overline{S}_{d+r}.$$

Thus

$$\overline{X_0^N \cdot {}^h U} = \sum_{i=1}^{mn} \lambda_i \overline{X_0^r A_i}, \quad \lambda_i \in k$$

so

$$X_0^N \cdot {}^h U = \sum_{i=1}^{mn} \lambda_i X_0^r A_i + BF + CG$$

where $B, C \in S$. Now we have

$$U = {}^a(X_0^N \cdot {}^h U) = \sum_{i=1}^{mn} \lambda_i \cdot {}^a A_i + {}^a B^a F + {}^a C^a G$$

and hence $u = \overline{U} = \sum_{i=1}^{mn} \lambda_i a_i$.

We finally show that the $a_i, i = 1, \ldots, mn$ are linearly independent. Indeed, assume that

$$\sum_{i=1}^{mn} \lambda_i a_i = 0 \quad \text{where not all } \lambda_i \text{ are } = 0,$$

so

$$\sum_{i=1}^{mn} \lambda_i \cdot {}^a A_i = B \cdot {}^a F + C \cdot {}^a G$$

where $B, C \in k[X, Y]$. Thus

$$^h \left(\sum_{i=1}^{mn} \lambda_i \cdot {}^a A_i \right) = {}^h(B \cdot {}^a F + C \cdot {}^a G)$$

i.e.

$$\sum_{i=1}^{mn} \lambda_i \cdot {}^h({}^a A_i) = {}^h B \cdot {}^h({}^a F) + {}^h C \cdot {}^h({}^a G).$$

By Proposition 4.29(vi), we have in general

$$A = X_0^r \cdot {}^h({}^a A)$$

for some integer $r \geq 0$, thus if we choose $N \gg 0$ we have

$$X_0^N \cdot {}^h({}^a A_i) = X_0^{m_i} A_i$$

$$X_0^N \cdot {}^h({}^a F) = X_0^r F$$

$$X_0^N \cdot {}^h({}^a G) = X_0^s G.$$

This gives

$$\sum_{i=1}^{mn} \lambda_i \cdot X_0^{m_i} A_i = {}^h B X_0^r F + {}^h C X_0^s G \in (F,G)S.$$

Since $(F,G)S$ is a homogeneous ideal, all homogeneous components of $\sum_{i=1}^{mn} \lambda_i \cdot X_0^{m_i} A_i$ is contained in it. The homogeneous component of lowest degree, i.e. the initial form, is

$$\lambda_{i_1} \cdot X_0^{m_{i_1}} A_{i_1} + \lambda_{i_2} \cdot X_0^{m_{i_2}} A_{i_2} + \cdots + \lambda_{i_p} \cdot X_0^{m_{i_p}} A_{i_p}$$

where all $\lambda_{i_1}, \lambda_{i_2}, \ldots, \lambda_{i_p}$ are non-zero. All A_i are homogeneous of the same degree d, thus all m_i are equal, say all $m_i = \mu$. Then we have the following relation in $\overline{S}_{d+\mu}$:

$$\lambda_{i_1} \cdot \overline{X_0^\mu A_{i_1}} + \lambda_{i_2} \cdot \overline{X_0^\mu A_{i_2}} + \cdots + \lambda_{i_p} \cdot \overline{X_0^\mu A_{i_p}} = 0.$$

However, the elements $\overline{X_0^\mu A_1}, \overline{X_0^\mu A_2}, \ldots, \overline{X_0^\mu A_{mn}}$ form a base for $\overline{S}_{d+\mu}$ as a k-vector space. Hence we have derived a contradiction, which proves that $a_1, a_2, \ldots, a_{mn}$ are linearly independent. This completes the proof of Lemma 4.33 and hence Theorem 4.30.   □

By the properties of intersection numbers for plane projective curves developed in Sect. 4.4, Theorem 4.30 implies the following:

Corollary 4.34 *Assume as before that k is algebraically closed. Then we have the following:*

1. *If two plane projective curves C_1 and C_1 of degrees m and n have no common component, then*

$$mn \geq \sum_P m_P(C_1) \cdot m_P(C_2).$$

2. *If these curves have exactly mn points in common, then these points of intersection are all smooth points on C_1 and C_2.*
3. *If the curves have more than mn points in common, then they have a common component.*
4. *A plane projective curve without singular points is irreducible.*

Proof 1. Is immediate by the theorem and Proposition 4.16.

2. Follows since if the number of common points is exactly mn, then equality must hold in 1, thus for all common points P we have $m_P(C_1) \cdot m_P(C_2) = 1$, thus $m_P(C_1) = m_P(C_2) = 1$.

3. Is immediate by 1.

Proof of 4. Let the curve C be given by the homogeneous polynomial $F(X_0, X_1, X_2)$. Suppose that

$$F(X_0, X_1, X_2) = F_1(X_0, X_1, X_2) \cdot F_2(X_0, X_1, X_2)$$

and let C_1 and C_2 be the corresponding projective curves. By the theorem $C_1 \cap C_2 \neq \emptyset$. Let $P \in C_1 \cap C_2$. After a change of projective coordinate system we may assume that $P = (1:0:0)$. Taking affine restrictions we get the curves K, K_1 and K_2 in $\mathbb{A}_k^2$, given by the polynomials

$$^aF(X,Y) = f(X,Y), \qquad ^aF_1(X,Y) = f_1(X,Y), \qquad ^aF_2(X,Y) = f_2(X,Y).$$

With

$$f_1(X,Y) = m_1(X,Y) + \text{higher terms}$$
$$f_2(X,Y) = m_2(X,Y) + \text{higher terms}$$

where $m_i(X,Y)$ denote the initial terms, we find

$$f(X,Y) = m(X,Y) + \text{higher terms}, \quad \text{where}$$
$$m(X,Y) = m_1(X,Y)m_2(X,Y).$$

Thus $m_P(C) = \deg(m(X,Y)) = \deg(m_1(X,Y)) + \deg(m_2(X,Y)) \geq 2$, and hence P is a singular point of C. $\qquad\square$

4.9 Algebraic Derivatives and the Jacobian

Let R be a commutative ring, and let $F(X) \in R[X]$, so

$$F(X) = a_N x^N + a_{N_1} X^{N-1} + \cdots + a_1 X + a_0$$

where all $a_i \in R$. We then *define* the derivative of $F(X)$ over R by

$$F'(X) = \frac{d}{dX} F(X) = N a_N x^{N-1} + (N-1)a_{N-1} X^{N-2} + \cdots + a_1.$$

For completeness we list the following

Proposition 4.35 *The operation of polynomial derivation over a commutative ring satisfies all the usual properties:*

1. *For all polynomials $F(X)$ and elements $r \in R$, $r' = 0$ and*

$$(rF)'(X) = rF'(X).$$

2. $(F+G)'(X) = F'(X) + G'(X).$
3. $(FG)'(X) = F'(X)G(X) + F(X)G'(X).$
4. $F(G(X))' = F'(G(X))G'(X).$

Proof This is known by elementary calculus when R is the field of real numbers, i.e. for $R = \mathbb{R}$. For polynomials over $\mathbb{R}$ purely algebraic manipulations therefore suffice to verify 1–4. As the algebraic rules employed also apply to any commutative ring, the claims in the proposition follow.   □

Remark 4.36 The same argument applies to derivation in the ring of formal power series $R[[X]]$ over R.

To determine whether a point P on the plane projective curve C is singular or not, we need only examine *the Jacobian matrix*, or strictly speaking, in this case just the Jacobian vector. We return to the concept of the Jacobian matrix in Sect. 5.1, when we move on to space curves, and later when we are dealing with algebraic geometry in higher dimensions.

Let the plane projective curve C be given by the following homogeneous polynomial, so the indexing set I is finite:

$$F(X_0, X_1, X_2) = \sum_{(i_0, i_1, i_2) \in I(F)} A_{(i_0, i_1, i_2)} X_0^{i_0} X_1^{i_1} X_2^{i_2}.$$

In accordance with the above remarks, we then define the partial derivative with respect to X_0 as follows:

$$\frac{\partial}{\partial X_0} F(X_0, X_1, X_2) = \sum_{(i_0, i_1, i_2) \in I(F)} i_0 A_{(i_0, i_1, i_2)} X_0^{i_0 - 1} X_1^{i_1} X_2^{i_2}$$

and similarly we define the operators $\frac{\partial}{\partial X_1}$ and $\frac{\partial}{\partial X_2}$.

These definitions, made for polynomials only, do of course agree with the partial derivatives encountered in calculus. Also, the result analogous to Theorem 4.35, including the multivariate chain rule, holds for partial derivatives by the same argument as in the proof of Proposition 4.35.

We now make the

Definition 4.9 The Jacobian matrix of the curve C given by the polynomial F is defined as

$$\left(\frac{\partial}{\partial X_0}(F(X_0, X_1, X_2)), \frac{\partial}{\partial X_1}(F(X_0, X_1, X_2)), \frac{\partial}{\partial X_2}(F(X_0, X_1, X_2)) \right).$$

Clearly this Jacobian matrix may only have the rank 0 or 1 at a point $P \in \mathbb{P}_k^2$. We have the following result:

Proposition 4.37 *Let C be a projective curve in $\mathbb{P}_k^2$, where k is algebraically closed. A point $P \in C$ is non singular if and only if the Jacobian matrix is of rank 1 at P.*

In this case the tangent line to C at P has the equation

$$\frac{\partial F}{\partial X_0}(P)X_0 + \frac{\partial F}{\partial X_1}(P)X_1 + \frac{\partial F}{\partial X_2}(P)X_2 = 0.$$

Proof Let the curve be given by the polynomial

$$F(X_0, X_1, X_2) = \sum_{(i_0, i_1, i_2) \in I(F)} A_{(i_0, i_1, i_2)} X_0^{i_0} X_1^{i_1} X_2^{i_2}$$

which is homogeneous of degree d, thus for all $(i_0, i_1, i_2) \in I(F)$, $i_0 + i_1 + i_2 = d$.

As usual we may assume that $P = (1 : 0 : 0)$. Indeed, the multivariate chain rule guarantees that formation of the Jacobian matrix is compatible with a change of projective coordinates: Let the change of projective coordinate system be given by

$$\overline{X}_0 = \alpha_{0,0} X_0 + \alpha_{0,1} X_1 + \alpha_{0,2} X_2$$
$$\overline{X}_1 = \alpha_{1,0} X_0 + \alpha_{1,1} X_1 + \alpha_{1,2} X_2$$
$$\overline{X}_2 = \alpha_{2,0} X_0 + \alpha_{2,1} X_1 + \alpha_{2,2} X_2$$

where of course the determinant is $\neq 0$:

$$\begin{vmatrix} \alpha_{0,0} & \alpha_{0,1} & \alpha_{0,2} \\ \alpha_{1,0} & \alpha_{1,1} & \alpha_{1,2} \\ \alpha_{2,0} & \alpha_{2,1} & \alpha_{2,2} \end{vmatrix} \neq 0.$$

We put

$$A = \begin{bmatrix} \alpha_{0,0} & \alpha_{0,1} & \alpha_{0,2} \\ \alpha_{1,0} & \alpha_{1,1} & \alpha_{1,2} \\ \alpha_{2,0} & \alpha_{2,1} & \alpha_{2,2} \end{bmatrix}.$$

We then find, by the chain rule, that

$$\left(\frac{\partial F}{\partial \overline{X}_0}, \frac{\partial F}{\partial \overline{X}_1}, \frac{\partial F}{\partial \overline{X}_2} \right) = A \cdot \left(\frac{\partial F}{\partial X_0}, \frac{\partial F}{\partial X_1}, \frac{\partial F}{\partial X_2} \right)^t.$$

Now take the affine restriction K of C to $\mathbb{A}_k^2 = D_+(X_0) \subset \mathbb{P}_k^2$, given by the polynomial

$$^a F(X, Y) = \sum_{(i_0, i_1, i_2) \in I(F)} A_{i_0, i_1, i_2} X^{i_1} Y^{i_2}.$$

Since $(1 : 0 : 0)$ lies on C, we have $A_{(d,0,0)} = 0$. $P \in K$ is a non singular point if and only if the initial form $\mathrm{in}(^a F(X, Y))$ is of degree 1, which is the case if and only if

$$L = A_{(d-1,1,0)} X + A_{(d-1,0,1)} Y$$

is not identically equal to 0, in which case it is equal to $\mathrm{in}(^a F(X, Y))$.

On the other hand the Jacobian matrix (vector) evaluated in $(1 : 0 : 0)$ is

$$(0, A_{(d-1,1,0)}, A_{(d-1,0,1)}).$$

We see that this matrix has rank 1 if and only if the coefficients of L do not both vanish.

In this case the tangent at P is given by $L = 0$, i.e. the initial form,

$$A_{(d-1,1,0)}X + A_{(d-1,0,1)}Y = 0$$

so the claim about the tangent at P follows for the new coordinate system. But then it also holds for all other coordinate systems.   □

In Chap. 2, Sect. 2.9 we defined the *tangent cone* of a projective plane curve for $k = \mathbb{R}$. The same notion is used for general k, although its definition is somewhat different from the one we gave in the case $k = \mathbb{R}$. The above argument justifies one way of defining this concept in the general case: If $P = (1:0:0)$ then the tangent cone to C at P is given by the homogeneous polynomial ${}^h(\text{in}({}^a F(X,Y)))$.

Indeed, from the properties of intersection numbers in Sect. 4.3 it follows that all but a finite number of lines through P will intersect C with intersection multiplicity equal to $m_P(C)$, the exceptions being the lines which constitute the components of $V_+({}^h(\text{in}({}^a F(X,Y))))$.

4.10 Simple Elimination Theory

Consider two polynomials in X with coefficients from a field k,

$$f(X) = a_0 + a_1 X + \cdots + a_m X^m, \quad a_m \neq 0,$$
$$g(X) = b_0 + b_1 X + \cdots + b_n X^n, \quad b_n \neq 0.$$

We then have the following result:

Theorem 4.38 *The polynomials $f(X)$ and $g(X)$ have a common non-constant factor $h(X)$ if and only if the following matrix is not invertible:*

$$\begin{bmatrix}
a_0 & a_1 & \cdots & a_m & & \cdots & \\
& a_0 & \cdots & a_{m-1} & a_m & \cdots & \\
& & \cdots & & & & \\
& & & a_0 & \cdots & & a_m \\
b_0 & b_1 & \cdots & & b_n & \cdots & \\
& b_0 & \cdots & & b_{n-1} & b_n & \cdots \\
& & \cdots & & & & \\
& & & b_0 & \cdots & & b_n
\end{bmatrix}.$$

In the statement of the theorem, the matrix has n rows of a's and m rows of b's, so the matrix is $(m+n) \times (m+n)$.

Proof of the theorem Assume first that the polynomials have a common non constant factor. Replacing k by a finite extension if necessary, we may assume that they have a common root, say x_1. We then have

$$
\begin{aligned}
a_0 + a_1 x_1 + \quad \cdots \quad + a_m x_1^m &= 0 \\
a_0 x_1 + \quad a_1 x_1^2 + \quad \cdots \quad + a_m x_1^{m+1} &= 0 \\
\cdots \quad \cdots \\
a_0 x_1^{n-1} + \ a_1 x_1^n + \quad \cdots \quad + a_m x_1^{m+n-1} &= 0 \\
b_0 + b_1 x_1 + \quad \cdots \quad + b_n x_1^n &= 0 \\
b_0 x_1 + \quad b_1 x_1^2 + \quad \cdots \quad + b_n x_1^{n+1} &= 0 \\
\cdots \quad \cdots \\
b_0 x_1^{m-1} + \ b_1 x_1^m + \quad \cdots \quad + b_n x_1^{m+n-1} &= 0.
\end{aligned}
$$

We see that the homogeneous linear system with the $(m+n) \times (m+n)$ matrix of the resultant has a non zero solution, hence the determinant is zero.

Conversely, assume that the matrix is not invertible, i.e., that its determinant vanishes. We then consider the linear system of equations with the *transposed* matrix to the above one as *its* coefficient matrix. This system also have a non zero set of solutions, thus there are elements α_i, $i = 1, \ldots, m$ and β_j, $j = 1, \ldots n$ in k, not all zero, such that

$$
\begin{aligned}
a_0 \beta_1 &= b_0 \alpha_1 \\
a_1 \beta_1 + a_0 \beta_2 &= b_1 \alpha_1 + b_0 \alpha_2 \\
\cdots \quad \cdots \quad \cdots \\
a_m \beta_n &= b_n \alpha_m.
\end{aligned}
$$

Thus there are non zero polynomials

$$\phi(X) = \alpha_1 + \alpha_2 X + \cdots + \alpha_m X^{m-1} \quad \text{and} \quad \psi(X) = \beta_1 + \beta_2 X + \cdots + \beta_n X^{n-1}$$

such that

$$\psi(X) f(X) = \phi(X) g(X).$$

Since $k[X]$ is a UFD, we may factor $g(X)$ in prime polynomials, and each of these prime factors must appear in the prime factorization of $\phi(X)$ or $f(X)$, and as $\phi(X)$ is of lower degree than m, which is the degree of $f(X)$, at least one of them, call it $h(X)$, must divide $f(X)$. Thus $f(X)$ and $g(X)$ have the common non-constant factor $h(X)$, and the proof is complete. $\square$

Note that we do not really use that k is a field, only that k is a UFD. In fact, for the first part of the proof when we wish to reduce to a common root, i.e. to $h(X) = X - x_1$, we first pass to the field of quotients. Furthermore, the theory of linear systems of equations hold over any integral domain, and we only needed the UFD property for $k[X]$. In particular we may apply the theorem to $F[X]$ where $F = k[X_1, \ldots, X_r]$ is a polynomial ring in any number of variables (or indeterminates) over a field k.

Fig. 4.1 The twisted cubic
curve

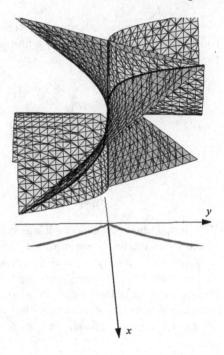

We make the following definition, which is especially useful in this case
with several variables:

Definition 4.10 The determinant R in Theorem 4.38 is called the resultant
of f and g with respect to X and denoted by $\mathrm{Res}(f, g, X)$.

4.11 An Application: The Twisted Cubic Curve

As an application of the simple elimination theory explained in Sect. 4.10 applied to several variables, we consider the curve of intersection of the following
two surfaces in $\mathbb{A}^3_{\mathbb{R}}$.

We consider the two surfaces defined by $F(x, y, z) = yz - x^2 = 0$ and
$G(x, y, z) = y - xz = 0$. The two surfaces are shown in Fig. 4.1, and the curve
of intersection is indicated fully drawn, the projection into the xy-plane being shown below it. As we see there is one curved space curve and another
component consisting of a straight line, namely the z-axis given by $x = y = 0$.
We now wish to project this curve of intersection down into the xy-plane.
Using MAPLE we get $\mathrm{Res}(F, G, z) = -x^3 + y^2$, thus this is the equation of the
projection as we have eliminated the variable z. This curve was introduced in
Sect. 2.4, it is a semi-cubic parabola. When we later need to treat algebraic
subsets more carefully using the full strength of commutative algebra, in the

framework of Grothendieck's theory of schemes, there will be one approach where the projection of the z-axis does not simply disappear, but will be present in the form of an *"embedded component"* of the projection at the origin, meaning some nilpotent structure located at the single point at the origin.

The twisted component of the curve of intersection has the parametric form

$$x = x(t) = t^2, \quad y = y(t) = t^3 \quad \text{and} \quad z = z(t) = t.$$

On Fig. 4.1 it is shown fully drawn, as is another component which is a straight line. The projection of the twisted component is shown below, it is a semi-cubic parabola as expected.

We of course verify immediately that this curve is contained in the intersection, and as it is an irreducible curve it must constitute one of the components.

4.12 Points of Inflexion and the Hessian

Definition 4.11 A non singular point P on the projective curve $C \subset \mathbb{P}^2_k$ is said to be a *point of inflexion*, or simply a flex for short, if the tangent $T_{C,x} = L$ to C at P satisfies

$$I(P, C \cap L) \geq 3.$$

If equality holds, then the flex is said to be an *ordinary flex*. When the inequality is strict, then we say that P is a *higher flex*.

If C has a line as one of its components, then a non singular point lying on this line-component is a flex, as the intersection number in question is then infinite. However, this degenerate case will be excluded in the following.

Now let C be given by the homogeneous polynomial $F(X_0, X_1, X_2)$, and form the 3×3 matrix of all second order derivatives

$$\mathcal{H}(F) = \begin{bmatrix} \dfrac{\partial^2 F}{\partial X_0 \partial X_0} & \dfrac{\partial^2 F}{\partial X_0 \partial X_1} & \dfrac{\partial^2 F}{\partial X_0 \partial X_2} \\[2mm] \dfrac{\partial^2 F}{\partial X_1 \partial X_0} & \dfrac{\partial^2 F}{\partial X_1 \partial X_1} & \dfrac{\partial^2 F}{\partial X_1 \partial X_2} \\[2mm] \dfrac{\partial^2 F}{\partial X_2 \partial X_0} & \dfrac{\partial^2 F}{\partial X_2 \partial X_1} & \dfrac{\partial^2 F}{\partial X_2 \partial X_2} \end{bmatrix}.$$

This matrix is referred to as the *Hessian matrix* of F, or of the curve C by abuse language. The Hessian polynomial $H(X_0, X_1, X_2)$ is the determinant of this matrix, and the Hessian curve $\text{Hess}(C)$ of C is the curve given by the Hessian polynomial. We have to check that this definition is independent of the coordinate system. Indeed, using the differential vector valued operator *gradient*

$$\nabla = \begin{bmatrix} \dfrac{\partial}{\partial X_0} \\[4pt] \dfrac{\partial}{\partial X_1} \\[4pt] \dfrac{\partial}{\partial X_2} \end{bmatrix}$$

the Hessian may be defined as a differential operator by the matrix product

$$\mathcal{H} = \nabla \cdot \nabla^t.$$

We introduce a new set of projective coordinates by

$$\overline{X}_0 = \alpha_{0,0}X_0 + \alpha_{0,1}X_1 + \alpha_{0,2}X_2$$
$$\overline{X}_1 = \alpha_{1,0}X_0 + \alpha_{1,1}X_1 + \alpha_{1,2}X_2$$
$$\overline{X}_2 = \alpha_{2,0}X_0 + \alpha_{2,1}X_1 + \alpha_{2,2}X_2$$

or equivalently

$$\begin{bmatrix} \overline{X}_0 \\ \overline{X}_1 \\ \overline{X}_2 \end{bmatrix} = \begin{bmatrix} \alpha_{0,0} & \alpha_{0,1} & \alpha_{0,2} \\ \alpha_{1,0} & \alpha_{1,1} & \alpha_{1,2} \\ \alpha_{2,0} & \alpha_{2,1} & \alpha_{2,2} \end{bmatrix} \begin{bmatrix} X_0 \\ X_1 \\ X_2 \end{bmatrix}$$

where

$$A = \begin{bmatrix} \alpha_{0,0} & \alpha_{0,1} & \alpha_{0,2} \\ \alpha_{1,0} & \alpha_{1,1} & \alpha_{1,2} \\ \alpha_{2,0} & \alpha_{2,1} & \alpha_{2,2} \end{bmatrix}$$

has determinant $\neq 0$. Denoting the gradient and Hessian with respect to $\overline{X}_0, \overline{X}_1, \overline{X}_2$ by $\overline{\nabla}$ and $\overline{\mathcal{H}}$, we obviously get

$$\nabla = A \cdot \overline{\nabla}$$
$$\mathcal{H} = A \cdot \overline{\nabla} \cdot (A \cdot \overline{\nabla})^t$$
$$= A \cdot \overline{\nabla} \cdot \overline{\nabla}^t A^t$$
$$= A \cdot \overline{\mathcal{H}} A^t$$

hence the two Hessian polynomials are equal up to a constant factor.

Moreover, we have the following result:

Proposition 4.39 *Assume that the ground field k is of characteristic 0. Then the intersection $C \cap \mathrm{Hess}(C)$ consists of all multiple points and flexes of C. Moreover, the intersection number of C and $\mathrm{Hess}(C)$ at a point P is 1 if and only if P is an ordinary flex of C.*

Proof Let $P \in C \cap \mathrm{Hess}(C)$. After a change of projective coordinates we may assume that $P = (1 : 0 : 0)$. We substitute $X_0, X_1, X_2 = 1, x, y$ in F and its partial derivatives, denoting the results by f, f_i and $f_{i,j}$. The affine restriction of $\mathrm{Hess}(C)$ is then given by the following polynomial in x, y:

$$h = \begin{vmatrix} f_{0,0} & f_{0,1} & f_{0,2} \\ f_{1,0} & f_{1,1} & f_{1,2} \\ f_{2,0} & f_{2,1} & f_{2,2} \end{vmatrix}.$$

Performing two row-operations and two column operations and using Euler's identity we find

$$h = \begin{vmatrix} d(d-1)f & (d-1)f_1 & (d-1)f_2 \\ (d-1)f_1 & f_{1,1} & f_{1,2} \\ (d-1)f_2 & f_{2,1} & f_{2,2} \end{vmatrix}$$

$$= (d-1)^2 \begin{vmatrix} \frac{d}{d-1}f & f_1 & f_2 \\ f_1 & f_{1,1} & f_{1,2} \\ f_2 & f_{2,1} & f_{2,2} \end{vmatrix} = (d-1)^2 \begin{vmatrix} \frac{d}{d-1}f & \frac{\partial f}{\partial x} & \frac{\partial f}{\partial y} \\ \frac{\partial f}{\partial x} & \frac{\partial^2 f}{\partial^2 x} & \frac{\partial^2 f}{\partial x \partial y} \\ \frac{\partial f}{\partial y} & \frac{\partial^2 f}{\partial y \partial x} & \frac{\partial^2 f}{\partial^2 y} \end{vmatrix}.$$

Since k is of characteristic 0, letting

$$g = \begin{vmatrix} 0 & \frac{\partial f}{\partial x} & \frac{\partial f}{\partial y} \\ \frac{\partial f}{\partial x} & \frac{\partial^2 f}{\partial^2 x} & \frac{\partial^2 f}{\partial x \partial y} \\ \frac{\partial f}{\partial y} & \frac{\partial^2 f}{\partial y \partial x} & \frac{\partial^2 f}{\partial^2 y} \end{vmatrix} = \left(\frac{\partial f}{\partial x}\right)^2 \frac{\partial^2 f}{\partial^2 y} + \left(\frac{\partial f}{\partial y}\right)^2 \frac{\partial^2 f}{\partial^2 x} - 2\frac{\partial f}{\partial x}\frac{\partial f}{\partial y}\frac{\partial^2 f}{\partial y \partial x}$$

we find

$$I(P, C \cap \mathrm{Hess}(C)) = I(P, C \cap H')$$

where H' is the curve given by g. If $P = (0,0)$ is a multiple point on C, then the partial derivatives of f have no constant terms, thus from the expression for g we see that $I(P, C \cap H') \geq 2$, in particular $P \in C \cap \mathrm{Hess}(C)$.

Assume that $P(=(0,0))$ is a simple point. We may assume that the tangent to C at this point has equation $y = 0$. Then

$$f(x,y) = y + Ax^2 + Bxy + Cy^2 + Dx^3 + Ex^2y + Fxy^2 + Gy^3 + \cdots.$$

If $\mathcal{O} = k[x,y]_{(x,y)}$ denotes the local ring of $\mathbb{A}_k^2$ at $P = (0,0)$, then

$$(y, f)\mathcal{O} = (y, Ax^2 + Dx^3 + Ex^4 + \cdots)\mathcal{O}$$

thus P is a flex if and only if $A = 0$, and an ordinary flex if and only if in addition $D \neq 0$. On the other hand we have

$$\frac{\partial f}{\partial x} = 2Ax + By + 3Dx^2 + 2Exy + Fy^2 + \cdots$$

$$\frac{\partial f}{\partial y} = 1 + Bx + 2Cy + Ex^2 + 2Fxy + 3Gy^2 + \cdots$$

$$\frac{\partial^2 f}{\partial^2 x} = 2A + 6Dx + 2Ey + \cdots$$

$$\frac{\partial^2 f}{\partial x \partial y} = B + 2Ex + 2Fy + \cdots$$

$$\frac{\partial^2 f}{\partial^2 y} = 2C + 2Fx + 6Gy + \cdots.$$

Substituting this into the expression for g, we obtain

$$g = 2A + 6Dx + (8AC - 2B^2 + 2E)y + \text{higher terms}$$

which shows that the intersection number $I(P, C \cap H')$ is 1 if and only if $A = 0$ and $D \neq 0$, i.e., if and only if P is an ordinary flex.   □

For the remainder of this section we assume that the ground field k is of characteristic 0. The last proposition has the following corollary:

Corollary 4.40 *A non singular curve C of degree ≥ 3 has a flex.*

Proof This follows by Bézout's Theorem since

$$\deg(\text{Hess}(C)) = 3(\deg(C) - 2).$$ □

For a non singular cubic we have more information:

Proposition 4.41 *Let C be a non singular cubic curve. Then after a suitable choice of projective coordinate system it has affine restriction to $\mathbb{A}_k^2 = D_+(X_0)$ defined by the equation*

$$y^2 - g(x) = 0,$$

where $g(x)$ is a cubic polynomial with three distinct roots. Moreover, C has nine flexes. These nine flexes have the property that a line joining two of them contains a third.

Proof By Corollary 4.40 C has at least one flex P, we choose the projective coordinate system such that $P = (0:0:1)$, and such that the tangent at that point is $V_+(X_0)$. In general, let the equation of C be given by the polynomial

$$f(X_0, X_1, X_2) = A_{300}X_0^3 + A_{210}X_0^2 X_1 + A_{201}X_0^2 X_2 + A_{030}X_1^3$$
$$+ A_{120}X_0 X_1^2 + A_{021}X_1^2 X_2 + A_{003}X_2^3$$
$$+ A_{102}X_0 X_2^2 + A_{012}X_1 X_2^2 + A_{111}X_0 X_1 X_2.$$

Now $f(0,0,1) = A_{003}$, thus $A_{003} = 0$. Moreover

$$\frac{\partial}{\partial X_0} f = 3A_{300}X_0^2 + 2A_{210}X_0X_1 + 2A_{201}X_0X_2 + A_{120}X_1^2$$

$$+ A_{102}X_2^2 + A_{111}X_1X_2$$

so $[\frac{\partial}{\partial X_0} f](0,0,1) = A_{102} \neq 0$, and similarly $[\frac{\partial}{\partial X_1} f](0,0,1) = A_{012} = 0$ and $[\frac{\partial}{\partial X_2} f](0,0,1) = 3A_{003} = 0$ since the tangent line at $(0:0:1)$ is $V_+(X_0)$. Thus the equation is given by the polynomial

$$f(X_0, X_1, X_2) = A_{300}X_0^3 + A_{210}X_0^2X_1 + A_{201}X_0^2X_2 + A_{030}X_1^3$$

$$+ A_{120}X_0X_1^2 + A_{021}X_1^2X_2 + A_{102}X_0X_2^2 + A_{111}X_0X_1X_2.$$

Now $f(0,0,1) = A_{003}$, thus $A_{003} = 0$. Moreover

$$\frac{\partial}{\partial X_0} f = 3A_{300}X_0^2 + 2A_{210}X_0X_1 + 2A_{201}X_0X_2 + A_{120}X_1^2$$

$$+ A_{102}X_2^2 + A_{111}X_1X_2$$

so $[\frac{\partial}{\partial X_0} f](0,0,1) = A_{102} \neq 0$, and similarly $[\frac{\partial}{\partial X_1} f](0,0,1) = A_{012} = 0$ and $[\frac{\partial}{\partial X_2} f](0,0,1) = 3A_{003} = 0$ since the tangent line at $(0:0:1)$ is $V_+(X_0)$. Thus the equation is given by the polynomial

$$f(X_0, X_1, X_2) = A_{300}X_0^3 + A_{210}X_0^2X_1 + A_{201}X_0^2X_2 + A_{030}X_1^3$$

$$+ A_{120}X_0X_1^2 + A_{021}X_1^2X_2 + A_{102}X_0X_2^2 + A_{111}X_0X_1X_2.$$

Thus the affine restriction $\overline{C}$ of C to $D_+(X_2)$ is given by the following polynomial, where $\overline{x} = \frac{X_0}{X_2}$ and $\overline{y} = \frac{X_1}{X_2}$:

$$g(\overline{x}, \overline{y}) = f(\overline{x}, \overline{y}, 1)$$

$$= A_{300}\overline{x}^3 + A_{210}\overline{x}^2\overline{y} + A_{201}\overline{x}^2 + A_{030}\overline{y}^3$$

$$+ A_{120}\overline{x}\overline{y}^2 + A_{021}\overline{y}^2 + A_{102}\overline{x} + A_{111}\overline{x}\overline{y}.$$

The point $(0,0)$ is a flex with inflectional tangent $V(\overline{x})$. But

$$I_{(0,0)}(g(\overline{x},\overline{y}),\overline{x}) = I_{(0,0)}(A_{030}\overline{y}^3 + A_{021}\overline{y}^2, \overline{x}) \geq 3 \quad \Longleftrightarrow \quad A_{021} = 0,$$

thus $A_{021} = 0$. Now the affine restriction of C to $D_+(X_0)$ is given by

$$f(1,x,y) = A_{210}x + A_{201}y + A_{030}x^3 + A_{120}x^2 + A_{102}y^2 + A_{111}xy.$$

Then since the curve C is irreducible of degree 3, $A_{030} \neq 0$, so we may assume $A_{030} = 1$. Hence

$$f(1,x,y) = x^3 + \alpha x + \beta y + \gamma x^2 + \delta y^2 + \epsilon xy,$$

where $\delta \neq 0$ since we found above that $A_{102} \neq 0$. We now solve the equation

$$\delta y^2 + y(\epsilon x + \beta) + x^3 + \gamma x^2 + \alpha x = 0$$

for y, and get

$$y = -\frac{\epsilon x + \beta}{2\delta} \pm \sqrt{\left(\frac{\epsilon x + \beta}{2\delta}\right)^2 - \frac{x^3 + \gamma x^2 + \alpha x}{\delta}} = ax + b \pm \sqrt{g(x)},$$

where

$$g(x) = \left(\frac{\epsilon x + \beta}{2\delta}\right)^2 - \frac{x^3 + \gamma x^2 + \alpha x}{\delta}.$$

Changing to a new coordinate system with the same x but $\bar{y} = y - ax - b$, we get the claimed form for the equation. It remains to show that $g(x)$ has no multiple roots. But if x_1 were a multiple root of $g(x) = 0$, then the point $(x_1, 0)$ would be a singularity on $C \cap D_+(X_0)$, thus $(1 : x_1 : 0)$ would be a singularity on C, against the assumption that C be non singular. Thus the proof of the first part of the proposition is complete.

Before we proceed, note that we may evidently assume that $g(x) = x^3 + ax^2 + bx$. The roots of the equation $g(x) = 0$ are $x = 0$ and $x = -\frac{a}{2} \pm \sqrt{(\frac{a}{2})^2 - b}$, and since these 3 roots are all different, it follows in particular that $(\frac{a}{2})^2 - b \neq 0$, that is $a^2 - 4b \neq 0$. Moreover, we also must have $b \neq 0$, thus we have the condition

$$b(a^2 - 4b) \neq 0.$$

To show that C has nine flexes, we note that the degree of Hess(C) is 3, so by Proposition 4.39 it suffices to show that there are 8 different flexes on $D_+(X_0) \cap C$.

Letting

$$\varphi(x, y) = y^2 - g(x)$$

we showed in the proof of Proposition 4.39 that these flexes are exactly the points in $V(\varphi) \cap V(h)$ where

$$h = \left(\frac{\partial \varphi}{\partial x}\right)^2 \frac{\partial^2 \varphi}{\partial^2 y} + \left(\frac{\partial \varphi}{\partial y}\right)^2 \frac{\partial^2 \varphi}{\partial^2 x} - 2 \frac{\partial \varphi}{\partial x} \frac{\partial \varphi}{\partial y} \frac{\partial^2 \varphi}{\partial y \partial x}.$$

Now we have that

$$\frac{\partial \varphi}{\partial x} = -3x^2 - 2ax - b$$

$$\frac{\partial^2 \varphi}{\partial^2 x} = -6x - 2a$$

$$\frac{\partial^2 \varphi}{\partial y \partial x} = 0$$

$$\frac{\partial \varphi}{\partial y} = 2y$$

$$\frac{\partial^2 \varphi}{\partial^2 y} = 2$$

and thus the equation

$$h(x,y) = (-3x^2 - 2ax - b)^2(2) + (2y)^2(-6x - 2a) = 0,$$

which together with $\varphi(x,y) = 0$ determines all the flexes of C in $D_+(X_0)$. Substituting $y^2 = g(x)$ in $h(x,y) = 0$ eliminates y, so we get the equation

$$(3x^2 + 2ax + b)^2 - 2(x^3 + ax^2 + bx)(6x + 2a) = 0$$

which after a short computation yields

$$k(x) = 3x^4 + 4ax^3 + 6bx^2 - b^2 = 0.$$

We claim that this equation has no multiple roots, that is, there are no common roots of this equation and

$$k'(x) = 12x^3 + 12ax^2 + 12bx = 0, \quad \text{i.e.,} \quad x^3 + ax^2 + bx = 0.$$

Now the claim follows by computing the resultant of $k(x)$ and $x^3 + ax^2 + bx$, it is found to be

$$b^4(a^2 - 4b)^2$$

after some computing, best performed using MAPLE or a similar system, and as we found above this is $\neq 0$.

Thus $k(x) = 0$ has four distinct roots, and to each of them correspond two values of y yielding flexes. Hence there are altogether eight flexes in $D_+(X_0)$. Since there is one flex at infinity, we thus have reached the maximum of nine possible flexes for a cubic curve.

Now choose any two flexes, we may assume that one of them is $(0:0:1)$ as above, and that the other one is $(1:x_0:y_0)$ where x_0 is one of the three distinct roots of $x^3 + ax^2 + bx = 0$. Then $y_0 = \pm\sqrt{g(x_0)}$, thus $(1:x_0:-y_0)$ is a third flex, and these three points are collinear as they lie on the line $V_+(x_0X_0 - X_1)$. $\square$

We have shown that if k is a field of characteristic zero then an elliptic curve in $\mathbb{P}^2_k$, that is a non singular cubic curve in $\mathbb{P}^2_k$, for a suitable choice of projective coordinate system will have an affine restriction to $D_+(X_0)$ which is on *Weierstrass Normal Form*:

Definition 4.12 An elliptic curve is on Weierstrass Normal Form if it is given by the equation

$$y^2 = x^3 + ax^2 + bx + c$$

where $x^3 + ax^2 + bx + c = 0$ do not have a multiple root in the algebraic closure of k.

The picture is completed by the following:

Proposition 4.42 *In the Weierstrass Normal form we may either assume that*

(i) $c = 0$

or that

(ii) $a = 0$.

If k *is algebraically closed, then we may assume that the form is* $y^2 = x(x-1)(x-\lambda)$ *where* $\lambda \in k$ *is a non zero element.*

Remark 4.43 In the case (ii) it is customary to write $y^2 = 4x^3 - g_1 x - g_2$.

Proof In the proof of Proposition 4.41 we may also assume that $(1 : 0 : 1) \in C$, thus that $c = 0$. Also, we may substitute $x = x' - \frac{a}{3}$ and the quadratic term vanishes while of course a constant term reappears.

When $c = 0$ we get

$$y^2 = x(x - x_1)(x - x_2)$$

as k is algebraically closed. Here $x_1 \neq 0$, and we obtain

$$\left(\frac{y}{\sqrt{x_1}}\right)^2 = \frac{x}{x_1}\left(\frac{x}{x_1} - 1\right)\left(\frac{x}{x_1} - \frac{x_2}{x_1}\right)$$

and introducing new variables the claim follows. $\square$

Chapter 5
Projective Varieties in $\mathbb{P}^N_k$

In this chapter we assume k to be an algebraically closed field, of any characteristic unless otherwise stated. Our aim is to give a summary of the most basic classical facts on higher dimensional projective algebraic geometry, as a preparation to Grothendieck's modern theory of schemes.

5.1 Subvarieties of $\mathbb{P}^N_k$

We now proceed in the higher dimensional case in the same way as we did for curves in Sect. 4.1.

Definition 5.1 We shall use the term *Projective Set* for a subset of some $\mathbb{P}^N_k$ which is the zero set of a collection of elements of, hence a homogeneous ideal in, some $k[X_0, \ldots, X_N]$. Since the set of zeroes is the same for a homogeneous ideal I and its radical $\sqrt{I}$, we shall normally associate a projective set with its radical homogeneous ideal.[1] A Projective Variety will designate a projective set for which the homogeneous radical ideal is prime.

As an example we first consider the projective closure of the twisted cubic curve in $\mathbb{P}^3_k$, as explained in Sect. 4.7 of Chap. 4. The curve is given on parametric form as

$$X = \left\{ (X_0 : X_1 : X_2 : X_3) = (u_0^3 : u_0^2 u_1 : u_0 u_1^2 : u_1^3) \mid (u_0 : u_1) \in \mathbb{P}^1_k \right\}$$

and thus it is the intersection of the following three surfaces in $\mathbb{P}^3_k$:

$$V_+(X_1 X_3 - X_2^2), \qquad V_+(X_0 X_2 - X_1^2), \qquad V_+(X_1 X_2 - X_0 X_3)$$

---

[1] If it becomes necessary to treat the case of non-radical ideals, we shall refer to the situation as *a projective set with nilpotent structure*.

A. Holme, *A Royal Road to Algebraic Geometry*,
DOI 10.1007/978-3-642-19225-8_5, © Springer-Verlag Berlin Heidelberg 2012

or

$$X = V_+(\mathfrak{J}) \quad \text{where}$$

$$\mathfrak{J} = (X_1 X_3 - X_2^2, X_0 X_2 - X_1^2, X_1 X_2 - X_0 X_3) k[X_0, X_1, X_2, X_3].$$

The *Jacobian matrix* of the set of polynomials

$$F_1 = X_1 X_3 - X_2^2, \qquad F_2 = X_0 X_2 - X_1^2, \qquad F_3 = X_1 X_2 - X_0 X_3$$

is

$$\mathfrak{J} = \begin{bmatrix} \frac{\partial}{\partial X_0} F_1 & \frac{\partial}{\partial X_1} F_1 & \frac{\partial}{\partial X_2} F_1 & \frac{\partial}{\partial X_3} F_1 \\ \frac{\partial}{\partial X_0} F_2 & \frac{\partial}{\partial X_1} F_2 & \frac{\partial}{\partial X_2} F_2 & \frac{\partial}{\partial X_2} F_2 \\ \frac{\partial}{\partial X_0} F_3 & \frac{\partial}{\partial X_1} F_3 & \frac{\partial}{\partial X_2} F_3 & \frac{\partial}{\partial X_3} F_3 \end{bmatrix} = \begin{bmatrix} 0 & X_3 & -2X_2 & X_1 \\ X_2 & -2X_1 & X_0 & 0 \\ -X_3 & X_2 & X_1 & -X_0 \end{bmatrix}.$$

Let us consider, e.g., the point $P = (1 : 1 : 1 : 1) \in \mathbb{P}_k^3$. We see immediately that $F_1(1 : 1 : 1 : 1) = F_2(1 : 1 : 1 : 1) = F_3(1 : 1 : 1 : 1) = 0$, and when substituted into the expression for the Jacobian, it yields

$$\mathfrak{J}(P) = \begin{bmatrix} 0 & 1 & -2 & 1 \\ 1 & -2 & 1 & 0 \\ -1 & 1 & 1 & -1 \end{bmatrix}.$$

When the second and third rows are added, a change of sign yields the first. Thus the rank of this matrix is 2, and the system of linear equations in X_0, X_1, X_2 and X_3 yields a line in $\mathbb{P}_k^3$, passing through the point $P = (1 : 1 : 1 : 1)$. This is the tangent line to the space curve at the point $P \in \mathbb{P}_k^3$.

We claim that the matrix $\mathfrak{J}$ has rank 2 at all points $P \in X$. Indeed, we know from the construction of the curve that these points are of the form

$$P = (u_0^3 : u_0^2 u_1 : u_0 u_1^2 : u_1^3)$$

where u_0, u_1 are not both 0. Then the Jacobian evaluated at this P yields

$$\mathfrak{J}(P) = \begin{bmatrix} 0 & u_1^3 & -2u_0 u_1^2 & u_0^2 u_1 \\ u_0 u_1^2 & -2u_0^2 u_1 & u_0^3 & 0 \\ -u_1^3 & u_0 u_1^2 & u_0^2 u_1 & -u_0^3 \end{bmatrix}.$$

Assume first that $u_0 = 0$. Then we may assume that $u_1 = 1$, and obtain

$$\mathfrak{J}(P) = \begin{bmatrix} 0 & 1 & 0 & 0 \\ 0 & 0 & 0 & 0 \\ -1 & 0 & 0 & 0 \end{bmatrix}.$$

which is of rank 2. If $u_1 = 0$ we may assume that $u_0 = 1$, and get

$$\mathcal{J}(P) = \begin{bmatrix} 0 & 0 & 0 & 0 \\ 0 & 0 & 1 & 0 \\ 0 & 0 & 0 & -1 \end{bmatrix}$$

and we find that the rank is 2 in this case as well.

Thus the Jacobian matrix is of rank $2 = \dim(\mathbb{P}^3_k) - \dim(X)$, and the tangent space to X in $\mathbb{P}^3_k$ should be given by the linear system of equations defined by $\mathcal{J}$, i.e., it should be cut out by the planes defined by the rows in the matrix.

Indeed, as the three equations define surfaces

$$Z_1 = V_+(F_1), \qquad Z_2 = V_+(F_2), \qquad Z_3 = V_+(F_3) \subset \mathbb{P}^3_k$$

which together cut out X, it should be reasonable to define the tangent space to X at the point $P \in X$ as the intersection of the tangent planes to these surfaces at the point P.

We therefore, as well as for some further good reasons, use the Jacobian matrix to define the (embedded) tangent space at non-singular points of the projective subvarieties $\mathbb{P}^N_k$ over an algebraically closer field k, in Sect. 5.4.

Let k be an algebraically closed field. An algebraic subset $X \subset \mathbb{P}^N_k$ is the zero set in $\mathbb{P}^N_k$ of a set of homogeneous polynomials

$$F_1(X_0, X_1, \ldots, X_N) = 0$$
$$F_2(X_0, X_1, \ldots, X_N) = 0$$
$$\vdots$$
$$F_m(X_0, X_1, \ldots, X_N) = 0.$$

If I is the homogeneous ideal generated by the above polynomials, then this X is denoted by $V_+(I)$. We shall refer to X as a *projective subvariety of* $\mathbb{P}^N_k$, or just as a *k-variety* when the context is clear, if the ideal generated by $F_1, F_2, \ldots, F_m$ in $k[X_0, X_1, \ldots, X_N]$ is a prime ideal. Furthermore, the *dimension of* X is defined as

$$\dim(X) = \dim(k[X_0, X_1, \ldots, X_N]/(F_1, F_2, \ldots F_m)k[X_0, X_1, \ldots, X_N]) - 1$$

where we have the *Krull dimension* to the right. Thus for instance, a plane projective curve defined by an irreducible $F(X_0, X_1, X_2) = 0$ will be of dimension

$$\dim(k[X_0, X_1, X_2]/(F(X_0, X_1, X_2)k[X_0, X_1, X_2]) - 1 = (3-1) - 1 = 1,$$

as it should.

The material explained of Sect. 4.1 in Chap. 4 applies also to the higher dimensional case. We shall develop it further in the general case.

So let X be a projective variety, and let $Y = X \cap D_+(X_0)$, an affine open subset of X. It is given by the polynomials in $R = k[\frac{X_1}{X_0}, \ldots, \frac{X_N}{X_0}] = k[Y_1, \ldots, Y_N]$:

$$F_1(1, Y_1, \ldots, Y_N) = 0$$
$$F_2(1, Y_1, \ldots, Y_N) = 0$$
$$\vdots$$
$$F_m(1, Y_1, \ldots, Y_N) = 0.$$

Then the quotient ring of R modulo the ideal generated by the polynomials above is referred to as the *affine coordinate ring* of the affine variety Y, and denoted by $\Gamma(Y) = k[y_1, \ldots, y_N]$. Similarly the ring

$$k[X_0, X_1, \ldots, X_N]/(F_1, \ldots, F_m)k[X_0, \ldots, X_N] = k[x_0, \ldots, x_N]$$

is denoted by $\Gamma_+(X)$ and referred to as *the homogeneous coordinate ring* of X.

The following result which we found for curves, holds in the higher dimensional case as well:

Theorem 5.1 1. *If $P = (p_1, \ldots, p_N)$ is a point on the affine variety Y, then it corresponds to a maximal ideal in the coordinate ring $\Gamma(Y)$ of Y in the following manner:*

$$Y \longrightarrow \mathrm{Max}(\Gamma(Y)) \text{ by } P \mapsto \mathfrak{m}(p_1, \ldots, p_N) = (y_1 - p_1, \ldots, y_N - p_N)\Gamma(Y).$$

This map is injective.
2. *If k is algebraically closed, then the map is bijective.*

The homogeneous coordinate ring $\Gamma_+(X)$ is a graded k-algebra,

$$\Gamma_+(X) = \bigoplus_{n=0}^{\infty} \Gamma_+(X)_n$$

where $\Gamma_+(X)_n$ denotes the elements of degree n. As in the case of curves we localize the ring $\Gamma_+(C)$ in the multiplicative subset

$$S = \{x_0^n \mid n = 0, \ldots\}$$

and obtain

$$\Gamma_+(X)_{x_0} = \left\{ \frac{P(x_0, \ldots, x_N)}{x_0^n} \;\middle|\; \begin{matrix} n = \\ 0, 1, 2, \ldots \end{matrix} \right\} = \bigoplus_{n=-\infty}^{\infty} (\Gamma_+(C)_{x_0})_n.$$

In particular we put

$$\Gamma_+(X)_{(x_0)} = (\Gamma_+(X)_{x_0})_0,$$

and note that we have

$$(\Gamma_+(X)_{x_0})_0 = k\left[\frac{x_1}{x_0}, \ldots, \frac{x_N}{x_0}\right].$$

Replacing $\frac{x_1}{x_0}, \ldots, \frac{x_N}{x_0}$ with $y_1, \ldots, y_N$, we thus obtain the affine coordinate ring of the affine restriction.

A similar argument applies to any linear form $L = a_0 X_0 + \cdots + a_N X_N$: Let ℓ be its image in $\Gamma_+(X)$. We proceed as before, and get

$$\Gamma_+(X)_\ell = \left\{ \frac{P(x_0, \ldots, x_N)}{\ell^n} \,\middle|\, \begin{matrix} n = \\ 0, 1, 2, \ldots \end{matrix} \right\} = \bigoplus_{n=-\infty}^{\infty} (\Gamma_+(C)_\ell)_n.$$

In particular we put

$$\Gamma_+(X)_{(\ell)} = (\Gamma_+(C)_\ell)_0.$$

We then have

$$(\Gamma_+(X)_\ell)_0 = k\left[\frac{x_0}{\ell}, \ldots, \frac{x_N}{\ell}\right],$$

and if $a_0 \neq 0$, then we may replace ℓ by x_0 and get

$$(\Gamma_+(X)_{x_0})_0 = k\left[\frac{x_1}{x_0}, \ldots, \frac{x_N}{x_0}\right].$$

Replacing $\frac{x_1}{x_0}, \ldots, \frac{y_N}{x_0}$ with $y_1, \ldots, y_N$, we again have the affine coordinate ring of the affine restriction to $D_+(x_0)$.

As for $\mathbb{P}_k^k$ we have that the points in $\mathbb{P}_k^N$ correspond to *homogeneous prime ideals* of height 1 in $k[X_0, \ldots, X_N]$, the only homogeneous maximal ideal in the ring being the irrelevant ideal generated by $X_0, \ldots, X_N$, this is also similar for $\Gamma_+(X)$, the only homogeneous maximal being the one generated by $x_0, \ldots, x_N$. The point $P = (p_0 : p_1 : \ldots : p_N) \in \mathbb{P}_k^N$ corresponds to the homogeneous prime ideal

$$\mathfrak{p}(P) = (\{p_i x_j - p_j x_i \mid 0 \leq i, j \leq N\})\Gamma(C).$$

Thus the set of homogeneous prime ideals of height 1 in $k[X_0, \ldots, X_N]$, respectively in $\Gamma_+(X)$, may be identified with the set of points of $\mathbb{P}_k^N$, respectively of X.

Proposition-Definition 5.2 *The field $k(X) = k(\frac{x_0}{\ell}, \frac{x_1}{\ell}, \ldots, \frac{x_N}{\ell})$ is independent of the linear form ℓ, and is referred to as the function field of the projective variety X.*

Proof Changing from one ℓ to another just amounts to a linear change of variables, and we may, without loss of generality, assume that $\ell = x_0$. □

Now let $P \in D_+(L) \cap X$, where $L = \alpha_0 X_0 + \cdots + \alpha_N X_N$, $\alpha_0, \ldots, \alpha_N \in k$. We have seen that $Y = X \cap D_+(\ell)$ is an affine variety, its coordinate ring is $(\Gamma_+(X)_\ell)_0 = k[y_0, \ldots, y_n]$ where $y_i = \frac{x_i}{\ell}$ and $\ell = \alpha_0 x_0 + \cdots + \alpha_N x_N$. We have the

Proposition-Definition 5.3 *The subset of $k(X)$*

$$\left\{ \frac{f}{g} \;\middle|\; \begin{array}{l} \text{where } f, g \in k[y_0, \ldots, y_N] \\ \text{and } g(P) \neq 0 \end{array} \right\}$$

is a subring of $k(X) = k(y_0, \ldots, y_N)$ which is a local ring. It is denoted by $\mathcal{O}_{X,P}$ and referred to as the local ring of X at P. Moreover, $\mathcal{O}_{X,P}$ is the localization of the subring $\Gamma(Y) = k[y_0, \ldots, y_N]$ of $k(X)$, the affine coordinate ring of $Y = X \cap D_+(X_0)$, in the maximal ideal $\mathfrak{m}(P)$.

The last assertion also holds for the coordinate ring of an affine variety of the form $D_+(H) \cap X$ where H is a homogeneous polynomial in $X_0, \ldots, X_N$ such that $P \in D_+(H)$. In particular $k(X)$ is the field of quotients of $\Gamma(Y)$, and may be described as the set of all fractions $\frac{F(x_0, \ldots, x_N)}{G(x_0, \ldots, x_N)}$ where numerator and denominator are in $\Gamma_+(X)$, homogeneous of the same degree.

Proof Without loss of generality we may assume that $L = X_0$ and that $P = (1 : 0 : \ldots : 0)$. Then $Y_i = \frac{X_i}{X_0}$ and $y_i = \frac{x_i}{x_0}$. The subset of $k(X)$

$$\left\{ \frac{f}{g} \;\middle|\; \begin{array}{l} \text{where } f, g \in k[y_1, \ldots, y_N] \\ \text{and } g(0, \ldots, 0) \neq 0 \end{array} \right\}$$

is evidently a subring of $k(X) = k(y_1, \ldots, y_N)$, and it is a local ring. It is also clear that $\mathcal{O}_{X,P}$ is the localization of the ring $\Gamma(Y) = k[\frac{x_1}{x_0}, \ldots, \frac{x_N}{x_0}]$ of $k(X)$ in the maximal ideal $\mathfrak{m}(0, \ldots, 0) = (y_1, \ldots, y_N)\Gamma(Y)$. Thus the first part of the claim follows. The last part is clear by the same argument as for L, subject to the need of replacing for instance $\frac{X_1}{L}$ by $\frac{X_1^{\deg(H)}}{H}$. $\qquad\square$

The concept of the *Zariski Topology* will be introduced and explained in Chap. 9, Sect. 9.1.

For now, we shall just say that a subset F of the projective variety X is *Zariski-closed* provided that it is a finite union

$$F = \bigcup_{i=1}^{m} F_i$$

where all the F_i are subsets of X which are themselves subvarieties of $\mathbb{P}_k^N$. The complement of a Zariski-closed subset of X is said to be Zariski-open. In Sect. 8.10 we also give the precise definition of the *structure sheaf* $\mathcal{O}_X$, but for now we provisionally define $\mathcal{O}_X(U)$ for the special Zariski-open subsets of

Fig. 5.1 Oscar Zariski.
Illustration by the author.
Drawing inspired by a
photo by George M.
Bergman. The
Oberwohlfach collection

the form $U = X \cap D_+(H) = D_+(h)$ where H is a homogeneous polynomial
as in Proposition-Definition 5.3, by

$$\mathcal{O}_X(D_+(h)) = (\Gamma_+(X)_h)_0 \subset k(X).$$

Here h denotes the image of H in S. In Sect. 8.10, we show how a definition
like this for a certain type of restricted class of open subsets may be extended
to all open subsets.

5.2 Projective Non-singular Curves

If X is of dimension 1, then we saw in Corollary 4.7, that for all non-singular
points $p \in X$ the local ring $\mathcal{O}_{X,p}$ is a discrete valuation ring of $k(X)$ over k.
But in fact, we have more, namely the

Theorem 5.4 *Assume that X is a projective non-singular curve. Then there is
a bijective correspondence between the points $p \in X$ and the discrete valuation
rings V of $k(X)$ over k, in the sense that $V = \mathcal{O}_{X,p}$ whenever V corresponds
to p.*

Proof A point $p \in X$ has a local ring $\mathcal{O}_{X,p}$ which is a discrete valuation ring
of $k(X)$, as recalled above. Conversely let the subring $V \subset k(X)$ be a discrete valuation ring. We have the affine covering of $X = \bigcup_{i=0}^{N} U_i$ where $U_i =
D_+(X_i) \cap X$, so the coordinate ring of U_i is $k[\frac{x_0}{x_i}, \frac{x_1}{x_i}, \ldots, \frac{x_N}{x_i}]$, all of these rings
are subrings of the function field $k(X)$. Thus $\frac{x_j}{x_i} \in k(X) = k(\frac{x_0}{x_i}, \frac{x_1}{x_i}, \ldots, \frac{x_N}{x_i})$
for all i and j, and if v_V denotes the valuation corresponding to the discrete
valuation ring V then we may we select $\frac{x_{j_0}}{x_{i_0}}$ among them so that

$$v_V\left(\frac{x_{j_0}}{x_{i_0}}\right) \geq v_V\left(\frac{x_j}{x_i}\right) \quad \text{for all } i \text{ and } j$$

since there are only a finite number of possibilities. Thus in particular

$$v_V\left(\frac{x_i}{x_{j_0}}\right) = v_V\left(\frac{x_{j_0}}{x_{i_0}}\frac{x_i}{x_{j_0}}\right) = v_V\left(\frac{x_{j_0}}{x_{i_0}}\right) - v_V\left(\frac{x_{j_0}}{x_i}\right) \geq 0$$

for all $i = 1, 2, \ldots, N$, and hence

$$k\left[\frac{x_0}{x_{j_0}}, \frac{x_1}{x_{j_0}}, \ldots, \frac{x_N}{x_{j_0}}\right] \subset V.$$

But then V is the localization of the ring $k[\frac{x_0}{x_{j_0}}, \frac{x_1}{x_{j_0}}, \ldots, \frac{x_N}{x_{j_0}}]$ in the maximal ideal

$$\mathfrak{m} = \mathfrak{m}_V \cap k\left[\frac{x_0}{x_{j_0}}, \frac{x_1}{x_{j_0}}, \ldots, \frac{x_N}{x_{j_0}}\right].$$

Now $\mathfrak{m}$ corresponds to a point $p \in U_{i_0} \cap X$, and the claim follows.   □

Remark 5.5 It is interesting to compare this proof with Theorem 15.3 of Chap. 15.2. In the language of proper morphisms and complete varieties, we have essentially shown here that projective varieties are complete varieties.

Corollary 5.6 *With X as in the theorem, all the affine coordinate rings $\Gamma(U)$ of open affine subsets $U \subset X$ are subrings of $k(X)$, and $\Gamma(U)$ is the intersection of all the discrete valuation rings $\mathcal{O}_{X,P}$ as P runs through U.*

Proof [2] In fact, for an integral domain R

$$R = \bigcap_{\text{All primes } \mathfrak{p} \subset R} R_{\mathfrak{p}}.$$

Assume $x, y \in R$, $y \neq 0$, and $\frac{x}{y} \in R_{\mathfrak{p}}$ for all prime ideals $\mathfrak{p}$. Let

$$\mathfrak{A} = (y):(x) = \left\{ z \in R \; \middle| \; \frac{x}{y} = \frac{w}{z} \text{ for some } w \in R \right\}.$$

Since $\frac{x}{y} \in R_{\mathfrak{p}}$, $\frac{x}{y}$ can be written $\frac{w}{z}$ with $w \in R$, $z \in R - \mathfrak{p}$. Therefore $\mathfrak{A} \not\subseteq \mathfrak{p}$. Therefore $\mathfrak{A}$ is not contained in a prime ideal, so $\mathfrak{A} = R$. This means that $1 \in \mathfrak{A}$, i.e. $\frac{x}{y} \in R$.   □

5.3 Divisors on a Projective Non-singular Curve

Let $X \subset \mathbb{P}_N^k$ be a projective non-singular curve, over the algebraically closed field k. Let $k(X)$ denote its function field. Recall that for all points $P \in X$, the

---

[2]We follow a footnote on p. 388 in [36].

local ring $\mathcal{O}_{X,P}$ is a discrete valuation ring of $k(X)$. Denote the corresponding valuation by $v_{X,P}$.

Definition 5.2 A divisor on the curve X is a an expression (a formal sum)

$$D = \sum_{P \in X} m_P P$$

where m_P is an integer and the $P = (a_0 : \ldots : a_N) \in \mathbb{P}_k^N$. m_P is equal to zero for all but finitely many points P. We put

$$\deg(D) = \sum_{P \in X} m_{D,P}.$$

The divisor D for which all $m_{D,P} = 0$ is called the *zero divisor* and denoted by 0. We add and subtract divisors by doing so point-wise, e.g.,

$$\sum m_P P + \sum m'_P P = \sum (m_P P + m'_P) P.$$

The set of divisors obviously form an Abelian group, which is (partially) ordered by

$$\sum m_P P \leq \sum m'_P P \iff m_P \leq m'_P \quad \text{for all } P \in X.$$

Now define

$$L(D) = \{ f \in k(X) \mid v_P(f) \geq -m_D \text{ for all } P \in X \}$$

which is a k-linear subspace of $k(X)$. We need the following

Proposition-Definition 5.7 *Let $z \in k(X)$. We say that the point $P \in X$ is a zero for z if $v_{X,P}(z) > 0$, and a pole if $v_{X,P}(z) < 0$. If $z \in k(X)$ is a non-zero element, then it has only a finite number of zeroes and poles. The order of the zero P is defined as $v_{X,P}(z)$, and the order of a pole P by $-v_{X,P}(z)$. The divisor of zeroes is defined as*

$$(z)_0 = \sum_{P \in X \ a \ zero \ of \ z} v_{X,P}(z) P$$

and the divisor of poles $(z)_\infty$ is defined as

$$(z)_\infty = \sum_{P \in X \ a \ pole \ of \ z} -v_{X,P}(z) P.$$

We have

$$\deg((z)_0) = \deg((z)_\infty).$$

The divisor (z) *is defined by*

$$(z) = (z)_0 - (z)_\infty$$

and referred to as a principal divisor.

Proof We first show that there is only a finite number of points $P \in X$ such that $v_{X,P}(z) \neq 0$. By a change of coordinate system we may assume that $X \not\subset V_+(X_0)$. Thus $X \cap V_+(X_0)$, being a Zariski closed subset of the curve X, consists of a finite number of points. They may be disregarded for the purpose of the present proof. Now $Y = X \cap D_+(X_0) \subset D_+(X_0) = \mathbb{A}_k^N$ and $z = \frac{f}{g}$ where $f, g \in k[1, \frac{x_1}{x_0}, \dots, \frac{x_N}{x_0}]$. Then $f = \frac{F(x_0,\dots,x_N)}{x_0^m}$ and $g = \frac{G(x_0,\dots,x_N)}{x_0^m}$ where $F(x_0,\dots,x_N), G(x_0,\dots,x_N) \in \Gamma_+(X)_m$ for the same m. Now a point $P = (a_0 : a_1 : \dots : a_N) \in D_+(X_0) \cap X$ such that $F(P)$ and $G(P) \neq 0$ is neither a zero nor a pole of X. Thus the zeroes and the poles are in $V(f, g) \subset X$, which is a proper Zariski-closed subset of X and thus finite. This proves the first part of the claim.

In Sect. 4.4 of Chap. 4 we introduced the intersection number $I(P, C_1 \cap C_2)$ of two curves C_1 and C_2 in the point P, and studied its basic properties. We also defined the intersection cycle of two curves as

$$C_1 \cdot C_2 = \sum_{P \in C_1 \cap C_2} I(P, C_1 \cap C_2) P$$

and showed as Bézout's Theorem (Theorem 4.30) that this cycle, if it exists, has degree

$$\sum_{P \in C_1 \cap C_2} I(P, C_1 \cap C_2) = \deg(C_1) \deg(C_2).$$

If now X were a plane curve, $X = V_+(H)$, then we would have that

$$(z)_0 = X \cdot V_+(F) \quad \text{and} \quad (z)_\infty = X \cdot V_+(G)$$

and thus by Bézout's theorem they are of the same degree, and the proof would be complete.

However, the general case presents no real difficulty. In fact, for a curve $X \subset \mathbb{P}_k^N$ and a hypersurface $S \subset \mathbb{P}_k^N$ we define the intersection cycle exactly as before, and observe that the proof of Bézout's theorem is still valid in this general setting, under the assumption that the curve X intersects the hypersurface S in a finite number of points, i.e., the intersection is *proper*, as we shall say later. As this assumption is obviously satisfied here, the proof is complete.   □

Definition 5.3 Two divisors D and D' are called linearly equivalent if $D - D' = (z)$ for some $z \in k(X)$.

5.4 Smoothness and Tangency in any Dimension

We now make the following definition:

Definition 5.4 Let X be a k-variety. We say that the point

$$P = (p_0 : p_1 : \ldots : p_N)$$

is a smooth, or non-singular, point of X if the Jacobian matrix evaluated at the point P

$$\mathcal{J}(P) = \begin{bmatrix} \frac{\partial F_1}{\partial X_0}(P), & \frac{\partial F_1}{\partial X_1}(P), & \ldots, & \frac{\partial F_1}{\partial X_N}(P) \\ \frac{\partial F_2}{\partial X_0}(P), & \frac{\partial F_2}{\partial X_1}(P), & \ldots, & \frac{\partial F_2}{\partial X_N}(P) \\ \vdots & \vdots & \ddots & \vdots \\ \frac{\partial F_m}{\partial X_0}(P), & \frac{\partial F_m}{\partial X_1}(P), & \ldots, & \frac{\partial F_m}{\partial X_N}(P) \end{bmatrix}$$

is of rank equal to $N - \dim(X)$. Otherwise the point is referred to as a singular point.

We may now define the tangent space at a smooth point of a projective subvariety of $\mathbb{P}_k^N$:

Definition 5.5 Let $X \subset \mathbb{P}_k^N$ be a projective variety. The tangent space $T_{X,P}$ of X in $\mathbb{P}_k^N$ at the point $P = (p_0 : \ldots : p_N)$ is the set of all points $Q = (q_0 : \ldots : q_N)$ such that

$$\begin{bmatrix} \frac{\partial F_1}{\partial X_0}(P), & \frac{\partial F_1}{\partial X_1}(P), & \ldots, & \frac{\partial F_1}{\partial X_N}(P) \\ \frac{\partial F_2}{\partial X_0}(P), & \frac{\partial F_2}{\partial X_1}(P), & \ldots, & \frac{\partial F_2}{\partial X_N}(P) \\ \vdots & \vdots & \ddots & \vdots \\ \frac{\partial F_m}{\partial X_0}(P), & \frac{\partial F_m}{\partial X_1}(P), & \ldots, & \frac{\partial F_m}{\partial X_N}(P) \end{bmatrix} \cdot \begin{bmatrix} q_0 \\ q_1 \\ \vdots \\ q_N \end{bmatrix} = \begin{bmatrix} 0 \\ 0 \\ \vdots \\ 0 \end{bmatrix}.$$

Note that when P is a non-singular point, then $\dim(T_{X,P}) = \dim(X)$. If P is a singular point, however, then this concept of tangent space is less interesting. We return to this later.

5.5 Hilbert Polynomial and Projective Invariants

We need an important theorem due to Hilbert and Serre, which we state and prove following the elegant treatment in [42], vol. II, Sect. 12, pp. 232–237.

Let k be a field. The below theorem is stated and proved in [42] for a ring A satisfying the descending chain condition, instead of for a field k. The general case is treated in essentially the same manner as we do here.

Fig. 5.2 David Hilbert

Let $R = k[X_1, \ldots, X_n]$, and let M be a finitely generated graded module over R, so $M = \bigoplus_{q=0}^{\infty} M_q$ where M_q is the k-space of elements of degree q. Then the vector space M_q has finite dimension which we denote by $\varphi_M(q)$. We then have the following

Theorem 5.8 (Hilbert-Serre) *For sufficiently large q, the function $\varphi_M(q)$ is given by a polynomial $P_M(q)$ with integral coefficients, whose degree in q is at most $n - 1$.*

Proof As M is a finitely generated R-module, it is also generated by a finite number of homogeneous elements, say $m_1, \ldots, m_r$. Then M_q is generated over k by

$$\left\{ X_1^{i_1} \cdots X_n^{i_n} m_p \mid i_1 + i_2 + \cdots + i_n + \deg(m_p) = q, p = 1, \ldots, r \right\}$$

for all q.

For $n = 0$ we have $R = k$, and by the above M_q is generated over k by those elements m_i which are of degree q. Thus $M_q = 0$ for all $q \gg 0$, and P_M is the zero polynomial in this case. The zero polynomial is assigned the degree -1, hence the assertion of the theorem holds for $n = 0$.

We now proceed by induction on n. Assume the claim for $n - 1$ variables. We consider the R-homomorphism given as multiplication by X_n, $M \longrightarrow M$, and complete to the exact sequence

$$0 \longrightarrow N \longrightarrow M \xrightarrow{\times X_n} M \longrightarrow P \longrightarrow 0$$

where $N = \{m \in M \mid X_n m = 0\}$ and $P = M/X_r M$. As both N and P are annihilated by X_n, they are actually modules over $k[X_1, \ldots, X_{n-1}]$. Moreover, it follows from this exact sequence that

$$\varphi_M(q + 1) - \varphi_M(q) = \varphi_P(q + 1) - \varphi_N(q).$$

By the induction assumption $\varphi_P(q+1)$ and $\varphi_N(q)$ are polynomials for $q \gg 0$, of degree at most $n-2$. Hence $\varphi_M(q+1) - \varphi_M(q)$ is a polynomial $f(q)$ for $q \gg 0$, of degree at most $n-2$.

We now claim that every polynomial $f(q)$ in q of degree d can be written in the form

$$f(q) = c_0 \binom{q}{d} + c_1 \binom{q}{d-1} + \cdots + c_{d-1} \binom{q}{1} + c_d$$

for some suitable coefficients c_i. Indeed, we have

$$q = \binom{q}{1}$$

$$q^2 = 2\binom{q}{2} + \binom{q}{1}$$

$$q^3 = 6\binom{q}{3} + 6\binom{q}{2} + \binom{q}{1}$$

etc., and obtain the claim by substituting these expressions for the powers of q into the polynomial $f(q)$.

We now show that in our situation, the coefficients are integers:

Lemma 5.9 *If $f(T)$ is a polynomial in T with rational coefficients such that*

$$f(q) = c_0 \binom{q}{d} + c_1 \binom{q}{d-1} + \cdots + c_{d-1}\binom{q}{1} + c_d$$

and $f(q) \in \mathbb{Z}$ for all $q \gg 0$, then all the coefficients $c_0, c_1, \ldots, c_d \in \mathbb{Z}$.

Proof Indeed, we proceed by induction on d. The case $d = 0$ is trivial, so assume the claim for $d - 1$. Now using the identity

$$\binom{h+1}{s} - \binom{h}{s} = \binom{h}{s-1}$$

we find

$$f(q+1) - f(q) = c_0 \binom{q}{d-1} + c_1 \binom{q}{d-2} + \cdots + c_{d-1}.$$

It then follows by the induction assumption that all $c_0, c_1, \ldots, c_{d-1}$ are integers since the difference is a polynomial in q of degree at most $d - 1$. Since the binomial coefficients $\binom{q}{s}$ are integers, it follows that also c_d is an integer, and the claim is proven. $\square$

We proceed with the proof of the theorem. We apply the lemma to the difference $\varphi_M(m) - \varphi_M(m-1)$, where m is a sufficiently big integer, say $m \geq N$, so that this difference is a polynomial of degree at most $n-2$ by what we saw above. By the lemma we get

$$\varphi_M(m) - \varphi_M(m-1) = a_0 \binom{m-1}{n-2} + a_1 \binom{m-1}{n-3} + \cdots + a_{n-2} \quad \text{for } m \geq N$$

where $a_0, \ldots, a_{n-2} \in \mathbb{Z}$. We also write

$$\varphi_M(m) - \varphi_M(m-1) = a_0 \binom{m-1}{n-2} + a_1 \binom{m-1}{n-3} + \cdots + a_{n-2} + c_m$$

$$\text{for } 2 \leq m \leq N-1$$

$$\varphi_M(1) = a_{n-2} + c_1$$

where $\binom{t}{s} = 0$ for $t < s$ and $c_i \in \mathbb{Q}$ are suitable correction terms. Note that $\varphi_M(0) = 0$.

We collect all these relations as follows, writing φ for φ_M:

$$\varphi(q) - \varphi(q-1) = \quad a_0\binom{q-1}{n-2} + a_1\binom{q-1}{n-3} + \cdots + a_i\binom{q-1}{n-2-i} + \cdots + a_{n-2}$$

$$\varphi(q-1) - \varphi(q-2) = a_0\binom{q-2}{n-2} + a_1\binom{q-2}{n-3} + \cdots + a_i\binom{q-2}{n-2-i} + \cdots + a_{n-2}$$

$$\cdots$$

$$\varphi(N) - \varphi(N-1) = a_0\binom{N-1}{n-2} + a_1\binom{N-1}{n-3} + \cdots + a_i\binom{N-1}{n-2-i} + \cdots + a_{n-2}$$

$$\varphi(N-1) - \varphi(N-2)$$
$$= a_0\binom{N-2}{n-2} + a_1\binom{N-2}{n-3} + \cdots + a_i\binom{N-2}{n-2-i} + \cdots + a_{n-2} + c_{N-1}$$

$$\varphi(N-2) - \varphi(N-3)$$
$$= a_0\binom{N-3}{n-2} + a_1\binom{N-3}{n-3} + \cdots + a_i\binom{N-3}{n-2-i} + \cdots + a_{n-2} + c_{N-2}$$

$$\cdots$$

$$\varphi(N-j) - \varphi(N-j-1)$$
$$= a_0\binom{N-j-1}{n-2} + a_1\binom{N-j-1}{n-3} + \cdots + a_i\binom{N-j-1}{n-2-i} + \cdots + a_{n-2} + c_{N-j}$$

$$\cdots$$

$$\varphi(n-2+1) - \varphi(n-2)$$
$$= a_0\binom{n-2}{n-2} + a_1\binom{n-2}{n-3} + \cdots + a_i\binom{n-2}{n-2-i} + \cdots + a_{n-2} + c_{n-2+1}$$

$$\cdots$$

$$\varphi(3) - \varphi(2) = \qquad a_{n-4}\binom{2}{2} + a_{n-3}\binom{2}{1} + a_{n-2} + c_3$$

$$\varphi(2) - \varphi(1) = \qquad a_{n-3}\binom{1}{1} + a_{n-2} + c_2$$

$$\varphi(1) = \qquad a_{n-2} + c_1.$$

Using the formula

$$\binom{q}{s} = \binom{q-1}{s-1} + \binom{q-2}{s-1} + \cdots + \binom{s}{s-1} + \binom{s-1}{s-1},$$

addition of the above relations yields that for all $q \geq N$,

$$\varphi_M(q) = a_0 \binom{q}{n-1} + a_1 \binom{q}{n-2} + \cdots + a_{n-2} \binom{q}{1} + a_{n-1}$$

where $a_{n-1} = \sum_{i=1}^{N-1} c_i$ is a constant, necessarily an integer since $a_0, \ldots,$ $a_{n-2} \in \mathbb{Z}$ and $\varphi_M(q) \in \mathbb{Z}$. Thus it follows that $\varphi_M(q)$ is a polynomial $P_M(q)$ for $q \gg 0$, and as a bonus we have shown that the polynomial $P_M(q)$ may be expressed as

$$\varphi_M(q) = a_0 \binom{q}{n-1} + a_1 \binom{q}{n-2} + \cdots + a_{n-2} \binom{q}{1} + a_{n-1}$$

where the coefficients $a_0, \ldots, a_{n-1}$ are integers.   □

We now make the following

Definition 5.6 Let $X = V_+(I)$ be a projective subset of $\mathbb{P}_k^N$. Put $\Gamma_+(X) = k[X_0, \ldots, X_N]/I$. We denote the polynomial $P_{\Gamma_+(X)}(T)$ by $P_X(T)$ and call it the Hilbert Polynomial of X.

We need the following:

Proposition 5.10 *The Hilbert Polynomial of the projective subset X has degree equal to $r = \dim(X)$. In particular*

$$P_X(t) = c_0 \binom{t}{r} + c_1 \binom{t}{r-1} + \cdots + c_{r-1} \binom{t}{1} + c_r$$

for integers $c_0, \ldots, c_r$, which are projective invariants of X.

Proof We shall prove this for the case when X is a variety over an algebraically closed field k, i.e., we assume that the ideal I be prime. The general case is shown by a straightforward extension of the proof we give here. We need an important result due to Emmy Noether, 1926:

Theorem 5.11 (Noether's Normalization Lemma) *Let $k[x_0, \ldots, x_N]$ be a finitely generated integral domain over the algebraically closed field k, and let $n+1$ be the transcendence degree of $k(x_0, \ldots, x_N)$ over k. Then there exist $r+1$ linear combinations in $x_0, \ldots, x_N$ with coefficients from k, $y_0, \ldots, y_n$, such that $k[x_0, \ldots, x_N]$ is a finite (graded) module over $k[y_0, \ldots, y_n]$.*

For a proof of this result we refer to [42], vol. I, Theorem 8 in Chap. V, Sect. 4 on p. 266.

Using this theorem together with Theorem 5.8 applied to $R = k[y_0, \ldots, y_r]$ and $M = k[x_0, \ldots, x_N]$, we find that the degree of $P_X(t)$ is at most $r + 1 - 1 = r$.

Assume that $\deg(P_X(t)) < r$. Let $Y = X \cap H$ be a *hyperplane section* of X, then Y is defined by the homogeneous ideal in S generated by the image $h \in k[x_0, \ldots, x_N]$ of the linear form in defining $H \subset \mathbb{P}_k^N$.

Let X be a projective set of dimension r, and $P_X(t) = a_0 t^t + a_1 t^{d-1} + \cdots + a_r$ its Hilbert Polynomial. We define the *degree* of X as $\deg(X) = r! a_0 = c_0$, with notations as above. We have the exact sequence

$$0 \longrightarrow (h)S \longrightarrow S \longrightarrow S/(h)S \longrightarrow 0$$

and as the leftmost arrow shifts the grading by 1, we get

$$P_X(q) - P_X(q-1) = P_{X_h}(q).$$

In particular the degree of the Hilbert Polynomial is reduced by 1 when cutting with a hyperplane. The dimension of X_h is $r - 1$, so repeating the procedure we eventually get a curve. By the assumption on $\deg(P_X(t))$ this curve will have a constant Hilbert polynomial, which is absurd. This completes the proof.   □

Now let X be a projective set, with Hilbert Polynomial

$$P_X(t) = c_0 \binom{t}{r} + c_1 \binom{t}{r-1} + \cdots + c_{r-1} \binom{t}{1} + c_r.$$

Definition 5.7 We define the degree $\deg(X)$ as c_0 in the above form of the Hilbert polynomial. The arithmetic genus of X is defined as

$$p_a(X) = (-1)^r (P_X(0) - 1) = (-1)^r (c_r - 1)$$

where r denotes the dimension of X. If X is non singular, the arithmetic genus is just called the genus of X and denoted by $g(X)$. We also note that the degree d is $r!$ times the leading coefficient of the polynomial $P_X(t)$.

If X is a hypersurface of degree d, defined by the degree d homogeneous polynomial f, then writing $S = k[X_0, \ldots, X_N]$ we have the exact sequence

$$0 \longrightarrow S(-d) \longrightarrow S \longrightarrow S/(f)S \longrightarrow 0.$$

Since the Hilbert polynomial of $\mathbb{P}_k^N$ is

$$P_{\mathbb{P}_k^N}(t) = \binom{t + N}{N},$$

we therefore find

$$P_X(t) = \binom{t+N}{N} - \binom{t-d+N}{N}$$

$$= \frac{1}{N!}((t+1)(t+2)\cdots(t+N)$$

$$- (t-d+1)(t-d+2)\cdots(t-d+N)).$$

Collecting the terms containing t^{N-1}, we need only to consider

$$\frac{1}{N!}\left(t^N + \frac{N(N+1)}{2}t^{N-1} - \left((t-d)^N + \frac{N(N+1)}{2}(t-d)^{N-1}\right)\right)$$

$$= \frac{1}{N!}\left(t^N + \frac{N(N+1)}{2}t^{N-1} - \left(t^N - dNt^{N-1} + \frac{N(N+1)}{2}t^{N-1}\right)\right)$$

$$= \frac{1}{N!}(dNt^{N-1}) = \frac{d}{(N-1)!}t^{N-1}.$$

Thus the concept of degree defined by the Hilbert polynomial for general algebraic sets coincide with the previous definition for hypersurfaces defined by the ordinary degree of a polynomial.

5.6 Emmy Noether, Her Family and Their Fate

At this point the name of *Noether* has appeared several times in our exposition. Emmy Noether and her father Max were mathematicians, as were several of Max' sons as well. Emmy was in a class way above her brothers and her father, although he certainly was also a very competent mathematician as well. When Adolf Hitler and his gang seised power in Germany, she was fired from her modest position in Göttingen, being a Jew. She came to America, where she did not get a job of the importance merited by her greatness as a mathematician. She died only a few years after being forced out of Germany. Emmy's father Max had the good fortune of passing away before all hell broke loose in Germany and Europe. But her younger brother Fritz, also a mathematician, made the wrong choice when he "escaped" to the Soviet Union. He got a position at the University of Tomsk, but in 1937 he was imprisoned as a "German spy". In 1941 ha was executed.

Emmy Noether's insights into the abstract nature of algebra were profound and immensely important. Nevertheless one could reasonably assert that her work in theoretical physics went at least as far as her algebra. Strangely, this formidable contribution is not so well known, Noether's name being overshadowed by that of Einstein.

Emmy's father Max also played a role in algebraic geometry, as we shall see below. He was born in 1844 and died in 1928.

Fig. 5.3 Emmy Noether.
Illustration by the author,
inspired by an image from
the Oberwohlfach
collection

Fig. 5.4 Max Noether.
Illustration by the author

5.7 The Riemann-Roch Theorem for Non-singular Curves

Let $X \subset \mathbb{P}_k^N$ be a non-singular projective curve over the algebraically closed field k. A formal sum

$$D = \sum_{i=1}^{m} n_{D,i} P_i$$

where $P_i \in X$ and $n_{D,i}$ are integers, is referred to as *a divisor* on X. Two divisors are regarded as equal if all their non-zero integers are the same. The set of all divisors on X form a group under addition in the obvious way, if $D = \sum_{i=1}^{m} n_{D,i} P_i$ and $D' = \sum_{i=1}^{m} n_{D',i} P_i$ then we put

$$D + D' = \sum_{i=1}^{m} (n_{D,i} + n_{D',i}) P_i.$$

A divisor D_0 is defined by all $n_{D_0,i} = 0$, we denote this divisor by 0. If no n_i occurring is negative, then D is called an *effective* divisor. We write $D \geq D'$ if for all i we have $n_{D,i} \geq n_{D',i}$. The set of all divisors $D \geq 0$ is a semigroup referred to as the *effective divisors* on X.

The degree of a divisor D, written $\deg(D)$ is the sum of its coefficients. If $X = V_+(F)$ is a plane non singular curve, and Y is another plane curve given by a polynomial G, not containing X as a component and where G may have multiple components, then we put

$$\text{Divisor}(G) = \sum_{P \in X \cap V_+(G)} I(P, X \cap V_+(G))P$$

where the multiple components of $V_+(G)$ are counted with their multiplicities. But as we have seen in the proof of Proposition-Definition 5.7, this definition may be extended to the case of a non singular curve $X \subset \mathbb{P}^N_k$ and a hypersurface $S = V_+(G) \subset \mathbb{P}^N_k$. Moreover, we saw that if $z = \frac{G(x_0,\ldots,x_N)}{H(x_0,\ldots,x_N)} \in k(X)$ then

$$(z) = \text{Divisor}(G) - \text{Divisor}(H).$$

Now let $\sum_{P \in X} n_P P$ be a divisor on the non singular curve X, so $n_P = 0$ for almost all P. Define a subspace of $k(X)$ by

$$L(D) = \{ f \in k(X) \mid v_{X,P}(f) \geq -n_P \text{ for all } P \text{ with } n_P \neq 0 \}$$

This is easily seen to be a k-vector subspace of $k(X)$. (Note that $0 \in L(D)$ since for all valuations v, $v(0) = \infty$.)

We have the

Lemma 5.12

$$\dim_k(L(D)) < \infty.$$

Rather than to produce a self contained *ad hoc* proof of this lemma at the present stage, we shall return to it later in the context of coherent sheaves and proper morphisms. We denote $\dim_k(L(D))$ by $\ell(D)$.

In the history of algebraic geometry there are several important grand themes running through it, of increasing complexity and beauty. The first such theme we encounter here is that of the *Riemann-Roch theorems*. The fascinating history of these themes would merit books in their own right, so this we will have to pass by. However, the starting point for Riemann-Roch theorems is the following result.

Theorem 5.13 (Riemann's Inequality) *Let g_X denote the genus of the non singular projective curve X, defined over the algebraically closed field k. Then for any divisor D on X,*

$$\ell(D) \geq \deg(D) + 1 - g_X.$$

Fig. 5.5 Georg Friedrich
Bernhard Riemann.
Illustration by the author

Fig. 5.6 Gustav Roch.
Illustration by the author

Initially this was proved by Riemann in 1857, and at this time it related the topological genus of a compact Riemann surface to algebraic properties of the surface. Here g_X is the topological *genus* of the surface X, roughly the number of holes in it: So it is 0 when X is a sphere, 1 for a torus etc. Finding an expression for the difference of the two sides of the inequality completed this topological result, this was achieved by *Gustav Roch* in 1865, only the year before his death in 1866 at the age of 27. Incidentally, this was also the year of death for his teacher Riemann. Like Niels Henrik Abel Gustav Roch died of tuberculosis, another great loss to mathematics. The theorem as it then stood was named the *Riemann-Roch theorem* by *Max Noether*, Emmy Noethers father. It has later been enormously generalized. Among other themes, we shall explain some of this in the remaining part of this book.

In Sect. 16.2 we introduce an important divisor on X, referred to as the *canonical divisor* and usually denoted by K_X. It has some amazing properties. One of these properties is that it *fills the gap* in Riemann's inequality above:

Theorem 5.14 (The Riemann-Roch Theorem for Curves) *Let g_X denote the genus of the non singular projective curve X, defined over the algebraically closed field k. Then for any divisor D on X,*

$$\ell(D) - \ell(K_X - D) = \deg(D) + 1 - g_X.$$

Again, we shall present a much more general result than this, with some indications of the proof. It uses heavier techniques than available at this stage. Some details are given in Chap. 20. For now, we use the statement of the theorem for curves as given here, to deduce some observations which would be key steps in a direct proof of the theorem just for curves at this point.

Since $L(0) = k$, it follows from the Riemann-Roch theorem by taking $D = 0$ that $\ell(K_X) = g_X$. Thus taking $D = K_X$ we find

$$\ell(K_X) - 1 = \deg(K_X) + 1 - g_X$$

so $\deg(K_X) = 2g_X - 2$.

Another simple observation, once we have the theorem, is the

Proposition 5.15 (Noether's Reduction Lemma) *If $\ell(D) > 0$ and $\ell(K_X - D - P) \neq \ell(K_X - D)$, then $\ell(D + P) = \ell(D)$.*

Indeed, by two applications of Riemann-Roch we get

$$\ell(D) - \ell(K_X - D) = \deg(D) + 1 - g_X$$
$$\ell(D + P) - \ell(K_X - D - P) = \deg(D) + 2 - g_X$$

and hence

$$(\ell(D + P) - \ell(D)) + (\ell(K_X - D) - \ell(K_X - D - P)) = 1.$$

Since the second difference greater than 0, while the first is non negative, the claim follows.

We also readily deduce the

Corollary 5.16 (1) *If $\deg(D) \geq 2g_X - 1$ then $\ell(D) = \deg(D) + 1 - g_X$.*
(2) *If $\deg(D) \geq 2g_X$, then $\ell(D - P) = \ell(D) - 1$ for all $P \in X$.*
(3) (Clifford's Theorem) *If $\ell(D) > 0$ and $\ell(K_X - D) > 0$ then $\ell(D) \leq \frac{1}{2}\deg(D) + 1$.*

In Part II, we turn to the important and pathbreaking theory of schemes, which is due to the remarkable mathematician *Alexander Grothendieck*.

Part II
Introduction to Grothendieck's Theory
of Schemes

Chapter 6
Categories and Functors

In this first chapter of Part 2 we give a general, rapid introduction to the required language from category theory.

6.1 Objects and Morphisms

A *category* $\mathcal{C}$ is defined by the following data:

1. A collection of *objects* denoted by $\mathrm{Obj}(\mathcal{C})$.
2. For any two objects $A, B \in \mathrm{Obj}(\mathcal{C})$ there is a *set* denoted by $\mathrm{Hom}_{\mathcal{C}}(A, B)$, and referred to as *the set of morphisms from A to B*.
3. For any three objects A, B and C there is a *rule of composition for morphisms*, that is to say, a mapping

$$\mathrm{Hom}_{\mathcal{C}}(A, B) \times \mathrm{Hom}_{\mathcal{C}}(B, C) \longrightarrow \mathrm{Hom}_{\mathcal{C}}(A, C)$$

denoted as

$$(\varphi, \psi) \mapsto \psi \circ \varphi.$$

In general the collection $\mathrm{Obj}(\mathcal{C})$ is not a set, in the technical sense of set theory. But the collection of *all possible sets* form a category, which we denote by $\mathcal{S}\mathrm{et}$. For two sets A and B the set $\mathrm{Hom}_{\mathcal{S}\mathrm{et}}(A, B)$ is the set of all mappings from A to B,

$$\mathrm{Hom}_{\mathcal{S}\mathrm{et}}(A, B) = \{\varphi \,|\, \varphi : A \longrightarrow B\}.$$

In the category $\mathcal{S}\mathrm{et}$ the composition of morphisms is nothing but the usual composition of mappings.

For a general category we impose some conditions on the rule of composition of morphisms, which ensures that all properties of mappings of sets, which are expressible in terms of diagrams, are valid for the rule of composition of morphisms in any category.

A. Holme, *A Royal Road to Algebraic Geometry*,
DOI 10.1007/978-3-642-19225-8_6, © Springer-Verlag Berlin Heidelberg 2012

In particular we have the following two fundamental conditions:

Condition 6.1 *There is a morphism* $\mathrm{id}_A \in \mathrm{Hom}_{\mathcal{C}}(A, A)$, *referred to as the identity morphism on* A, *such that for all* $\varphi \in \mathrm{Hom}_{\mathcal{C}}(A, B)$ *we have* $\varphi \circ \mathrm{id}_A = \varphi$, *and for all* $\psi \in \mathrm{Hom}_{\mathcal{C}}(C, A)$ *we have* $\mathrm{id}_A \circ \psi = \psi$

and

Condition 6.2 *Composition of morphisms is associative, in the sense that whenever one side in the below equality is defined, so is the other and equality holds:*

$$(\varphi \circ \psi) \circ \xi = \varphi \circ (\psi \circ \xi).$$

6.2 Examples of Categories

A category $\mathcal{S}$ such that $\mathrm{Obj}(\mathcal{S})$ is a set is called a *small category*. Such categories are important in certain general constructions which we will come to later.

So far the most important category we have seen is the category Set. But examples abound, they are literally everywhere. We list some others below.

Example 6.1 The collection of all groups form a category, the morphisms being the group-homomorphisms. This category is denoted by $\mathcal{G}$rp.

Example 6.2 The collection of all Abelian groups form a category, the morphisms being the group-homomorphisms. This category is denoted by $\mathcal{A}$b.

Example 6.3 The collection of all rings form a category, the morphisms being the ring-homomorphisms. This category is denoted by $\mathcal{R}$ing.

Example 6.4 The collection of all commutative rings with 1 form a category, the morphisms being the ring-homomorphisms which map 1 to 1. This category is denoted by $\mathcal{C}$omm. Note the important condition of the unit element being mapped to the unit element. Note that there are ring-morphisms between commutative rings with 1 which are not morphisms in the category $\mathcal{C}$omm, only in the category $\mathcal{R}$ing.

Example 6.5 Let A be a commutative ring with 1. The collection of all A-modules form a category, the morphisms being the A-homomorphisms. This category is denoted by $\mathcal{M}$od$_A$.

Example 6.6 The class of all topological spaces, together with the continuous mappings, from a category which we denote by $\mathcal{T}op$.

6.3   The Dual Category

If $\mathcal{C}$ is a category, then we get another category $\mathcal{C}^*$ by keeping the objects, but putting

$$\operatorname{Hom}_{\mathcal{C}^*}(A, B) = \operatorname{Hom}_{\mathcal{C}}(B, A).$$

It is a trivial exercise to verify that $\mathcal{C}^*$ is then a category. It is referred to as the *dual category of* $\mathcal{C}$.

Instead of writing $\varphi \in \operatorname{Hom}_{\mathcal{C}}(A, B)$, we employ the notation

$$\varphi : A \longrightarrow B,$$

which is more in line with our usual thinking. If we have the situation

$$
\begin{array}{ccc}
A & \xrightarrow{\varphi} & B \\
f\downarrow & & \downarrow g \\
C & \xrightarrow{\psi} & D
\end{array}
$$

and the two compositions are the same, then we say that *the diagram commutes*. This language is also used for diagrams of different shapes, such as triangular ones, with the obvious modification. A complex diagram consisting of several sub-diagrams is called commutative if all the subdiagrams commute, and this is so if all the subdiagrams commute in the diagram obtained by adding in some (or all) compositions: Thus for instance the diagram

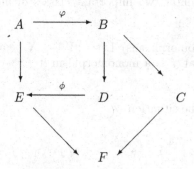

commutes if and only if all the subdiagrams in

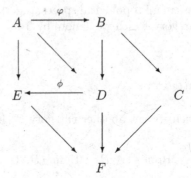

commute.

If A is an object in the category $\mathcal{C}$, then we define the *fiber category over* A, denoted by $\mathcal{C}_A$, by taking as objects

$$\{(B,\varphi) \mid \varphi : B \longrightarrow A\}$$

and letting

$$\mathrm{Hom}_{\mathcal{C}_A}((B,\varphi),(C,\psi)) = \{f \in \mathrm{Hom}_{\mathcal{C}}(B,C) \mid \psi \circ f = \varphi\}.$$

6.4 The Topology on a Topological Space Viewed as a Category

Let X be a topological space. We define a category $\mathcal{T}op(X)$ by letting the objects be the open subsets of X, and for two open subsets U and V we let $\mathrm{Hom}(U,V)$ be the set whose only element is the inclusion mapping if $U \subseteq V$, and $\emptyset$ otherwise. This is a category, as is easily verified. If $U \subseteq X$ is an open subset, then the category $\mathcal{T}op(X)_U$ is nothing but the category $\mathcal{T}op(U)$.

6.5 Monomorphisms and Epimorphisms

We frequently encounter two important classes of morphisms in a general category:

Definition 6.1 (Monomorphisms) Let $f : Y \longrightarrow X$ be a morphism in the category $\mathcal{C}$. We say that f is a monomorphism if $f \circ \psi_1 = f \circ \psi_2$ implies that $\psi_1 = \psi_2$.

In other words, the situation

$$Z \underset{\psi_2}{\overset{\psi_1}{\rightrightarrows}} Y \overset{f}{\longrightarrow} X \quad \text{where } f \circ \psi_1 = f \circ \psi_2$$

implies that $\psi_1 = \psi_2$.

Thus, to say that $f : Y \longrightarrow X$ is a monomorphism is equivalent to asserting that for all Z the mapping

$$\mathrm{Hom}_{\mathcal{C}}(Z,Y) \overset{\mathrm{Hom}_{\mathcal{C}}(Z,f)}{\longrightarrow} \mathrm{Hom}_{\mathcal{C}}(Z,X) \quad \text{where } \psi \mapsto f \circ \psi$$

is an injective mapping of sets. The following proposition is easily verified:

Proposition 6.1 1. *The composition of two monomorphisms is again a monomorphism.*
2. *If $f \circ g$ is a monomorphism then so is g.*

The dual concept to a monomorphism is that of an *epimorphism*:

Definition 6.2 (Epimorphisms) Let $f : X \longrightarrow Y$ be a morphism in the category $\mathcal{C}$. We say that f is an epimorphism if

$$\psi_1 \circ f = \psi_2 \circ f \quad \Longrightarrow \quad \psi_1 = \psi_2.$$

In other words, the situation

$$X \overset{f}{\longrightarrow} Y \overset{\psi_1}{\underset{\psi_2}{\rightrightarrows}} Z \quad \text{where } \psi_1 \circ f = \psi_2 \circ f$$

implies that $\psi_1 = \psi_2$. To say that $f : X \longrightarrow Y$ is an epimorphism is equivalent to asserting that for all Z the mapping

$$\mathrm{Hom}_{\mathcal{C}}(Y,Z) \overset{\mathrm{Hom}_{\mathcal{C}}(f,Z)}{\longrightarrow} \mathrm{Hom}_{\mathcal{C}}(X,Z) \quad \text{where } \psi \mapsto \psi \circ f$$

is an injective mapping of sets.

Proposition 6.2 1. *The composition of two epimorphisms is again an epimorphism.*
2. *If $f \circ g$ is an epimorphism then so is f.*

For some of the categories we most frequently encounter, monomorphisms are injective mappings, while the epimorphisms are surjective mappings. This is the case for Set, as well as for the category Mod_R of R-modules over a ring R. But for topological spaces a morphism (i.e. a continuous mapping) is an epimorphism if and only if the image of the source space is dense in the target space. Monomorphisms are the injective, continuous mappings, however.

At any rate, this phenomenon motivates the usual practice of referring to epimorphisms as *surjections* and monomorphisms as *injections*. A morphism which is both is said to be *bijective*. But this concept must not be confused with that of *an isomorphism*. The latter is always bijective, but the former need not always be an isomorphism.

6.6 Isomorphisms

A morphism

$$\varphi : A \longrightarrow B$$

is said, as in the examples cited above, to be an *isomorphism* if there is a morphism

$$\psi : B \longrightarrow A$$

such that the two compositions are the two identity morphisms of A and B, respectively. In this case we say that A and B are *isomorphic* objects, and as is easily seen the relation of being isomorphic is an equivalence relation on the class $\mathrm{Obj}(\mathcal{C})$. We write, as usual, $A \cong B$. A category such that the collection of isomorphism classes of objects is *a set*, is referred to as *an essentially small category*.

If $\varphi : A \longrightarrow B$ is an isomorphism, then the inverse $\psi : B \longrightarrow A$ is uniquely determined: Indeed, assume that

$$\psi \circ \varphi = \psi' \circ \varphi = \mathrm{id}_A \quad \text{and} \quad \varphi \circ \psi = \varphi \circ \psi' = \mathrm{id}_B$$

then multiplying the first relation to the right with ψ' and using associativity we get $\psi = \psi'$. We put, as usual, $\varphi^{-1} = \psi$.

6.7 Covariant and Contravariant Functors

Given two categories $\mathcal{C}$ and $\mathcal{D}$. A *covariant functor* from $\mathcal{C}$ to $\mathcal{D}$ is a mapping

$$F : \mathcal{C} \longrightarrow \mathcal{D}$$

and for any two objects A and B in $\mathcal{C}$ a mapping, by abuse of notation also denoted by F,

$$F : \mathrm{Hom}_{\mathcal{C}}(A, B) \longrightarrow \mathrm{Hom}_{\mathcal{D}}(F(A), F(B)),$$

which maps identity morphisms to identity morphisms and is compatible with the composition, namely $F(\varphi \circ \psi) = F(\varphi) \circ F(\psi)$.

We shall refer to the category $\mathcal{C}$ as the *source* category for the functor F, and to $\mathcal{D}$ as the *target* category.

As is easily seen the composition of two covariant functors is again a covariant functor.

A *contravariant* functor is defined in the same way, except that it *reverses the morphisms*. Another way of expressing this is to define a contravariant functor

$$T : \mathcal{C} \longrightarrow \mathcal{D}$$

as a covariant functor

$$T : \mathcal{C} \longrightarrow \mathcal{D}^*,$$

or equivalently as a covariant functor

$$T : \mathcal{C}^* \longrightarrow \mathcal{D}.$$

In particular the *identity mapping* of objects and morphisms from $\mathcal{C}$ to itself is a covariant functor, referred to as the *identity functor on* $\mathcal{C}$.

Example 6.7 Let B be an A-algebra and let

$$\varphi : A \longrightarrow B$$

be the homomorphism given by $\varphi(a) = a1_B$. The assignment

$$\mathrm{Mod}_A \longrightarrow \mathrm{Mod}_B$$

which to an A-module assigns a B-module by

$$T_B : M \mapsto M \otimes_A B$$

is a covariant functor, referred to as base extension. The assignment which to a B-module N assigns an A-module $N_{[\varphi]}$ defined by $an = \varphi(a)n$ is a covariant functor referred to as base restriction.

Example 6.8 The assignment

$$\mathrm{Mod}_A \longrightarrow \mathrm{Mod}_A$$
$$h_N : M \mapsto \mathrm{Hom}_A(M, N)$$

where N is a fixed A-module, is a contravariant functor.

Example 6.9 The assignment

$$\mathrm{Mod}_A \longrightarrow \mathrm{Mod}_A$$
$$h^N : M \mapsto \mathrm{Hom}_A(N, M)$$

where N is a fixed A-module, is a covariant functor.

Example 6.10 The assignment

$$F : \mathcal{A}\mathrm{b} \longrightarrow \mathcal{G}\mathrm{rp}$$

which merely regards an Abelian group as a general group, is a covariant functor. This is an example of so-called *forgetful functors*, to be treated below.

6.8 Forgetful Functors

The functor

$$T : \mathcal{A}b \longrightarrow \mathcal{S}et$$

which to an Abelian group assigns the underlying set, is called *a forgetful* functor. Similarly we have forgetful functors between many categories, where the effect of the functor merely is do disregard part of the structure of the objects in the source category. Thus for instance, we gave forgetful functors into the category $\mathcal{S}et$ from $\mathcal{T}op$, $\mathcal{C}omm$, etc., and from $\mathcal{M}od_A$ to $\mathcal{A}b$, and so on. Base restriction is also a forgetful functor.

6.9 The Category of Functors $\mathcal{F}un(\mathcal{C}, \mathcal{D})$

The category of covariant functors $\mathcal{F}un(\mathcal{C}, \mathcal{D})$ from the category $\mathcal{C}$ to the category $\mathcal{D}$ is defined by letting the objects be the covariant functors from $\mathcal{C}$ to $\mathcal{D}$, and for two such functors T and S we let

$$\text{Hom}_{\mathcal{F}un(\mathcal{C},\mathcal{D})}(S, T)$$

be collections

$$\{\Psi_A\}_{A \in \text{Obj}(\mathcal{C})}$$

of morphisms

$$\Psi_A : S(A) \longrightarrow T(A),$$

such that whenever $\varphi : A \longrightarrow B$ is a morphism in $\mathcal{C}$, then the following diagram commutes:

$$
\begin{array}{ccc}
S(A) & \xrightarrow{\ \Psi_A\ } & T(A) \\
{\scriptstyle S(\varphi)}\downarrow & & \downarrow{\scriptstyle T(\varphi)} \\
S(B) & \xrightarrow[\ \Psi_B\ ]{} & T(B)
\end{array}
$$

Morphisms of functors are often referred to as *natural transformations*. The commutative diagram above is then called *the naturality condition*.

6.10 Functors of Several Variables

We may also define a functor of n "variables", i.e. an assignment T which to a tuple of objects $(A_1, A_2, \ldots, A_n)$ from categories $\mathcal{C}_i$, $i = 1, 2, \ldots, n$ assigns an object $T(A_1, A_2, \ldots, A_n)$ of a category $\mathcal{D}$, and which is covariant in some of the variables, contravariant in others, and such that the obvious generalization of the naturality condition holds. In particular we speak of *bifunctors* when there are two source categories. The details are left to the reader.

6.11 Isomorphic and Equivalent Categories

We may regard the categories themselves as a category, the objects then being the categories and the morphisms being the covariant functors. Strictly speaking this "category" violates the requirement that $\mathrm{Hom}_{\mathcal{C}}(A, B)$ be a set, so the language "the categories of all categories" should be viewed as an informal way of speaking. We then, in particular, get the notion of *isomorphic categories*: Explicitly, two categories $\mathcal{C}$ and $\mathcal{D}$ are isomorphic if there are covariant functors $S : \mathcal{C} \longrightarrow \mathcal{D}$, and $T : \mathcal{D} \longrightarrow \mathcal{C}$ such that $S \circ T = \mathrm{id}_{\mathcal{D}}$ and $T \circ S = \mathrm{id}_{\mathcal{C}}$.

The requirement of having an equal sign in the relations above is so strong as to render the concept of limited usefulness. But bearing in mind that the functors from $\mathcal{C}$ to $\mathcal{D}$ do form a category, we may amend the definition by requiring only that the two composite functors above be *isomorphic* to the respective identity functors. We get the following important notion:

Definition 6.3 (Equivalence of Categories) Two categories $\mathcal{C}$ and $\mathcal{D}$ are equivalent if there are covariant functors

$$S : \mathcal{C} \longrightarrow \mathcal{D} \quad \text{and} \quad T : \mathcal{D} \longrightarrow \mathcal{C}$$

such that there are isomorphisms Ψ and Φ of covariant functors

$$\Psi : S \circ T \xrightarrow{\cong} \mathrm{id}_{\mathcal{D}}$$

and

$$\Phi : T \circ S \xrightarrow{\cong} \mathrm{id}_{\mathcal{C}},$$

such that

$$S \circ \Phi = \Psi \circ S$$

in the sense that

$$S(\Phi_A) = \Psi_{S(A)} \quad \text{for all objects } A \text{ in } \mathcal{C}$$

and moreover,

$$T \circ \Psi = \Phi \circ T$$

in the sense that

$$T(\Psi_B) = \Phi_{T(B)} \quad \text{for all objects } B \text{ in } \mathcal{D}.$$

The functors are then referred to as *equivalences of categories*, and the two categories are said to be *equivalent*.

We express the two compatibility conditions by saying that S and T commute with Φ, Ψ. If a functor Φ or a functor Ψ belong to such a pair, then the functor is referred to as an equivalence of categories, or just as *an equivalence* when no confusion is possible.

Remark 6.3 In arguments which merely involve analyzing diagrams and properties of morphisms involving only diagrams, such as being isomorphisms or inverse to one another, one may simplify by "pretending" an equivalence of categories to be an isomorphism. The resulting "bogus" proof may be updated to a fully valid proof by adding the notational complexity required by introducing isomorphisms of functors of the type

$$\Psi : S \circ T \xrightarrow{\cong} \mathrm{id}_{\mathcal{D}}$$

and

$$\Phi : T \circ S \xrightarrow{\cong} \mathrm{id}_{\mathcal{C}},$$

at the various places where they are required, rather than just writing

$$S \circ T = \mathrm{id}_{\mathcal{D}} \quad \text{and} \quad T \circ S = \mathrm{id}_{\mathcal{C}}.$$

Proposition 6.4 *A covariant functor*

$$S : \mathcal{C} \longrightarrow \mathcal{D}$$

is an equivalence of categories if and only if the following two conditions are satisfied:

1. *For all $A_1, A_2 \in \mathrm{Obj}(\mathcal{C})$, S induces a bijection*

$$s : \mathrm{Hom}_{\mathcal{C}}(A_1, A_2) \longrightarrow \mathrm{Hom}_{\mathcal{D}}(S(A_1), S(A_2)) \quad f \mapsto S(f).$$

2. *For all $B \in \mathrm{Obj}(\mathcal{D})$ there exists $A \in \mathrm{Obj}(\mathcal{C})$ such that $B \cong S(A)$.*

Proof Assume first that $S : \mathcal{C} \longrightarrow \mathcal{D}$ is an equivalence. Let T be the functor going the other way as in Definition 6.3. Then for all $B \in \mathrm{Obj}(\mathcal{D})$ we have the isomorphism $\Phi_B : S(T(B)) \longrightarrow B$, hence the condition 2. is satisfied.

To prove 1., we use Remark 6.3. We construct an inverse t to the mapping s

$$s : \mathrm{Hom}_{\mathcal{C}}(A_1, A_2) \longrightarrow \mathrm{Hom}_{\mathcal{D}}(S(A_1), S(A_2))$$

as follows: For all $g : \mathrm{Hom}_{\mathcal{D}}(S(A_1), S(A_2))$ the morphism $t(g)$ is identified with $T(g)$ via the diagram:

$$
\begin{array}{ccc}
T(S(A_1)) & \xrightarrow{\;T(g)\;} & T(S(A_2)) \\
\Big\downarrow{=} & & {=}\Big\downarrow \\
A_1 & \xrightarrow[t(g)]{} & A_2
\end{array}
$$

the functorial isomorphisms $\Psi_{S(A_i)}$ being replaced by equal signs. Then $T(S(f)) = f$ since the diagram below commutes:

$$
\begin{array}{ccc}
T(S(A_1)) & \xrightarrow{\;T(S(f))\;} & T(S(A_2)) \\
\Big\downarrow{=} & & {=}\Big\downarrow \\
A_1 & \xrightarrow[f]{} & A_2
\end{array}
$$

Conversely, let $g : \mathrm{Hom}_{\mathcal{D}}(S(A_1), S(A_2))$. Then the diagram below commutes

$$
\begin{array}{ccc}
S(T(S(A_1))) & \xrightarrow{\;S(T(g))\;} & S(T(S(A_2))) \\
\Big\downarrow{=} & & {=}\Big\downarrow \\
S(A_1) & \xrightarrow[g]{} & S(A_2)
\end{array}
$$

Thus

$$
s(t(g)) = g.
$$

It remains to prove the sufficiency of the conditions 1. and 2., in other words to prove that these two conditions together imply the existence of the functor T making the pair (S, T) into an equivalence of categories in the sense of Definition 6.3. For this we can of course not simplify by using Remark 6.3.

We proceed as follows: For each object $B \in \mathrm{Obj}(\mathcal{D})$ we use condition 2. to fix an object $A_B \in \mathrm{Obj}(\mathcal{C})$ and an isomorphism $\beta_B : B \xrightarrow{\cong} S(A_B)$. We let $T(B) = A_B$ and for a morphism $\varphi : B_1 \longrightarrow B_2$ we define $T(\varphi) = \psi$ as the unique morphism which makes the following diagram commutative:

$$
\begin{array}{ccc}
B_1 & \xrightarrow{\;\varphi\;} & B_2 \\
{\cong}\Big\downarrow{\beta_{B_1}} & & \beta_{B_2}\Big\downarrow{\cong} \\
S(A_{B_1}) & \xrightarrow[\psi]{} & S(A_{B_2})
\end{array}
$$

T defined in this way is a functor. This verification is simple and is left to the reader.

To complete the proof we have to define isomorphisms of functors, commuting with S and T,

$$\Psi : S \circ T \xrightarrow{\cong} \mathrm{id}_{\mathcal{D}}$$

and

$$\Phi : T \circ S \xrightarrow{\cong} \mathrm{id}_{\mathcal{C}} .$$

First, the isomorphisms

$$\beta_B : B \xrightarrow{\cong} S(A_B)$$

yield an isomorphism of functors

$$\beta_{\mathcal{D}} \longrightarrow S \circ T$$

and we let Ψ be its inverse.

The details of the remainder of the proof is left to the reader: Let A be an object in $\mathcal{C}$, put $S(A) = B$, so $A \cong A_B$. Keeping track of the isomorphisms involved, this yields the isomorphism

$$\Phi : T \circ S \longrightarrow \mathrm{id}_{\mathcal{C}},$$

which commutes with Ψ as required. This completes the proof. $\square$

6.12 When are two Functors Isomorphic?

It is useful to be able to determine when two functors

$$S, T : \mathcal{C} \longrightarrow \mathcal{D},$$

are isomorphic. If they are, then it follows that for all objects A in $\mathcal{C}$, $S(A) \cong T(A)$. But the existence of isomorphisms $S(A) \cong T(A)$ for all objects A does not imply that the functors S and T are isomorphic. Instead, we have the following result:

Proposition 6.5 *A morphism of functors $\Gamma : S \longrightarrow T$ for the functors $S, T :$ $\mathcal{C} \longrightarrow \mathcal{D}$ is an isomorphism if and only if all Γ_A are isomorphisms.*

Proof One way is by definition. We need to show that if all Γ_A are isomorphisms in $\mathcal{D}$, then Γ is an isomorphism of functors. We let $\Delta_A = \Gamma_A^{-1}$. To show is that this defines a morphism of functors

$$\Delta : T \longrightarrow S,$$

which is then automatically inverse to Γ.

We need to verify that the following diagram commutes, for all $\varphi : A \longrightarrow B$:

$$
\begin{array}{ccc}
T(A) & \xrightarrow{\ \Delta_A\ } & S(A) \\
{\scriptstyle T(\varphi)}\big\downarrow & & \big\downarrow {\scriptstyle S(\varphi)} \\
T(B) & \xrightarrow[\ \Delta_B\ ]{} & S(B)
\end{array}
$$

In fact, we have the commutative diagram

$$
\begin{array}{ccc}
S(A) & \xrightarrow{\ \Gamma_A\ } & T(A) \\
{\scriptstyle S(\varphi)}\big\downarrow & & \big\downarrow {\scriptstyle T(\varphi)} \\
S(B) & \xrightarrow[\ \Gamma_B\ ]{} & T(B)
\end{array}
$$

or

$$
T(\varphi) \circ \Gamma_A = \Gamma_B \circ S(\varphi).
$$

This implies that

$$
\Delta_B \circ (T(\varphi) \circ \Gamma_A) \circ \Delta_A = \Delta_B \circ (\Gamma_B \circ S(\varphi)) \circ \Delta_A,
$$

from which the claim follows by associativity of composition. $\square$

6.13 Left and Right Adjoint Functors

Let $\mathcal{A}$ and $\mathcal{B}$ be two categories and let

$$
F : \mathcal{A} \longrightarrow \mathcal{B} \quad \text{and} \quad G : \mathcal{B} \longrightarrow \mathcal{A}
$$

be covariant functors. Assume that for all objects

$$
A \in \mathrm{Obj}(\mathcal{A}) \quad \text{and} \quad B \in \mathrm{Obj}(\mathcal{B})
$$

there are given bijections

$$
\Phi_{A,B} : \mathrm{Hom}_{\mathcal{B}}(F(A), B) \longrightarrow \mathrm{Hom}_{\mathcal{A}}(A, G(B))
$$

which are functorial in A and in B. We then say that F is left adjoint to G, and is G right adjoint to F. Less precisely we say simply that F and G are adjoint functors.

For contravariant functors the definition is analogous, or it may be reduced to the covariant case by passage to the dual category of $\mathcal{A}$ or $\mathcal{B}$.

Example 6.11 Let $\varphi : R \longrightarrow S$ be a homomorphism of commutative rings. Recall that if M is an S-module, then we define an R-module denoted by $M_{[\varphi]}$ by putting $rm = \varphi(r)m$ whenever $r \in R$ and $m \in M$. The covariant functor

$$S\text{-modules} \longrightarrow R\text{-modules}$$

$$M \mapsto M_{[\varphi]}$$

is called the Reduction of Structure-functor. This functor has a left adjoint called Base extension or the Extension of Structure-functor

$$R\text{-modules} \longrightarrow S\text{-modules}$$

$$N \mapsto N \otimes_R S.$$

In a more fancy language one may express the definition above by saying that

Definition 6.4 The functor F is left adjoint to the functor G, or G is right adjoint to F, where

$$\mathcal{A} \underset{G}{\overset{F}{\underset{\longleftarrow}{\longrightarrow}}} \mathcal{B}$$

provided that there is an isomorphism of bifunctors

$$\Phi : \mathrm{Hom}_{\mathcal{B}}(F(\ ),\ ) \overset{\cong}{\longrightarrow} \mathrm{Hom}_{\mathcal{A}}(\ , G(\ )).$$

Whenever we have a *morphism* of bifunctors as above, i.e. functorial mappings

$$\Phi_{A,B} : \mathrm{Hom}_{\mathcal{B}}(F(A), B) \longrightarrow \mathrm{Hom}_{\mathcal{A}}(A, G(B)),$$

then the morphism $\mathrm{id}_{F(A)}$ is mapped to a morphism $\varphi_A : A \longrightarrow G(F(A))$. We then obtain a morphism of functors

$$\varphi : \mathrm{id}_{\mathcal{A}} \longrightarrow G \circ F.$$

We then have that $\Phi_{A,B}$ is given by

$$(F(A) \longrightarrow B) \mapsto (A \overset{\varphi_A}{\longrightarrow} G(F(A)) \longrightarrow G(B)).$$

Similarly, a morphism of bifunctors Ψ in the opposite direction yields a morphism of functors

$$\psi : F \circ G \longrightarrow \mathrm{id}_{\mathcal{B}},$$

and $\Psi_{A,B}$ is then given by

$$(A \longrightarrow G(B)) \mapsto (F(A) \longrightarrow F(G(B)) \xrightarrow{\psi_B} B).$$

The assertion that Φ and Ψ are inverse to one another may then be expressed solely in terms of commutative diagrams, involving F, G, φ and ψ. We do not pursue this line of thought further here.

6.14 Representable Functors

In the following sections we turn to the very important and useful notion of a *representable functor*. Because we shall mainly use this in the contravariant case, we shall take that approach here, although of course the contravariant and the covariant cases are essentially equivalent by the usual trick of passing to the dual category.

So let $\mathcal{C}$ be a category, and let $X \in \text{Obj}(\mathcal{C})$. We define a contravariant functor

$$h_X : \mathcal{C} \longrightarrow \text{Set}$$

putting

$$h_X(Y) = \text{Hom}_{\mathcal{C}}(Y, X)$$

for any object Y in $\mathcal{C}$. For a morphisms $\varphi : Y_1 \longrightarrow Y_2$ we let

$$h_X(\varphi) : \text{Hom}_{\mathcal{C}}(Y_2, X) \longrightarrow \text{Hom}_{\mathcal{C}}(Y_1, X)$$

be given by

$$\psi \mapsto \psi \circ \varphi.$$

It is easily verified that h_X so defined is a contravariant functor. We shall extend a notation from algebraic geometry, and refer to the functor h_X as *the functor of points of the object X*. We also shall refer to the set $h_X(Y) = \text{Hom}_{\mathcal{C}}(Y, X)$ as the set of Y-valued points of the object X in $\mathcal{C}$.

We are now ready for the following important definition:

Definition 6.5 A contravariant functor

$$F : \mathcal{C} \longrightarrow \text{Set}$$

is said to be representable by the object X of $\mathcal{C}$ if there is an isomorphism of functors

$$\Psi : h_X \longrightarrow F.$$

Clearly all properties of the object X in $\mathcal{C}$ which may be formulated in category-theoretical terms are reflected in a functors representing it. In particular we have the

Proposition 6.6 *Two objects X and Y in the category $\mathcal{C}$ are isomorphic if and only if the functors h_X and h_Y are isomorphic.*

Proof An isomorphism $\varphi : X \longrightarrow Y$ yields, for all objects Z, functorial isomorphisms

$$\varphi_Z : h_Y(Z) = \operatorname{Hom}_{\mathcal{C}}(Y, Z) \longrightarrow \operatorname{Hom}_{\mathcal{C}}(X, Z)$$

by composition. As is easily seen this yields an isomorphism of functors

$$\Phi : h_Y \xrightarrow{\cong} h_X.$$

Conversely, given an isomorphism Φ as above, we get a morphism $f : Y \to X$ which corresponds to id_X, and it is easily checked that this is an isomorphism. The details are left to the reader as a simple exercise. $\square$

For the covariant case we define

$$h^X : \mathcal{C} \longrightarrow \operatorname{Set}$$

by

$$h^X(Y) = \operatorname{Hom}_{\mathcal{C}}(X, Y),$$

which is a covariant functor. We then similarly get the notion of a representable covariant functor $\mathcal{C} \longrightarrow \operatorname{Set}$. The details are left to the reader. Of course this amounts to applying the contravariant case to the dual category $\mathcal{C}^*$.

The functors of points play important roles in many situations.

6.15 Representable Functors and Universal Properties: Yoneda's Lemma

A vast number of constructions in mathematics are best understood as representing an appropriate functor. The key to a unified understanding of this lies in the theorem below.

Given a contravariant functor

$$F : \mathcal{C} \longrightarrow \operatorname{Set}.$$

Let X be an object in $\mathcal{C}$, and let $\xi \in F(X)$. For all objects Y of $\mathcal{C}$ we then define a mapping as follows:

$$\Phi_Y : h_X(Y) = \operatorname{Hom}_{\mathcal{C}}(Y, X) \longrightarrow F(Y)$$

$$\varphi \mapsto F(\varphi)(\xi).$$

It is an easy exercise to verify that this is a *morphism of contravariant functors*,

$$\Phi : h_X \longrightarrow F.$$

We now have the

Theorem 6.7 (Yoneda's Lemma) *The functor F is representable by the object X if and only if there exists an element $\xi \in F(X)$ such that the corresponding Φ is an isomorphism of contravariant functors. This is the case if and only if all Φ_Y are bijective.*

Proof By Proposition 6.5 Φ is an isomorphism if and only if all Φ_Y are bijective. Thus, if all Φ_Y are bijective then F is representable via the isomorphism Φ.

On the other hand, if F is representable, then there is an isomorphism of functors

$$\Psi : h_X \xrightarrow{\cong} F.$$

Put $\xi = \Psi_X(\mathrm{id}_X)$ and let $\varphi \in h_X(Y)$. For all objects Y we get the commutative diagram

$$
\begin{array}{ccc}
h_X(X) & \xrightarrow{\ \Psi_X\ } & F(X) \\
{\scriptstyle \varphi \circ (\ )} \downarrow & & \downarrow {\scriptstyle F(\varphi)} \\
h_X(Y) & \xrightarrow[\ \Psi_Y\ ]{} & F(Y)
\end{array}
$$

Noting what happens to id_X in this commutative diagram, we find the relation

$$F(\varphi)(\xi) = \Psi_Y(\varphi),$$

thus $\Psi_Y = \Phi_Y$ which is therefore bijective. $\qquad\qquad\qquad\Box$

We say that the object X represents the functor F and that the element $\xi \in F(X)$ is the *universal element*. This language is tied to the following *Universal Mapping Property*:

The *Universal Mapping Property* of the pair (X, ξ) representing the contravariant functor F is formulated as follows:

For all elements $\eta \in F(Y)$ there exists a unique morphism

$$\varphi : Y \longrightarrow X$$

such that

$$F(\varphi)(\xi) = \eta.$$

In fact, this is nothing but a direct translation of the assertion that Φ_Y be bijective.

Another remark to be made here, is that two objects representing the same functor are isomorphic by a unique isomorphism. The universal elements correspond under the mapping induced by this isomorphism. The proof of this observation is left to the reader.

Chapter 7
Constructions and Representable Functors

This chapter proceeds to treat products and coproducts in the usual framework of representable functors.

7.1 Products and Coproducts

Let $\mathcal{C}$ be a category. Let B_i, $i = 1, 2$ be two objects in $\mathcal{C}$. Define a functor

$$F : \mathcal{C} \longrightarrow \mathrm{Set}$$

by

$$B \mapsto \{(\psi_1, \psi_2) \mid \psi_i \in \mathrm{Hom}_{\mathcal{C}}(B, B_i), i = 1, 2\}.$$

If this functor is representable, then the representing object, unique up to a unique isomorphism, is denoted by $B_1 \times B_2$, and referred to as the *product* of B_1 and B_2. The universal element (p_1, p_2) is of course a pair of two morphisms, from $B_1 \times B_2$ to B_1, respectively B_2:

$$B_1 \times B_2 \xrightarrow{\;p_2\;} B_2$$
$$\downarrow{\scriptstyle p_1}$$
$$B_1$$

The morphism p_1 and p_2 are called the first and second projection, respectively. The product $B_1 \times B_2$ and the projections solve the following so called *universal problem*: For all morphisms f_1 and f_2 as below, there exists a unique morphism h such that the triangular diagrams commute:

A. Holme, *A Royal Road to Algebraic Geometry*,
DOI 10.1007/978-3-642-19225-8_7, © Springer-Verlag Berlin Heidelberg 2012

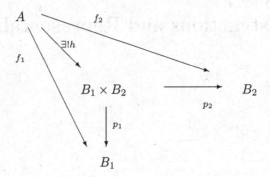

We obtain, as always, a dual notion by applying the above to the dual category $\mathcal{C}^*$. Specifically, we consider the functor

$$G : \mathcal{C} \longrightarrow \mathrm{Set}$$

defined by

$$B \mapsto \{(\ell_1, \ell_2) \mid \ell_i \in \mathrm{Hom}_{\mathcal{C}}(B_i, B), i = 1, 2\}.$$

Whenever this functor is representable, the representing object is denoted by $B_1 \coprod B_2$ and referred to as the *coproduct* of B_1 and B_2. The morphisms η_i, for $i = 1, 2$, are called the canonical injections:

$$
\begin{array}{c}
B_1 \\
\downarrow{\scriptstyle \eta_1} \\
B_1 \coprod B_2 \xleftarrow{\ \eta_2\ } B_2
\end{array}
$$

We similarly define products and coproducts of sets of objects, in particular infinite sets of objects. For a set of morphisms

$$\varphi_i : B \longrightarrow B_i, \quad i \in I,$$

we get

$$(\varphi_i \mid i \in I) : B \longrightarrow \prod_{i \in I} B_i$$

such that all the appropriate diagrams commute.

Further, if $\psi_i : A_i \longrightarrow B_i$ are morphisms for all $i \in I$, then we get a morphism

$$\psi = \prod_{i \in I} \psi_i : \prod_{i \in I} A_i \longrightarrow \prod_{i \in I} B_i$$

uniquely determined by making all of the following diagrams commutative:

$$
\begin{array}{ccc}
\prod_{i\in I} A_i & \xrightarrow{\ \psi\ } & \prod_{i\in I} B_i \\
{\scriptstyle \mathrm{pr}_i}\downarrow & & \downarrow{\scriptstyle \mathrm{pr}_i} \\
A_i & \xrightarrow[\psi_i]{} & B_i
\end{array}
$$

In the category Set products exist, and are nothing but the usual set-theoretic product:

$$
\prod_{i\in I} A_i = \{(a_i | i \in I) \,|\, a_i \in A_i\}.
$$

The coproduct is the *disjoint union* of all the sets:

$$
\coprod_{i\in I} A_i = \{(a_i, i) \,|\, i \in I, a_i \in A_i\}.
$$

Adding the index as a second coordinate only serves to make the union disjoint.

7.2 Fibered Products and Coproducts

When we apply the above concepts to the categories $\mathcal{C}_A$, respectively $\mathcal{C}^A$, then we get the notions of *fibered* products and coproducts, respectively. We go over this version in detail, as it is important in algebraic geometry.

Let A be an object in the category $\mathcal{C}$. Let (B_i, φ_i), $i = 1, 2$ be two objects in $\mathcal{C}_A$. Define a functor

$$
F : \mathcal{C}_A \longrightarrow \text{Set}
$$

by

$$
(B, \varphi) \mapsto \{(\psi_1, \psi_2) \mid \psi_i \in \operatorname{Hom}_{\mathcal{C}_A}((B, \varphi), (B_i, \varphi_i)), i = 1, 2\}.
$$

If this functor is representable, then the representing object, unique up to a unique isomorphism, is denoted by $B_1 \times_A B_2$, and referred to as the *fibered product* of B_1 and B_2 over A. The universal element (p_1, p_2) is of course a pair of two morphisms, from $B_1 \times_A B_2$ to B_1, respectively B_2 such that the following diagram commutes:

$$
\begin{array}{ccc}
B_1 \times_A B_2 & \xrightarrow{\ p_2\ } & B_2 \\
{\scriptstyle p_1}\downarrow & & \downarrow{\scriptstyle \varphi_2} \\
B_1 & \xrightarrow[\varphi_1]{} & A
\end{array}
$$

The morphisms p_1 and p_2 are called the first and second projection, respectively. We may illustrate the universal property of the fibered product as follows:

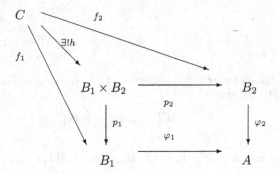

where all the triangular diagrams are commutative.

Again we get the dual notion by applying the above to the dual category $\mathcal{C}^*$. Specifically, we consider the functor

$$G : \mathcal{C}^A \longrightarrow \mathfrak{Set}$$

by

$$(\tau, B) \mapsto \{(\ell_1, \ell_2) \mid \ell_i \in \mathrm{Hom}_{\mathcal{C}^A}((\tau, B), (\tau_i, B_i)), i = 1, 2\}.$$

Whenever this functor is representable, the representing object is denoted by $B_1 \coprod_A B_2$ and referred to as the *fibered coproduct* of B_1 and B_2. The morphisms ℓ_i, $i = 1, 2$ are called the canonical injections, and the following diagram commutes:

$$
\begin{array}{ccc}
A & \xrightarrow{\ \tau_1\ } & B_1 \\
\tau_2 \downarrow & & \downarrow \ell_1 \\
B_2 & \xrightarrow{\ \ell_2\ } & B_1 \coprod_A B_2
\end{array}
$$

Chapter 8
Abelian Categories

This chapter deals with Abelian categories, Grothendieck topologies, presheaves and sheaves.

8.1 Definitions

In the category of Abelian groups, or more generally the category of modules over a commutative ring A, the product and the coproduct of two objects always exist. And moreover, they are isomorphic. In fact, it is easily seen that the direct sum $M_1 \oplus M_2$ of two A-modules satisfies the universal properties of both $M_1 \times M_2$ and $M_1 \coprod M_2$. The universal elements of the appropriate functors are given by the A-homomorphisms

$$p_1(m_1, m_2) = m_1, \qquad p_2(m_1, m_2) = m_2,$$
$$\ell_1(m_1) = (m_1, 0), \qquad \ell_2(m_2) = (0, m_2).$$

This also applies to finite products and coproducts in $\mathcal{A}$b: They exist, and are equal.[1]

Another characteristic feature of this category is that $\mathrm{Hom}_{\mathcal{C}}(M, N)$ is an Abelian group.

. The category of A-modules is an example of an *Abelian category*. The property that finite products and coproducts always exist and are equal is one of the defining properties of this concept. Another important part of the definition of an Abelian category, is the existence of a *zero object*. For the category of A-modules this is the A-module consisting of the element 0

---

[1] But although infinite products and coproducts do exist, they are not equal: The direct sum of a family of Abelian groups $\{A_i\}_{i \in \mathbb{N}}$ is the subset of the direct product $A_1 \times A_2 \times \cdots \times A_n \times \cdots$ consisting of all tuples such that only a finite number of coordinates are different from the zero element in the respective A_i's.

A. Holme, *A Royal Road to Algebraic Geometry*,
DOI 10.1007/978-3-642-19225-8_8, © Springer-Verlag Berlin Heidelberg 2012

alone. It is denoted by $\mathbf{0}$, and has the property that for all objects X there is a unique morphism from it to X, and a unique morphism from X to $\mathbf{0}$. We say that $\mathbf{0}$ is both *final and cofinal* in $\mathcal{C}$.

8.2 Product and Coproduct in the Non Abelian Category of Commutative Rings

An important category from commutative algebra is *not Abelian*, however. Namely, the category Comm. In Comm, the product of two objects, of two commutative rings with 1 A and B, is the ring $A \times B$ consisting of all pairs (a, b) with $a \in A$ and $b \in B$. The *coproduct*, however, is the tensor product $A \otimes B$. In the category Comm_R of R-algebras the coproduct is $A \otimes_R B$.

8.3 Localization as Representing a Functor

Let A be a commutative ring with 1, and $S \subset A$ a multiplicatively closed subset. We say that an element $a \in A$ is invertible on the A-module N provided that the A-homomorphism

$$\mu_a : N \longrightarrow N \quad \text{given by } \mu_a(n) = an,$$

is invertible, i.e. is a bijective mapping on N. The subset S of A is said to be invertible on N if all elements in S are.

We now let $\mathcal{C}$ be the subcategory of Mod_A of modules where S is invertible, and let M be an A-module. Define $F : \mathcal{C} \longrightarrow \text{Set}$ by

$$F(N) = \text{Hom}_A(M, N).$$

In introductory courses in commutative algebra we construct the localization of M in S, the A-module $S^{-1}M$, with the canonical A-homomorphism $\tau : M \longrightarrow S^{-1}M$. It is a simple exercise to show that the pair $(S^{-1}M, \tau)$ represents the functor F.

8.4 Kernel and Cokernel of Two Morphisms

Let

$$A \underset{\varphi_2}{\overset{\varphi_1}{\rightrightarrows}} B$$

be two morphisms in the category $\mathcal{C}$.

We consider the functor

$$F : \mathcal{C} \longrightarrow \mathrm{Set}$$

given by

$$F(X) = \{\varphi \in \mathrm{Hom}_{\mathcal{C}}(X, A) \mid \varphi_1 \circ \varphi = \varphi_2 \circ \varphi\}.$$

This functor is contravariant. If it is representable, then we refer to the representing object as the *kernel* of the pair φ_1, φ_2.

Thus an object N together with a morphism $\iota : N \longrightarrow A$ is called a *kernel for* φ_1 *and* φ_2 if the following two conditions are satisfied:

1. We have $\varphi_1 \circ \iota = \varphi_2 \circ \iota$.
2. If $\varphi : X \longrightarrow A$ is a morphism such that $\varphi_1 \circ \varphi = \varphi_2 \circ \varphi$, then there exists a unique morphism $\psi : X \longrightarrow N$ such that $\iota \circ \psi = \varphi$.

The dual concept is that of a *cokernel:* $\tau : B \longrightarrow M$ is a cokernel for the morphisms $\varphi_1, \varphi_2 : A \longrightarrow B$ if the following universal property holds:

1. We have $\tau \circ \varphi_1 = \tau \circ \varphi_2$.
2. If $\varphi : B \longrightarrow X$ is a morphism such that $\varphi \circ \varphi_1 = \varphi \circ \varphi_2$, then there exists a unique morphism $\psi : M \longrightarrow X$ such that $\psi \circ \tau = \varphi$.

We write

$$\ker(f, g) \quad \text{and} \quad \mathrm{coker}(f, g)$$

for the kernel, respectively the cokernel, of the pair (f, g).

It is easily verified that kernels and cokernels exist in the categories $\mathcal{A}b$, $\mathcal{M}od_R$, $\mathcal{T}op$ and $\mathcal{S}et$.

Two mappings of sets,

$$A \overset{f}{\underset{g}{\longrightarrow}} B$$

have a kernel and a cokernel: The kernel is defined as

$$K = \{a \in A \mid f(a) = g(a)\}$$

and the map η is the obvious inclusion. The cokernel is defined as $C = B/\sim$, where $\sim$ is the equivalence relation on B generated by the relation ρ given below:[2]

$$b_1 \rho b_2$$
$$\Updownarrow$$
$$\exists a \in A \text{ such that } f(a) = b_1, g(a) = b_2.$$

As is immediately seen, this is a cokernel for (f, g).

---

[2] A relation ρ on a set B generates an equivalence relation $\sim$ by putting $a \sim b$ if either $a = b$, or there is a sequence $a \sim a_1 \sim a_2 \sim \cdots \sim a_n = b$, or a sequence $b = b_1 \sim b_2 \sim \cdots \sim b_m = a$.

For the category $\mathfrak{Top}$, these sets carry a natural topology: Namely the induced topology from the space $A \times B$ in the former case, the quotient topology in the latter. We get a kernel and a cokernel in $\mathfrak{Top}$ with this choice of topology on the set-theoretic versions.

If f and g are morphisms in $\mathfrak{Mod}_R$, then the set theoretic kernel is automatically an R-module. The same is true for the cokernel, since the relation $\sim$ is actually a *congruence relation* for the operations, that is to say, it is compatible with the operations.[3]

The diagrams

$$X \overset{\longrightarrow}{\underset{\longrightarrow}{}} Y \longrightarrow Z$$

or

$$X \longrightarrow Y \overset{\longrightarrow}{\underset{\longrightarrow}{}} Z$$

are said to be *exact* if the former is a cokernel-diagram or the latter a kernel-diagram, respectively.

In an Abelian category we have the usual concept of kernel and cokernel of a single morphism. The link between this and the case of a pair of morphisms are the definitions

$$\ker(f) = \ker(f, 0) \quad \text{and} \quad \operatorname{coker}(f) = \operatorname{coker}(f, 0),$$

where 0 denotes the zero morphism.

8.5 Inductive and Projective Limits

Let $\mathfrak{J}$ and $\mathcal{C}$ be two categories, where $\mathfrak{J}$ is small, and let

$$F : \mathfrak{J} \longrightarrow \mathcal{C}$$

be a covariant functor. We define a functor

$$L : \mathcal{C} \longrightarrow \mathfrak{Set}$$

by

$$L(A) = \left\{ \{v_X\}_{X \in \mathrm{Obj}(\mathfrak{J})} \,\middle|\, \begin{array}{l} v_X : A \longrightarrow F(X) \text{ is a morphism in } \mathcal{C} \\ \text{such that for all morphisms} \\ \alpha : X \longrightarrow Y \text{ we have } F(\alpha) \circ v_X = v_Y \end{array} \right\}.$$

---

[3]Hence addition and multiplication with an element in R may be defined on the set of equivalence classes by performing the operations on elements representing the classes and taking the resulting classes. The details are left to the reader.

When this functor is representable, we call the representing object *the projective limit of the functor F*.

Similarly we define the *inductive limit* of a covariant functor: We define a functor

$$S : \mathcal{C} \longrightarrow \mathrm{Set}$$

by

$$S(A) = \left\{ \{v_X\}_{X \in \mathrm{Obj}(\mathcal{I})} \ \middle| \ \begin{array}{l} v_X : F(X) \longrightarrow A \text{ is a morphism in } \mathcal{C} \\ \text{such that for all morphisms} \\ \alpha : X \longrightarrow Y \text{ we have } v_Y \circ F(\alpha) = v_X \end{array} \right\}.$$

These limits of the functor F, the projective and the inductive, will be denoted respectively by and $\varprojlim_{X \in \mathrm{Obj}(\mathcal{I})}(F)$ and $\varinjlim_{X \in \mathrm{Obj}(\mathcal{I})}(F)$.

By Yoneda's Lemma we have the following universal properties for the two limits introduced above:

Universal property of the projective limit For all objects A of $\mathcal{C}$ and objects X of $\mathcal{I}$ with morphisms $v_X : A \longrightarrow F(X)$ compatible with morphisms in $\mathcal{I}$, there exists a unique morphism $A \longrightarrow \varprojlim(F)$ such that the following composition is v_X,

$$A \longrightarrow \varprojlim_{X \in \mathrm{Obj}(\mathcal{I})}(F) \longrightarrow F(X).$$

Universal property of the inductive limit For all objects A of $\mathcal{C}$ and objects X of $\mathcal{I}$ with morphisms $v_X : F(X) \longrightarrow A$ compatible with morphisms in $\mathcal{I}$, there exists a unique morphism $\varinjlim(F) \longrightarrow A$ such that the following composition is v_X,

$$F(X) \longrightarrow \varinjlim_{X \in \mathrm{Obj}(\mathcal{I})}(F) \longrightarrow A.$$

The subscript $X \in \mathrm{Obj}(\mathcal{I})$ is deleted when no ambiguity is possible.

We define the inductive and the projective limit for contravariant functors similarly, or rely on the definition for the covariant case by regarding a contravariant functor

$$F : \mathcal{I} \longrightarrow \mathcal{C}$$

as a covariant functor

$$G : \mathcal{I}^* \longrightarrow \mathcal{C}.$$

Then $\varinjlim(F) = \varinjlim(G)$ and $\varprojlim(F) = \varprojlim(G)$.[4] Another name for the inductive limit is *direct limit*, while the projective limit is called *inverse limit*.

---

[4]Note that if we turn the contravariant functor into a covariant one as $H : \mathcal{I} \longrightarrow \mathcal{C}^*$, then the two kinds of limits are interchanged.

To sum up, for a covariant functor we have for all morphisms $\varphi : X \longrightarrow Y$ in $\mathfrak{I}$:

$$\varprojlim(F) \longrightarrow F(X) \xrightarrow{F(\varphi)} F(Y) \longrightarrow \varinjlim(F),$$

and for a contravariant functor we have for all morphisms $\varphi : X \longrightarrow Y$ in $\mathfrak{I}$:

$$\varprojlim(F) \longrightarrow F(Y) \xrightarrow{F(\varphi)} F(X) \longrightarrow \varinjlim(F).$$

8.6 Important Special Case: Projective and Inductive Systems and Their Limits

We now turn to the most important special case of the situation in the preceding section.

Let I be a partially ordered set, that is to say a set I where there is given an ordering-relation $\leq$ such that

1. $i \leq i$
2. $i \leq j$ and $j \leq k \Longrightarrow i \leq k$.

We regard the set I as the set of objects in a category $\mathfrak{Ind}(I)$, necessarily *small*, and let the set of morphisms between the objects i and j be

$$\mathrm{Hom}_{\mathfrak{Ind}(I)}(i,j) = \begin{cases} \emptyset & \text{if } i \not\leq j, \\ \iota_{i,j} & \text{one element if } i \leq j. \end{cases}$$

As we see, the category $\mathfrak{Top}(X)$ is the result of applying this to the partially ordered set of open subsets in the topological space X. Thus the present point of view represents a generalization of the concept of a topological space. When we wish to distinguish between I viewed as just a partially ordered set and this category we write some times $\mathfrak{Ind}(I)$ instead of just I.

An *inductive system* in a category $\mathcal{C}$ over I is by definition a covariant functor

$$F : \mathfrak{Ind}(I) \longrightarrow \mathcal{C},$$

and a *projective system* in $\mathcal{C}$ over I is a contravariant functor between the same categories, regarded as a covariant functor

$$F : \mathfrak{Ind}(I)^* \longrightarrow \mathcal{C}.$$

Usually we write F_i instead of $F(i)$ in the above situations, and refer to $\{F_i\}_{i \in I}$ as an inductive, respectively projective, system.

Note that if we give I the partial ordering $\preceq$ by letting $i \preceq j \Leftrightarrow j \leq i$ and denote the resulting partially ordered set by I^*, then $\mathfrak{Ind}(I)^* = \mathfrak{Ind}(I^*)$.

We let $\{F_i\}_{i \in I}$ be an inductive system, and define a functor

$$L : \mathcal{C} \longrightarrow \text{Set}$$

as follows:

$$L(X) = \left\{ \{\varphi_i\}_{i \in I} \mid \varphi_i \in \text{Hom}_{\mathcal{C}}(F_i, X) \text{ and if } i \leq j \text{ then } \varphi_i = \varphi_j \circ \iota_{i,j} \right\}.$$

If this functor happens to be representable, then we denote the representing object by

$$\varinjlim_{i \in I} F_i,$$

and note that the universal element is a collection of morphisms, compatible with the inductive structure,

$$F_i \longrightarrow \varinjlim_{i \in I} F_i.$$

The universal property amounts to that whenever we have such a set of compatible morphisms,

$$F_i \longrightarrow Y,$$

then they factor uniquely through a morphism

$$\varinjlim_{i \in I} F_i \longrightarrow Y.$$

Similarly we define the *projective limit*, denoted by $\varprojlim_{i \in I} F_i$.

As in the general case we may sum this up as follows, for an inductive system $\{F_i\}_{i \in I}$ over a partially ordered set I, where $i \leq j$:

$$\varprojlim_{i \in I} F_i \longrightarrow F_i \longrightarrow F_j \longrightarrow \varinjlim_{i \in I} F_i.$$

8.7 On the Existence of Projective and Inductive Limits

There are several results on the existence of inductive and projective limits. The most general theorem is the following, which we prove following [4], pp. 54–56:

Theorem 8.1 *Let $\mathcal{C}$ be a category. Then every covariant functor from a small category $\mathcal{A}$*

$$F : \mathcal{A} \longrightarrow \mathcal{C}$$

has an inductive limit if and only if $\mathcal{C}$ has infinite coproducts and cokernels always exist in $\mathcal{C}$, and every functor as above has a projective limit if and only if $\mathcal{C}$ has infinite products and kernels always exist in $\mathcal{C}$.

Proof We prove the assertion for projective limits, noting that the inductive case follows by replacing the target category by its dual.

To show that the condition is necessary, we note that the product of a family of objects in the category $\mathcal{C}$ may be viewed as a projective limit: Indeed, let $\{C_i\}_{i \in I}$ denote a set of objects from $\mathcal{C}$. Let $\mathcal{A}$ denote the category defined by $\mathrm{Obj}(\mathcal{A}) = I$ and $\mathrm{Hom}_{\mathcal{A}}(i,j) = \{\mathrm{id}_i\}$ if $i = j$, empty otherwise. Define a functor

$$F : \mathcal{A} \longrightarrow \mathcal{C}$$

$$i \mapsto C_i.$$

As is immediately seen, the assertion that $\varprojlim_F C_i$ exists is equivalent to the assertion that $\prod_{i \in I} C_i$ exists. To show the necessity of the last part of the condition, let

$$f_1, f_2 : C_1 \longrightarrow C_2$$

be two morphisms in $\mathcal{C}$. Let $\mathcal{A}$ be the category consisting of two objects denoted by 1 and 2, and such that $\mathrm{Hom}_{\mathcal{A}}(1,2) = \{\varphi_1, \varphi_2\}$. Apart from the identity morphisms, these are the only morphisms in $\mathcal{A}$. Define the functor F by

$$F : \mathcal{A} \longrightarrow \mathcal{C}$$

$$i \mapsto C_i \quad \text{for } i = 1, 2,$$

$$\varphi_i \mapsto f_i \quad \text{for } i = 1, 2.$$

The kernel of f_1 and f_2 is the projective limit $\varprojlim_F C_i$.

To prove the sufficiency, put

$$\Pi = \prod_{X \in \mathrm{Obj}(\mathcal{A})} F(X),$$

and let

$$\mathrm{pr}_X : \Pi \longrightarrow F(X)$$

denote the projections. Further, let

$$\Upsilon = \prod_{Y \in \mathrm{Obj}(\mathcal{A}), \alpha \in \mathrm{Hom}_{\mathcal{A}}(X,Y)} F(Y)_\alpha,$$

where $F(Y)_\alpha = F(Y)$ for all α and let

$$\mathrm{pr}_{Y,\alpha} : \Upsilon \longrightarrow F(Y)$$

denote the projections.

Now each morphism $\alpha : X \longrightarrow Y$ yields a morphism

$$F(\alpha) \circ \mathrm{pr}_X : \Pi \longrightarrow F(Y),$$

hence by the universal property of the product Υ there is a unique morphism v which makes the following diagrams commutative:

$$
\begin{array}{ccc}
\Pi & \xrightarrow{\;v\;} & \Upsilon \\
{\scriptstyle \mathrm{pr}_X}\downarrow & & \downarrow{\scriptstyle \mathrm{pr}_{Y,\alpha}} \\
F(X) & \xrightarrow[F(\alpha)]{} & F(Y)
\end{array}
$$

We also have a morphism $\mathrm{pr}_Y : \Pi \longrightarrow F(Y)$, for a given Y and morphism $\alpha : X \longrightarrow Y$. This yields a morphism w such that the following diagrams are commutative:

$$
\begin{array}{ccc}
\Pi & \xrightarrow{\;w\;} & \Upsilon \\
{\scriptstyle \mathrm{pr}_Y}\downarrow & & \downarrow{\scriptstyle \mathrm{pr}_{Y,\alpha}} \\
F(Y) & \xrightarrow[=]{} & F(Y)
\end{array}
$$

Let

$$\ell : L \longrightarrow \Pi$$

be the kernel of (v, w), so in particular $v \circ \ell = w \circ \ell$. Let $\ell_X = \mathrm{pr}_X \circ \ell$, and consider the system

$$\ell_X : L \longrightarrow F(X).$$

We claim that this is the projective limit of the functor F. First of all, we have to show that the compositions behave right, namely that whenever

$$\alpha : X \longrightarrow Y$$

is a morphism in $\mathcal{A}$, then

$$\ell_Y : L \xrightarrow{\ell_X} F(X) \xrightarrow{F(\alpha)} F(Y).$$

Indeed, we have

$$F(\alpha) \circ \ell_X = F(\alpha) \circ \mathrm{pr}_X \circ \ell = \mathrm{pr}_{Y,\alpha} \circ v \circ \ell = \mathrm{pr}_{Y,\alpha} \circ w \circ \ell = \mathrm{pr}_Y \circ \ell = \ell_Y.$$

We finally show that the universal property of the projective limit is satisfied. So let the family of morphisms

$$s_X : S \longrightarrow F(X), \quad X \in \mathrm{Obj}(\mathcal{A})$$

be such that whenever $\alpha : X \longrightarrow Y$ is a morphism in $\mathcal{A}$, then $F(\alpha) \circ s_X = s_Y$.
In particular we obtain a unique morphism $\sigma : S \longrightarrow \Pi$, such that $s_X = \mathrm{pr}_X \circ \sigma$. We now have

$$\mathrm{pr}_{Y,\alpha} \circ v \circ \sigma = F(\alpha) \circ \mathrm{pr}_X \circ \sigma = F(\alpha) \circ s_X = s_Y$$

and

$$\mathrm{pr}_{Y,\alpha} \circ w \circ \sigma = \mathrm{pr}_Y \circ \sigma = s_Y.$$

Thus by the universal property of the product Υ it follows that

$$v \circ \sigma = w \circ \sigma,$$

and hence σ factors uniquely through the kernel L: There is a unique $S \xrightarrow{s} L$
such that $\sigma = \ell \circ s$. Thus $\ell_X \circ s = \mathrm{pr}_X \circ \ell \circ s = \mathrm{pr}_X \circ \sigma = s_X$, and we are done. $\square$

We note that we are now guaranteed the existence of inductive and projective limits, in a rather general setting, in the categories Set, Top, and Mod_A.
It is, however, useful in our practical work to have a good description of these limits. This is particularly important in the case of some inductive limits we encounter in *sheaf theory*. This is the subject of the next section.

8.8 The Stalk of a Presheaf on a Topological Space

Let X be a topological space, and let $x \in X$ be a point. Let $\mathcal{B}$ be a set consisting of open subsets of X containing x, such that the following condition holds:

For any two open subsets U and V containing x there is an open subset $W \in \mathcal{B}$ contained in $U \cap V$. We say that $\mathcal{B}$ is a basis for the system of open neighborhoods around $x \in X$. Let $\mathcal{C}$ be one of the categories Set or Mod_A, and let

$$\mathcal{F} : \mathsf{Top}_X \longrightarrow \mathcal{C}$$

be a contravariant functor. We refer to $\mathcal{F}$ as a *presheaf* of $\mathcal{C}$ on the topological space X. If $\iota_{U,V} : U \hookrightarrow V$ is the inclusion mapping of U into V, then $\mathcal{F}(\iota_{U,V}) : \mathcal{F}(V) \longrightarrow \mathcal{F}(U)$ is denoted by $\rho^{\mathcal{F}}_{V,U}$ and referred to as the *restriction morphism from V to U*. $\mathcal{F}$ is deleted from the notation when no ambiguity is possible. The image of an element f under the restriction morphism $\rho_{V,U}$ is referred to as the restriction of f from V to U.

Then we get the inductive limit $\varinjlim_{V \in \mathcal{B}} \mathcal{F}(V)$ as follows: We form the disjoint union of all $\mathcal{F}(V)$:

$$M(x) = \coprod_{V \in \mathcal{B}} \mathcal{F}(V).$$

We define an equivalence relation in $M(x)$ by putting $f \sim g$ for $f \in \mathcal{F}(U)$ and $g \in \mathcal{F}(V)$ provided that they have the same restriction to a smaller open subset in $\mathcal{B}$. We then have

$$\varinjlim_{V \in \mathcal{B}} \mathcal{F}(V) = M(x)/\sim.$$

In fact, for the category Mod_A we find well defined addition and scalar multiplication which makes this set into an A-module by putting

$$[f_U] + [g_V] = [\rho_{U,W}(f_U) + \rho_{V,W}(g_V)]$$

where f_U and g_V are elements in $\mathcal{F}(U)$ and $\mathcal{F}(V)$, respectively. We also let

$$a[g_V] = [ag_V],$$

here the equivalence class is indicated by brackets. It is easy to see that there are canonical isomorphisms between $\varinjlim_{V \in \mathcal{B}} \mathcal{F}(V)$ and $\varinjlim_{V \in \mathcal{D}} \mathcal{F}(V)$ when $\mathcal{B}$ and $\mathcal{D}$ are two bases for the neighborhood system at x. In particular we may take all the open subsets containing x: Indeed, we consider first the case when $\mathcal{B}$ is arbitrary and $\mathcal{D}$ is the set of all open subsets containing x. Then $\mathcal{D} \supset \mathcal{B}$ induces a morphism

$$\varinjlim_{V \in \mathcal{B}} \mathcal{F}(V) \longrightarrow \varinjlim_{V \in \mathcal{D}} \mathcal{F}(V),$$

which is an isomorphism since whenever U is an open subset containing x, there is an open subset V in $\mathcal{B}$ contained in U. Since we may do this for all bases for the neighborhood system around x, the claim follows.

The inductive limit defined above is denoted by $\mathcal{F}_x$, and referred to as *the stalk* of the presheaf $\mathcal{F}$ at the point $x \in X$. For an open subset V containing the point x we have, in particular, a mapping of sets or an A-homomorphism

$$\iota_{V,x} : \mathcal{F}(V) \longrightarrow \mathcal{F}_x$$

of $\mathcal{F}(V)$ into the stalk at the point x. Frequently we write f_x instead of $\iota_{V,x}(f)$.

8.9 Grothendieck Topologies, Sheaves and Presheaves

A Grothendieck topology $\mathcal{G}$ consists of a category $\mathrm{Cat}(\mathcal{G})$ and a set $\mathrm{Cov}(\mathcal{G})$, called *coverings*, of families of morphisms

$$\{\varphi_i : U_i \longrightarrow U\}_{i \in I}$$

in $\mathrm{Cat}(\mathcal{G})$ such that

1. All sets consisting of one isomorphism are coverings.
2. If $\{\varphi_i : U_i \longrightarrow U\}_{i \in I}$ and $\{\varphi_{i,j} : U_{i,j} \longrightarrow U_i\}_{j \in I_i}$ are coverings, then so is the set consisting of all the compositions

$$\{\psi_{i,j} : U_{i,j} \longrightarrow U\}_{i \in I, j \in I_i}.$$

3. If $\{\varphi_i : U_i \longrightarrow U\}_{i \in I}$ is a covering, and $V \longrightarrow U$ is a morphism in $\mathrm{Cat}(\mathcal{G})$, then the products $U_i \times_U V$ exist for all $i \in I$ and the projections $\{U_i \times_U V \longrightarrow V = U \times_U V\}_{i \in I}$ is a covering.

At this point we offer one example only, namely the following: Let X be a topological space, and let $\mathrm{Cat}(\mathcal{G})$ be $\mathrm{Top}(X)$. Whenever U is an open subset, a covering is given as the set of all open injections of the open subsets in an open covering in the usual sense:

$$\{\varphi_i : U_i \longrightarrow U\}_{i \in I} \in \mathrm{Cov}(\mathcal{G}) \quad \Longleftrightarrow \quad U = \bigcup_{i \in I} U_i.$$

The verification that this is a Grothendieck topology is simple, perhaps modulo the following hint: If V and W are open subsets of the open set U, then $V \times_U W = V \cap W$ in the category $\mathrm{Top}(X)$. The concepts of presheaf, sheaf etc. are extended from usual topology to Grothendieck topology in rather obvious ways, in particular we have the

Definition 8.1 *Let $\mathcal{G}$ be a Grothendieck topology and $\mathcal{C}$ be a category with infinite products. A presheaf of $\mathcal{G}$ on $\mathcal{C}$ is defined as a contravariant functor*

$$\mathcal{F} : \mathrm{Cat}(\mathcal{G}) \longrightarrow \mathcal{C}.$$

If $\varphi : U \longrightarrow V$ is a morphism in $\mathrm{Cat}(\mathcal{G})$, then we say, as for a topological space, that $\mathcal{F}(\varphi)$ is the restriction morphism from $\mathcal{F}(V)$ to $\mathcal{F}(U)$, and when the objects of $\mathcal{C}$ have an underlying set, then we refer to the image of individual elements $s \in \mathcal{F}(V)$ as the restriction of s from V to U.

$\mathcal{F}$ is said to be a *sheaf* if it satisfies the following condition:

Condition 8.1 (Sheaf Condition) *If $\{\varphi_i : U_i \longrightarrow U\}_{i \in I}$ is a covering, then the diagram below is exact:*

$$\mathcal{F}(U) \xrightarrow{\alpha} \prod_{i \in I} \mathcal{F}(U_i) \overset{\beta}{\underset{\gamma}{\rightrightarrows}} \prod_{i,j \in I} \mathcal{F}(U_i \times_U U_j).$$

Here $\alpha = (\mathcal{F}(\varphi_i) | i \in i)$ is the canonical morphism which is determined by the universal property of the product, and β, γ also come from the universal property of the product by means of the two sets of morphisms:

$$\mathcal{F}(U_i) \xrightarrow{\mathcal{F}(\mathrm{pr}_1) = \beta_{i,j}} \mathcal{F}(U_i \times_U U_j)$$

$$\mathcal{F}(U_j) \xrightarrow{\mathcal{F}(\mathrm{pr}_2) = \gamma_{j,i}} \mathcal{F}(U_i \times_U U_j)$$

so

$$\beta = (\beta_j | j \in I) \quad \text{where } \beta_j = \prod_{i \in I} \beta_{i,j}$$

and

$$\gamma = (\gamma_j | j \in I) \quad \text{where } \gamma_j = \prod_{i \in I} \gamma_{j,i}.$$

We have a simple but clarifying result on sheaves of $\mathfrak{Set}$ and sheaves of $\mathfrak{Mod}_A$:

Proposition 8.2 *Let*

$$\mathcal{F} : \mathfrak{Cat}(\mathcal{G}) \longrightarrow \mathfrak{Mod}_A$$

be a presheaf on the Grothendieck topology $\mathcal{G}$, *and*

$$T : \mathfrak{Mod}_A \longrightarrow \mathfrak{Set}$$

be the forgetful functor. Then $\mathcal{F}$ *is a sheaf if and only if* $T \circ \mathcal{F}$ *is a sheaf.*

Proof Kernels are the same in the two categories.   □

For presheaves of $\mathfrak{Set}$ and $\mathfrak{Mod}_A$ the sheaf-condition takes on a concrete form. We have the

Proposition 8.3 *A presheaf* $\mathcal{F}$ *of* $\mathfrak{Set}$ *or* $\mathfrak{Mod}_A$ *on the Grothendieck topology* $\mathcal{G}$ *is a sheaf if and only if the following two conditions are satisfied:*

Sheaf Condition 1: *If* $\{\varphi_i : U_i \longrightarrow U\}$ *is a covering, and if* s' *and* $s'' \in \mathcal{F}(U)$ *have the same restrictions to* U_i *for all* $i \in I$, *then they are equal.*

Sheaf Condition 2: *If* $\{\varphi_i : U_i \longrightarrow U\}$ *is a covering, and if there is given* $s_i \in \mathcal{F}(U_i)$ *for each* $i \in I$ *such that* s_i *and* s_j *have the same restrictions to* $U_i \cap U_j = U_i \times_U U_j$, *then there exists* $s \in \mathcal{F}(U)$ *such that the restriction of* s *to* U_i *is equal to* s_i.

Proof Immediate from the description of kernels in the categories $\mathfrak{Set}$ and $\mathfrak{Mod}_R$.   □

The following example contains much of the geometric intuition behind the concept of a sheaf:

Example 8.1 (Continuous Mappings) Let X be a topological space, and let $\mathcal{C}^0(U)$ denote the set of all continuous functions from the open subset U to the field of real numbers $\mathbb{R}$, with the usual topology given by the metric $d(r,s) = |s - r|$. Then $\mathcal{C}^0$ is a sheaf of $\mathcal{Ab}$ on $\mathfrak{Top}_X$.

If X is an open subset of $\mathbb{R}^N$ for some N, then we let $\mathcal{C}^m(U)$ denote the set of all functions on U which are m times differentiable. This is also a sheaf of $\mathcal{A}b$. We use this notation for $m = \infty$ as well.

We may replace $\mathbb{R}$ by the field of complex numbers, also with the usual topology.

The category of presheaves of $\mathcal{C}$ on the Grothendieck topology $\mathcal{G}$ is the category of contravariant functors from $\mathcal{C}at(\mathcal{G})$ to $\mathcal{C}$. The *sheaves* form a *subcategory*, where we keep the morphisms but subject the objects to the additional sheaf-condition. We say that the sheaves form a *full subcategory* of the category of presheaves.

8.10 The Sheaf Associated to a Presheaf

We have the following general fact, valid for presheaves on any Grothendieck topology $\mathcal{G}$:

Proposition 8.4 *Let $\mathcal{F}$ be a presheaf of Set or Mod_R on a Grothendieck topology $\mathcal{G}$. Letting $\mathrm{Sheaves}_{\mathcal{G}}$ denote the category of sheaves of Set or Mod_R, as the case may be, the following functor is representable:*

$$\mathrm{Sheaves}_{\mathcal{G}} \longrightarrow \mathrm{Set}$$

$$\mathcal{H} \mapsto \mathrm{Hom}(\mathcal{F}, \mathcal{H}).$$

In other words, the presheaf $\mathcal{F}$ determines uniquely a sheaf $[\mathcal{F}]$ and a morphism $\tau_{\mathcal{F}}$

$$\mathcal{F} \xrightarrow{\tau_{\mathcal{F}}} [\mathcal{F}]$$

such that whenever $\mathcal{F} \longrightarrow \mathcal{H}$ is a morphism from the presheaf $\mathcal{F}$ to the sheaf $\mathcal{H}$, then there exists a unique morphism

$$[\mathcal{F}] \longrightarrow \mathcal{H},$$

such that the appropriate diagram commutes. The sheaf $[\mathcal{F}]$ is referred to as the sheaf associated to the presheaf $\mathcal{F}$.

It also follows from the proof that we may start from a base $\mathcal{B}$ for the topology and a presheaf $\mathcal{F}$ which is defined just on $U \in \mathcal{B}$, and construct the sheaf $[\mathcal{F}]$ on all open subsets.

Proof For simplicity we consider the case when $\mathcal{G}$ is the usual topology on a topological space. We only treat the case of a presheaf of Mod_R, as the case of Set is an obvious modification. We make the following definition:

$$[\mathcal{F}](U) = \left\{ (\xi_x)_{x \in U} \in \prod_{x \in U} \mathcal{F}_x \,\middle|\, \begin{array}{l} \forall x \in U \exists V \subset U \text{ such that } x \in V \\ \text{and such that } \exists \eta_V \in \mathcal{F}(V) \text{ with} \\ \iota_{V,y}(\eta_V) = (\eta_V)_y = \xi_y \text{ for all } y \in V \end{array} \right\}.$$

The definition of the restriction map from U to some $W \subset U$ is obvious, it is denoted by $\rho_{U,W}^{[\mathcal{F}]}$. Likewise, it is an immediate exercise to check that this is a sheaf of Mod_A on the topological space X. Also, the definition of $\tau_{\mathcal{F}}$ is obvious:

$$\tau_{\mathcal{F}} : \mathcal{F}(U) \longrightarrow [\mathcal{F}](U)$$

$$f \mapsto (f_x | x \in U).$$

Clearly $\tau_{\mathcal{F},x}$ is an isomorphism for all $x \in X$.

To verify the universal property, let

$$\varphi : \mathcal{F} \longrightarrow \mathcal{H}$$

be a morphism from $\mathcal{F}$ to a sheaf $\mathcal{H}$. We have to define a morphism

$$[\varphi] : [\mathcal{F}] \longrightarrow \mathcal{H}$$

which makes the appropriate diagram commutative. Now φ yields, for all $x \in X$,

$$\varphi_x : \mathcal{F}_x \longrightarrow \mathcal{H}_x.$$

Thus we also have

$$\psi(x) : [\mathcal{F}]_x \longrightarrow \mathcal{H}_x,$$

and to show is that there is a morphism of sheaves

$$\psi : [\mathcal{F}] \longrightarrow \mathcal{H}$$

such that $\psi(x) = \psi_x$. Let U be an open subset of X, and let $\xi_U = (\xi_x)_{x \in U} \in [\mathcal{F}](U)$, where of course the ξ_x's satisfy the condition in the definition of $[\mathcal{F}](U)$. In particular there is an open covering of U by open subsets V where there are $\eta_V \in \mathcal{F}(V)$ such that for all $y \in V$ we have $\xi_y = (\eta_V)_y$. We put $\varphi_V(\eta_V) = \zeta_V \in \mathcal{H}(V)$. Then it is easy to see that by the sheaf condition of $\mathcal{H}$ these elements ζ_V may be glued to an element $\zeta_U \in \mathcal{H}(U)$. As is easily verified, putting $\psi_U(\xi_U) = \zeta_U$ gives a morphism of sheaves $\psi : [\mathcal{F}] \longrightarrow \mathcal{H}$, and we put $\psi = [\varphi]$.

Uniqueness is a consequence of the following lemma, the proof of which is left to the reader:

Lemma 8.5 *Given two morphisms of sheaves on the topological space* X,

$$\psi, \phi : \mathcal{A} \longrightarrow \mathcal{B}$$

such that for all $x \in X$

$$\psi_x = \phi_x : \mathcal{A}_x \longrightarrow \mathcal{B}_x.$$

Then $\psi = \phi$.

The verification that the appropriate diagram commutes is also straight-forward.   □

We finally note that the assignment

$$\mathcal{F} \mapsto [\mathcal{F}]$$

defines a covariant functor

$$\text{Presheaves}_{\mathcal{G}} \longrightarrow \text{Seaves}_{\mathcal{G}}.$$

8.11 The Category of Abelian Sheaves

We conclude this introductory section by summarizing the basic properties of the category of sheaves of Abelian groups on a topological space X. This category is denoted by $\mathcal{A}b_X$. It is commonly refereed to as the *category of Abelian sheaves on X*. All of this is valid in more general settings, say for modules over commutative rings, etc.

The direct sum of two Abelian sheaves $\mathcal{A}$ and $\mathcal{B}$ is defined by

$$(\mathcal{A} \oplus \mathcal{B})(U) = \mathcal{A}(U) \oplus \mathcal{B}(U),$$

which does indeed define an Abelian sheaf on X. For a morphism of Abelian sheaves,

$$\varphi : \mathcal{A} \longrightarrow \mathcal{B},$$

we define the Abelian sheaf $\ker(\varphi)$ by

$$\ker(\varphi)(U) = \ker(\varphi_U),$$

and let the restriction homomorphisms be the restrictions of the corresponding ones for the sheaf $\mathcal{A}$. It is a simple exercise to verify that $\ker(\varphi)$ so defined is an Abelian sheaf. For the definition of $\text{coker}(\varphi)$, however, the situation is different: In this case we only get a *presheaf* by

$$U \mapsto \text{coker}(\varphi_U).$$

It is important to reflect on the significance of this difference. We define $\text{coker}(\varphi)$ by taking the associated sheaf to the above presheaf. Similarly we have to define the Abelian sheaf $\text{im}(\varphi)$, by first defining the obvious presheaf, then taking the associated sheaf.

Proposition 8.6 φ *is a monomorphism if and only if* $\ker(\varphi)$ *is the zero sheaf,* **0**. *Moreover,* $\text{coker}(\varphi) = \mathbf{0}$ *if and only if f is an epimorphism.*

Remark The terms injective, respectively surjective, are also used.

The *proof* of the proposition a is simple routine exercise, and is left to the reader.

We also have the following simple result, the proof of which is likewise left to the reader as an exercise:

Proposition 8.7 *Let*

$$\varphi : \mathcal{A} \longrightarrow \mathcal{B}$$

be a morphism of Abelian sheaves. The following are equivalent:

1. φ *is an isomorphism.*
2. *All* φ_U *are bijective.*
3. *All* φ_x *are bijective.*
4. φ *is a monomorphism and an epimorphism.*

Remark Thus we have another example where "isomorphism" and "bijection" is the same thing. As we know, this is not always the case in general categories.

If $\iota : \mathcal{S} \hookrightarrow \mathcal{F}$ is the inclusion of a *subsheaf* into the Abelian sheaf $\mathcal{F}$, then $\mathrm{coker}(\iota)$ is denoted by $\mathcal{F}/\mathcal{S}$.

A sequence of Abelian sheaves

$$\cdots \longrightarrow \mathcal{A}_{i-1} \xrightarrow{\varphi_{i-1}} \mathcal{A}_i \xrightarrow{\varphi_i} \mathcal{A}_{i+1} \longrightarrow \cdots$$

is said to be *exact* at $\mathcal{A}_i$ if $\mathrm{im}(\varphi_{i-1}) = \ker(\varphi_i)$.

In the category of Abelian sheaves $\mathrm{Hom}(\mathcal{A}, \mathcal{B})$ is always an Abelian group with the obvious definition of addition. The category is, in fact, an Abelian category. Functors compatible with the additive structure on the Hom-sets are called additive functors. Here are two examples:

Let $f : X \longrightarrow Y$ be a continuous mapping of topological spaces. We define a functor referred to as *the direct image under f*,

$$f_* : \mathcal{A}b_X \longrightarrow \mathcal{A}b_Y$$

by putting

$$f_*(\mathcal{F})(U) = \mathcal{F}(f^{-1}(U)).$$

As is easily seen, this defines an Abelian sheaf on Y, and moreover, $f_*(\ )$ is a covariant additive functor from $\mathcal{A}b_X$ to $\mathcal{A}b_Y$.

The fiber $f_*(\mathcal{F})_{f(x)}$ is related to $\mathcal{F}_x$ in the following manner: By definition

$$f_*(\mathcal{F})_{f(x)} = \varinjlim_{\{V \subset Y | f(x) \in V\}} \mathcal{F}(f^{-1}(V)) \xrightarrow{\text{canonical}} \mathcal{F}_x$$

where the homomorphism labelled "canonical" is the one coming from forming $\varinjlim$ over an inductive system and over a subsystem.

We also define an *"inverse image functor"* for a continuous mapping $f :$ $X \longrightarrow Y$

$$f^* : \mathcal{A}b_Y \longrightarrow \mathcal{A}b_X$$

by first defining a presheaf

$$f^{-1}(\mathcal{G})(U) = \lim_{\longrightarrow \{V | f(U) \subseteq V\}} \mathcal{G}(V),$$

and then taking the associated sheaf. This is also a covariant, additive functor.

Remark The notation f^* is used in a variety of different situations. Here f is a continuous mapping and the categories are categories of sheaves of $\mathcal{A}b$ on topological spaces. When f is a morphism of schemes, as encountered in algebraic geometry later in this book, and the categories are categories of modules on these schemes, then f^* will have a different meaning.

We have

$$f^*(\mathcal{G})_x = \mathcal{G}_{f(x)}.$$

Indeed, we show that $f^{-1}(\mathcal{G})_x = \mathcal{G}_{f(x)}$. We have by the definition

$$f^*(\mathcal{G})_x = f^{-1}(\mathcal{G})_x = \lim_{\longrightarrow \{U | x \in U\}} f^{-1}(\mathcal{G})(U)$$

$$= \lim_{\longrightarrow \{U | x \in U\}} \left(\lim_{\longrightarrow \{V | f(U) \subseteq V\}} \mathcal{G}(V) \right) \overset{\text{canonical}}{\longrightarrow} \lim_{\longrightarrow \{V | f(x) \in V\}} \mathcal{G}(V) = \mathcal{G}_{f(x)}$$

where again the homomorphism labelled "canonical" is the one coming from forming $\lim_{\longrightarrow}$ over an inductive system and over a subsystem. In this case this homomorphism is an isomorphism, however, since the "subsystem" in question is actually the whole system: In fact, take $f(x) \in V \subset Y$, and put $U = f^{-1}(V)$. Then $x \in U$ and $f(U) \subset V$.

Whenever X is a subspace of Y and f is the natural injection, we write $\mathcal{F}|X$ instead of $f^*(\mathcal{F})$. If f is an open embedding, this is nothing but the obvious restriction to open subsets contained in U.

For two Abelian sheaves $\mathcal{A}$ and $\mathcal{B}$ we define the sheaf $\mathcal{H}om(\mathcal{A}, \mathcal{B})$, referred to as *the Sheaf Hom* of $\mathcal{A}$ and $\mathcal{B}$, as the associated sheaf of the presheaf

$$U \mapsto \text{Hom}(\mathcal{A}|U, \mathcal{B}|U).$$

The category $\mathcal{A}b_X$ plays an important role in algebraic geometry, and we will return to it as we need more specialized or advanced features.

We use the material from Sect. 6.13 to study the pair of functors

$$\mathcal{A}b_X \overset{f_*}{\underset{f^*}{\rightleftarrows}} \mathcal{A}b_Y.$$

They are adjoint functors, as we shall now explain.

For all Abelian sheaves $\mathcal{G}$ on Y we define the functorial morphism of Abelian sheaves

$$\rho_{\mathcal{G}} : \mathcal{G} \longrightarrow f_*(f^*(\mathcal{G}))$$

by letting

$$\rho_{\mathcal{G},V} : \mathcal{G}(V) \overset{\text{canonical}}{\longrightarrow} f^{-1}(\mathcal{G})(f^{-1}(V)) = f_*(f^{-1}(\mathcal{G}))(V) \overset{\tau_{f^{-1}(V)}}{\longrightarrow} f_*(f^*(\mathcal{G}))(V)$$

where the last homomorphism is the one coming from the morphism of a presheaf to its associated sheaf.[5]

We next define functorial homomorphisms

$$\sigma_{\mathcal{F},U} : f^*(f_*(\mathcal{F}))(U) \longrightarrow \mathcal{F}(U)$$

as follows:

$$f^{-1}(f_*(\mathcal{F}))(U) = \varinjlim_{\{V \subset Y | f(U) \subset V\}} f_*(\mathcal{F})(V)$$

$$= \varinjlim_{\{V \subset Y | f(U) \subset V\}} \mathcal{F}(f^{-1}(V)) \longrightarrow \mathcal{F}(U),$$

where the last homomorphism comes from the restrictions from $f^{-1}(V)$ to U. We thus obtain a morphism of presheaves $f^{-1}(f_*(\mathcal{F})) \longrightarrow \mathcal{F}$ and hence by the universal property of the associated sheaf, a morphism of sheaves $f^*(f_*(\mathcal{F})) \longrightarrow \mathcal{F}$ as claimed.

We now have the following result:

Proposition 8.8 *The morphism of functors defined above* $\rho : \mathrm{id}_{\mathcal{A}b_Y} \longrightarrow f_* \circ f^*$ *defines an isomorphism of bifunctors*

$$\Phi : \mathrm{Hom}_X(f^*(\ ),\ ) \longrightarrow \mathrm{Hom}_Y(\ , f_*(\ )),$$

thus f_* *is right adjoint to* f^*. *The inverse functor* Ψ *of* Φ *is given by the morphism* $\sigma : f^* \circ f_* \longrightarrow \mathrm{id}_{\mathcal{A}b_X}$ *defined above.*

The remaining details of the proof are left to the reader. Hint: See [15], vol. I, pp. 30–33.

We introduce the following notation: The image of $\mu : f^*(\mathcal{G}) \longrightarrow \mathcal{F}$ under $\Phi_{\mathcal{G},\mathcal{F}}$ is denoted by $\mu^\flat : \mathcal{G} \longrightarrow f_*(\mathcal{F})$, whereas the preimage of $\nu : \mathcal{G} \longrightarrow f_*(\mathcal{F})$ is denoted by $\nu^\sharp : f^*(\mathcal{G}) \longrightarrow \mathcal{F}$.

---

[5]One readily verifies that direct image f_* of a presheaf commutes with forming the associated sheaf.

Chapter 9
The Concept of Spec(A)

The Spec(A) for a commutative ring A is constructed in this chapter. Ample algebraic details are given, and some of the standard examples are discussed.

9.1 The Affine Spectrum of a Commutative Ring

We start out by reminding the reader of a simple but fundamental concept from general topology:

Given a topological space X, recall that a *basis for the topology* on X is a collection of open subsets $\mathcal{B}$ of X such that

1. For all $x \in X$ there exists $B \in \mathcal{B}$ such that $x \in B$.
2. For all B_1 and B_2 in $\mathcal{B}$ and all $x \in B_1 \cap B_2$ there exists $B_3 \in \mathcal{B}$ contained in $B_1 \cap B_2$ such that $x \in B_3$.

Conversely, a collection $\mathcal{B}$ of subsets of a set X defines a topology by letting the open subsets be all possible unions of sets from $\mathcal{B}$ *if and only if* the two conditions above are satisfied. This is easily verified.

We define a topology, referred to as *the Zariski topology*, on the set of prime ideals of a commutative ring with 1 as follows: Let A be such a ring. We consider the set of all prime ideals in A, that is to say all ideals $\mathfrak{p} \neq A$ such that

$$ab \in \mathfrak{p} \quad \text{and} \quad a \notin \mathfrak{p} \implies b \in \mathfrak{p}$$

and denote the set of all such prime ideals by $\mathrm{Spec}(A)$. For $a \in A$ we define the subset $D(a) \subseteq \mathrm{Spec}(A)$ by

$$D(a) = \{\mathfrak{p} \in \mathrm{Spec}(A) \,|\, a \notin \mathfrak{p}\}$$

and we put

$$V(a) = \{\mathfrak{p} \in \mathrm{Spec}(A) \,|\, a \in \mathfrak{p}\}.$$

A. Holme, *A Royal Road to Algebraic Geometry*,
DOI 10.1007/978-3-642-19225-8_9, © Springer-Verlag Berlin Heidelberg 2012

Clearly $D(1) = \mathrm{Spec}(A)$, and as is easily seen,

$$D(a) \cap D(b) = D(ab),$$

hence all the subsets $D(a)$ as $a \in A$ constitute a *basis for a topology* on $\mathrm{Spec}(A)$.

Definition 9.1 (The Zariski Topology) The topology referred to above is called the Zariski topology on $\mathrm{Spec}(A)$.

It is easily seen that the closed subsets in this topology are given as

$$\mathcal{F} = \{F \mid F = V(S)\}$$

where $S \subset A$ is a subset, and

$$V(S) = \{\mathfrak{p} \mid \mathfrak{p} \supseteq S\}.$$

Evidently $V(S) = V((S)A)$, thus the closed subsets of $\mathrm{Spec}(A)$ are described by the ideals in A in this manner. Note that $V(A) = \emptyset$.

We similarly have that all the open subsets of $\mathrm{Spec}(A)$ are described as

$$U = D(S)$$

where

$$D(S) = \{\mathfrak{p} \mid \mathfrak{p} \not\supseteq S\}.$$

We note that $D(S) = D((S)A)$.

This establishes an important relation between the closed subsets of the topological space $\mathrm{Spec}(A)$ and the ideals in the ring A. We summarize this as follows:

Proposition 9.1 1. *Let $\mathfrak{a}$ and $\mathfrak{b}$ be two ideals in A. Then*

$$V(\mathfrak{a} \cap \mathfrak{b}) = V(\mathfrak{a}\mathfrak{b}) = V(\mathfrak{a}) \cup V(\mathfrak{b}).$$

2. *Let $\{\mathfrak{a}_i\}_{i \in I}$ be a family if ideals in A. Then*

$$V\left(\sum_{i \in I} \mathfrak{a}_i\right) = \bigcap_{i \in I} V(\mathfrak{a}_i).$$

3. *We have for all ideals $\mathfrak{a}$ that $V(\mathfrak{a}) = V(\sqrt{\mathfrak{a}})$.*

4. *V establishes a bijective correspondence between the radical ideals in A and the closed subsets of $\mathrm{Spec}(A)$.*

Proof 1. Is a direct consequence of the well known fact from commutative algebra, that if $\mathfrak{p}$ is a prime ideal then for any two ideals $\mathfrak{a}$ and $\mathfrak{b}$

$$\mathfrak{ab} \subseteq \mathfrak{p} \quad \text{and} \quad \mathfrak{b} \not\subseteq \mathfrak{p} \implies \mathfrak{a} \subseteq \mathfrak{p}.$$

2. For an deal $\mathfrak{I}$, in particular for a prime ideal, it is true that it contains all the ideals $\mathfrak{a}_i$ if and only if it contains their sum.

3. If $\mathfrak{p} \supseteq \sqrt{\mathfrak{a}}$, then in particular $\mathfrak{p} \supseteq \mathfrak{a}$. On the other hand if $\mathfrak{p} \supseteq \mathfrak{a}$, and if $a \in \sqrt{\mathfrak{a}}$, then for some integer N we have $a^N \in \mathfrak{a}$, thus $a^N \in \mathfrak{p}$, thus $a \in \mathfrak{p}$. Hence $\mathfrak{p} \supseteq \sqrt{\mathfrak{a}}$.

4. This assertion follows already from the previous ones, but we note the inverse mapping to V: Namely, letting

$$I(F) = \bigcap_{\mathfrak{p} \in F} \mathfrak{p},$$

we get a radical ideal such that $V(I(F)) = F$. The details of this simple verification is left to the reader.   □

Example 9.1 If k is a field, then $\mathrm{Spec}(k)$ consists of a single point.

Example 9.2 Let $\mathbb{Z}$ be the ring if integers. Then $\mathrm{Spec}(\mathbb{Z})$ is the set

$$\{0, 2, 3, 5, 7, \dots\}$$

consisting of the set of all prime numbers and the number 0. The closure of the set consisting of 0 alone is all of $\mathrm{Spec}(\mathbb{Z})$, while the closure of any other point is the point itself. The point 0 is referred to as *the generic point* of $\mathrm{Spec}(\mathbb{Z})$, while the others are closed points.

Recall that if Δ is a multiplicatively closed subset of A, then there is a bijective correspondence between the prime ideals in A which do not intersect Δ, and the prime ideals in the ring $\Delta^{-1}A$, A localized by Δ, given by

$$\mathfrak{p} \mapsto \mathfrak{P} = (\mathfrak{p})\Delta^{-1}A.$$

As usual we let A_a denote the localization of A in the multiplicatively closed set $S = \{1, a, a^2, a^3, \dots\}$ of all powers of a. In particular, if $a \notin \mathfrak{p}$ and $\mathfrak{P} = (\mathfrak{p})A_a$, then $\mathfrak{P}$ is a prime in A_a, and all primes of A_a are obtained in this manner. We then get that $\mathrm{Spec}(A_a)$ homeomorphic to the open subset $D(a)$ with the topology induced from $\mathrm{Spec}(A)$.

Example 9.3 Let A be a commutative ring, and let $a \in A$. Then $\mathrm{Spec}(A/(a)A)$ is homeomorphic as a topological space with the subspace $V(a)$ of $\mathrm{Spec}(A)$.

We now come to the structure sheaf on $\mathrm{Spec}(A)$. The complement of a set-theoretic union of prime ideals in a commutative ring with 1 is a multiplicatively closed subset. Indeed, let $\{\mathfrak{p}_i\}_{i \in I}$ be a set of prime ideals, and let Δ be the complement in A of the set $\bigcup_{i \in I} \mathfrak{p}_i$. Then if $a, b \in \Delta$ we have $a, b \notin \mathfrak{p}_i \ \forall i \in I$, thus $ab \notin \mathfrak{p}_i \ \forall i \in I$, thus $ab \in \Delta$.

Now for all open $U \subseteq \mathrm{Spec}(A)$ let $\Delta(U)$ denote the multiplicatively closed subset of A given by the complement of the union of all primes $\mathfrak{p} \in U$. Note that for two open subsets U and V of $\mathrm{Spec}(A)$ we have

$$U \subseteq V \implies \Delta(U) \supseteq \Delta(V).$$

Also note that $s \in \Delta(D(s))$. We define a presheaf of Comm $\mathcal{O}'$ on the topological space $\mathrm{Spec}(A)$ by

$$\mathcal{O}'(U) = \Delta(U)^{-1}A$$

and for two open subsets $U \subset V$ we define the restriction map by

$$\rho_{V,U}^{\mathcal{O}'} : \Delta(V)^{-1}A \longrightarrow \Delta(U)^{-1}A$$
$$\frac{a}{s} \mapsto \frac{a}{s},$$

which makes sense as $\Delta(U) \supseteq \Delta(V)$.[1]

Definition 9.2 We denote the associated sheaf of the presheaf $\mathcal{O}'$ by $\mathcal{O}_{\mathrm{Spec}(A)}$, or just $\mathcal{O}$ when no ambiguity is possible. We refer to it as the structure sheaf of the pair $(\mathrm{Spec}(A), \mathcal{O})$. The pair itself is called the affine spectrum associated to the commutative ring A, or also the spectrum of the ring A. From now on $\mathrm{Spec}(A)$ will denote this pair, rather than just the underlying topological space. The commutative ring $\mathcal{O}(U)$ is also denoted by $\Gamma(U, \mathcal{O})$.

Let $\mathfrak{U}(x)$ be the set of all open subsets in $\mathrm{Spec}(A)$ containing the point $x \in \mathrm{Spec}(A)$, corresponding to the prime ideal $\mathfrak{p}_x \subset A$. Then for all $U \in \mathfrak{U}(x)$,

$$\Delta(U) \subset \Delta(x) = \{s \in A \mid s \notin \mathfrak{p}_x\}.$$

This inclusion induces a homomorphism in Comm,

$$\varphi_{U,x} : \mathcal{O}'(U) \longrightarrow A_{\mathfrak{p}_x},$$

and as these homomorphisms are compatible with the restriction homomorphisms of $\mathcal{O}'$, we obtain a homomorphism of commutative rings with 1,

$$\varphi_x : \mathcal{O}'_x \longrightarrow A_{\mathfrak{p}_x}.$$

We have the following

---

[1] Since we adhere to the requirement that for an object A of Comm we always have $1_A \neq 0_A$, the category on which this presheaf is defined is strictly speaking not $\mathcal{T}\mathrm{op}_{\mathrm{Spec}(A)}$, but rather the category obtained from it by deleting the empty set from the objects.

Lemma 9.2 φ_x *is an isomorphism.*

Proof To show is that φ_x is bijective.

1. φ_x is surjective: Let $\alpha \in A_{\mathfrak{p}_x}$. Then $\alpha = \frac{a}{s}$, where $s \notin \mathfrak{p}_x$. Thus $\alpha = \varphi_{D(s),x}(\frac{a}{s})$, the latter fraction now to be understood as an element in the ring $\Delta(D(s))^{-1}A$. Then the image of this element in the inductive limit $\mathcal{O}'_x$ is mapped to α by φ_x. Thus φ_x is onto.

2. φ_x is injective: We show that $\ker(\varphi_x) = 0$. Suppose that $\varphi_x(\beta) = 0$. We wish to show that $\beta = 0$. There is an open subset $U \ni x$ and $s \in \Delta(U)$ and an element $b \in A$ such that $\beta = [\frac{b}{s}]$, in the notation we used describing the stalks. We may assume that $U = D(s)$.

It suffices to show that the restriction of $\frac{b}{s}$ to some smaller open neighborhood containing x is zero. Now $\varphi_{D(s),x}(\frac{b}{s}) = \varphi_x(\beta) = 0$. Hence there exists $t \in \Delta(x)$ such that $tb = 0$. But then the restriction of $\frac{b}{s}$ to $U \cap D(t) = D(s) \cap D(t) = D(st)$ is zero. $\square$

For all non empty open subsets $U \subset \mathrm{Spec}(A)$ we have the homomorphism of $\mathcal{C}$omm coming from the canonical morphism from a presheaf to its associated sheaf, which of course is compatible with restriction to a smaller open subset,

$$\tau_U : \Delta(U)^{-1}A \longrightarrow \mathcal{O}_{\mathrm{Spec}(A)}(U).$$

Moreover, for a not nilpotent and $U = D(a)$ we have

$$\{1, a, a^2, a^3, \dots\} \subset \Delta(D(a)),$$

which defines a homomorphism

$$\varsigma_a : A_a \longrightarrow \Delta(D(a))^{-1}A,$$

by

$$\frac{b}{a^n} \mapsto \frac{b}{a^n}.$$

Now we have the following:

Proposition 9.3 1. $c \in \Delta(D(a)) \Leftrightarrow$ *For some m there exists $r \in A$ such that* $rc = a^m$.

2. *For all a not nilpotent ς_a is an isomorphism.*

3. *For all a not nilpotent $\tau_{D(a)}$ is an isomorphism.*

Remark In particular there is a canonical isomorphism

$$\Gamma(\mathrm{Spec}(A), \mathcal{O}_{\mathrm{Spec}(A)}) \cong A.$$

Proof We first show 1. We have that $c \in \Delta(D(a)) \Leftrightarrow c \notin \cup\{\mathfrak{p}|a \notin \mathfrak{p}\}$ This condition on c amounts to

$$a \notin \mathfrak{p} \implies c \notin \mathfrak{p} \quad \text{or equivalently} \quad c \in \mathfrak{p} \implies a \in \mathfrak{p}.$$

As the radical of an ideal is the intersection of all prime ideals containing it, the latter condition is again equivalent to

$$\sqrt{(a)A} \subseteq \sqrt{(c)A}$$

which again is equivalent to

$$a \in \sqrt{(c)A}.$$

That is to say, $a^m \in (c)A$ for some m. Thus 1. follows.

To show 2., we prove that ς_a is injective and surjective. So suppose that $\varsigma_a(\frac{b}{a^n}) = 0$. Then there is an element $c \in \Delta(D(a))$ such that $cb = 0$. But by 1. there exists $r \in A$ such that $rc = a^m$ for some m. Thus we also have $a^m b = 0$, hence $\frac{b}{a^n} = 0$, and η_a is injective. Next, let $\frac{b}{c} \in \Delta(D(a))^{-1}A$. As above we find $m \in \mathbb{N}$ and $r \in A$ such that $a^m = rc$. As clearly $r \in \Delta(D(a))$, we thus find $\frac{b}{c} = \frac{rb}{a^m}$, which is in the image of ς_a. Thus 2. is proven.

Proof of 3.: By 2. it suffices to show that the composition

$$\eta_a : A_a \xrightarrow{\varsigma_a} \Delta(D(a))^{-1}A \longrightarrow \mathcal{O}_{\mathrm{Spec}(A)}(D(a))$$

is an isomorphism. We write, for the canonical homomorphism from A_a to $A_{\mathfrak{p}}$ where $a \notin \mathfrak{p}$,

$$A_a \longrightarrow A_{\mathfrak{p}}$$

$$\frac{b}{a^n} \mapsto \left(\frac{b}{a^n}\right)_{\mathfrak{p}}.$$

Then

$$\eta_a\left(\frac{b}{a^n}\right) = \left(\left.\left(\frac{b}{a^n}\right)_{\mathfrak{p}}\right| \mathfrak{p} \in D(a)\right).$$

The homomorphism η_a for $a \in A$ is the same as the homomorphism η_1 for $1 \in A_a$. Thus it suffices to show that in general $\eta_1 = \eta$ is bijective:

$$\eta : A \longrightarrow \mathcal{O}(\mathrm{Spec}(A))$$

$$b \mapsto (b_{\mathfrak{p}}|\mathfrak{p} \in \mathrm{Spec}(A)).$$

We show that η is injective: Suppose that $\eta(b) = 0$. Then for all prime ideals $\mathfrak{p}$ of A there is $p \notin \mathfrak{p}$ such that $sb = 0$. Hence $\mathfrak{a} = \mathrm{Ann}(b)$ is contained in no prime ideal, thus $1 \in \mathrm{Ann}(b)$, so $b = 0$.

We finally show that η is surjective: Recall that

$$\mathcal{O}(\mathrm{Spec}(A)) = \left\{ (s_{\mathfrak{p}} | \mathfrak{p} \in \mathrm{Spec}(A)) \left| \begin{array}{l} \text{For all } \mathfrak{p} \in \mathrm{Spec}(A) \text{ there exists} \\ \text{an open } V \subset \mathrm{Spec}(A) \text{ containing } \mathfrak{p} \text{ such} \\ \text{that there exists } s_V \in \Delta(V)^{-1}A \text{ with} \\ (s_V)_{\mathfrak{q}} = s_{\mathfrak{q}} \text{ for all } \mathfrak{q} \in V \end{array} \right. \right\}.$$

Clearly we may assume that all the open subsets V are of the form $D(a_i)$ as i runs through some indexing set I. We then have

$$\mathrm{Spec}(A) = \bigcup_{i \in I} D(a_i),$$

thus

$$\bigcap_{i \in I} V(a_i) = V((a_i | i \in I)A) = \emptyset.$$

Hence

$$(\{a_i | i \in I\})A = A,$$

in particular we have for some indices $i_1, i_2, \ldots, i_r$

$$c_{i_1} a_{i_1} + \cdots + c_{i_r} a_{i_r} = 1,$$

and we may assume that $I = \{1, 2, \ldots, r\}$.

Now $s_{D(a_i)} \in \Delta(D(a_i))^{-1}A$, thus by the lemma

$$s_{D(a_i)} = \frac{b_i}{a_i^{n_i}}.$$

However, since $D(a_i) = D(a_i^{n_i})$ for all i, and the localizations are the same as well, we may assume that all $n_i = 1$. Thus

$$s_{D(a_i)} = \frac{b_i}{a_i}.$$

To compare this for different values of i, consider the canonical homomorphisms

$$A_{a_i} \xrightarrow{\varphi_i} A_{a_i a_j} \xleftarrow{\varphi_j} A_{a_j}.$$

We show that

$$\varphi_i\left(\frac{b_i}{a_i}\right) = \varphi_j\left(\frac{b_j}{a_j}\right):$$

Indeed, letting $\varphi_i(\frac{b_i}{a_i}) - \varphi_j(\frac{b_j}{a_j}) = b_{i,j}$, the image of $\beta_{i,j}$ in $(A_{a_i a_j})_{\mathfrak{P}}$ is zero for all prime ideals $\mathfrak{P}$ of $(A_{a_i a_j})$. Thus as above, $\beta_{i,j} = 0$.

Hence we have the identity

$$\frac{b_i a_j}{a_i a_j} = \frac{b_j a_i}{a_j a_i}$$

in $A_{a_i a_j}$, and thus there are non negative integers $m_{i,j}$ such that

$$(a_i a_j)^{m_{i,j}} (b_i a_j - b_j a_i) = 0.$$

Being finite in number, we may assume that these integers are equal, say to M, and get the relation

$$a_i^M a_j^{M+1} b_i = a_j^M a_i^{M+1} b_j.$$

As

$$\frac{b_i}{a_i} = \frac{a_i^M b_i}{a_i^{M+1}},$$

we may replace b_i by $a_i^M b_i$ and a_i by a_i^{M+1}, and finally obtain the simple relation [2]

$$a_i b_j = a_j b_i.$$

Using the $c_1, \ldots, c_r$ which we found above with the property that

$$1 = c_1 a_1 + \cdots + c_r a_r,$$

we let

$$b = b_1 c_1 + \cdots + b_r c_r.$$

We claim that in A_{a_i},

$$\frac{b}{1} = \frac{b_i}{a_i}.$$

Indeed,

$$b a_i = \sum_{j=1}^{r} c_j b_j a_i = \sum_{j=1}^{r} c_j b_i a_j = b_i.$$

Thus η is surjective and the proof is complete. $\square$

---

[2]The argument would be much simpler if A were an integral domain. However, an important aspect of scheme-theory is to have a theory which is valid in the presence of zero-divisors and even nilpotent elements.

9.2 Very first Examples of Affine Spectra

The simplest possible cases are the affine spectra of fields: if k is a field, then $\mathrm{Spec}(k)$ has an underlying topological space consisting of one point, $X = \{s\}$ where s corresponds to the zero ideal of k. The structure sheaf is simply given by $\mathcal{O}(s) = k$.

$\mathrm{Spec}(\mathbb{Z})$ has as underlying topological space the set

$$\{0, 2, 3, 5, \ldots, p, \ldots\},$$

the set of 0 and all prime numbers. The topology is given by the open sets being the whole space as well as the empty set and the complements of all finite sets of prime numbers. The structure sheaf has $\mathbb{Q}$ as stalk in the point 0, called the generic point, and at a prime number the stalk is $\mathbb{Z}$ localized at that prime.

We consider $\mathrm{Spec}(k[X_1, X_2, \ldots, X_n])$, the affine spectrum of the polynomial ring in n variables over the field k. It is referred to as the *scheme theoretic* affine n-space over the field k. It is denoted by $\mathbb{A}_k^n$. Note that we distinguish between this and k^n, which is identified with a special set of closed points in $\mathbb{A}_k^n$, namely those corresponding to maximal ideals of the type

$$\mathfrak{m} = (X_1 - a_1, X_2 - a_2, \ldots, X_n - a_n).$$

If k is not algebraically closed, there are of course other closed points than these: Namely, all maximal ideals are closed points, and to capture these as points of the above type we have to *extend the base* to the algebraic closure K of k. Note that it is definitely *not true* that $\mathbb{A}_k^n \subset \mathbb{A}_K^n$. The reader should take a few moments to contemplate this phenomenon.

Let $\mathfrak{a}$ be an ideal in $\mathrm{Spec}(k[X_1, X_2, \ldots, X_n])$, the polynomial ring in n variables. Let $B = k[X_1, X_2, \ldots, X_n]/\mathfrak{a}$. Then $\mathrm{Spec}(B)$ has as underlying topological space a closed subset of the affine n-space over k. An affine spectrum of this kind is called *an affine spectrum of finite type over k*. They constitute the class of closed subschemes of $\mathbb{A}_k^n$. We return to this later.

The construction of $\mathcal{O}_{\mathrm{Spec}(A)}$ has an important generalization:

Definition 9.3 Let M be an A-module. Then the sheaf $\widetilde{M}$ on $\mathrm{Spec}(A)$ is the sheaf associated to the presheaf $\mathcal{M}$ defined by $\mathcal{M}(U) = \Delta(U)^{-1}M$, with the restriction maps being the canonical ones induced from localization:

$$U \subset V \quad \Longrightarrow \quad \Delta(V)^{-1}M \longrightarrow \Delta(U)^{-1}M, \quad \frac{m}{s} \mapsto \frac{m}{s}.$$

We immediately observe that for all open subsets $U \subset \mathrm{Spec}(A)$, $\widetilde{M}(U)$ is a module over the ring $\mathcal{O}_{\mathrm{Spec}(A)}(U)$. Moreover, if $V \supset U$ then the restriction map

$$\rho_{V,U}^{\widetilde{M}} : \widetilde{M}(V) \longrightarrow \widetilde{M}(U)$$

is an $\mathcal{O}_{\mathrm{Spec}(A)}(V), \mathcal{O}_{\mathrm{Spec}(A)}(U)$ homomorphism.

Definition 9.4 If M and N are modules over A and B, respectively, and if $\varphi : A \longrightarrow B$ is a homomorphism of rings, then a mapping $f : M \longrightarrow N$ is called an A, B-homomorphism if it is additive and $f(am) = \varphi(a)f(m)$.

The following observations are proved in exactly the same fashion as the corresponding ones for the sheaf $\mathcal{O}_{\mathrm{Spec}(A)}$:

Proposition 9.4 1. *The canonical homomorphism for* $\mathfrak{p} \in U$

$$\Delta(U)^{-1}M \longrightarrow M_{\mathfrak{p}}$$
$$\frac{m}{s} \mapsto \frac{m}{s}$$

induces an isomorphism

$$\varphi_x : \mathcal{M}_x \xrightarrow{\cong} M_{\mathfrak{p}_x}$$

where x *corresponds to (is equal to) the prime ideal* $\mathfrak{p}_x$.
2. *The canonical homomorphism*

$$\Delta(D(a))^{-1}M \longrightarrow M_a$$
$$\frac{m}{s} \mapsto \frac{m}{s}$$

is an isomorphism.
3. *The morphism which maps a presheaf to its associated sheaf induces a homomorphism over the basis open sets* $D(a)$

$$\tau_{D(a)} : \Delta(D(a))^{-1}M \longrightarrow \widetilde{M}(D(a))$$

which is an isomorphism.

We observe that for all non-empty open subsets $U \subset \mathrm{Spec}(A)$, $\Delta(U)^{-1}M$ is a module over $\Delta(U)^{-1}A$, and that the restriction mappings are bi-homomorphisms as defined in Definition 9.4. Thus we have the same situation for the associated sheaves:

$\widetilde{M}(U)$ is an $\mathcal{O}_{\mathrm{Spec}(A)}(U)$-module, and restrictions of $\widetilde{M}$ are bi-homomorphisms for the corresponding restrictions of $\mathcal{O}_{\mathrm{Spec}(A)}$.

Definition 9.5 A sheaf $\mathcal{M}$ of modules satisfying the above is called an $\mathcal{O}_X$-module on $X = \mathrm{Spec}(A)$. A morphism $f : \mathcal{M} \longrightarrow \mathcal{N}$ of sheaves between two $\mathcal{O}_X$-modules on X is called an $\mathring{\mathcal{O}}_X$-homomorphism if all f_U are $\mathcal{O}_X(U)$-homomorphisms.

If $f : M \longrightarrow N$ is a homomorphism of A-modules, then we have a $\mathcal{O}_X$-homomorphism $\widetilde{f} : \widetilde{M} \longrightarrow \widetilde{N}$. Thus $M \mapsto \widetilde{M}$ is a covariant functor from the category of A-modules to the category of $\mathcal{O}_X$-modules on $X = \mathrm{Spec}(A)$.

Chapter 10
The Category of Schemes

In this chapter we introduce the categories of preschemes and schemes, and explore some of their basic properties.

10.1 First Approximation: The Category of Ringed Spaces

A *ringed space* is a pair $(X, \mathcal{O}_X)$ consisting of a topological space X and a sheaf $\mathcal{O}_X$ of Comm on X, defined for all non empty open subsets of X. By abuse of notation the pair $(X, \mathcal{O}_X)$ is also denoted by X. The topological space is referred to as *the underlying topological space*, while the sheaf $\mathcal{O}_X$ is called the *structure sheaf* of X.

A morphism from the ringed space $(X, \mathcal{O}_X)$ to the ringed space $(Y, \mathcal{O}_Y)$

$$(f, \theta) : (X, \mathcal{O}_X) \longrightarrow (Y, \mathcal{O}_Y)$$

is a pair consisting of a continuous mapping $f : X \longrightarrow Y$ and a homomorphism of sheaves of Comm,

$$\theta : \mathcal{O}_Y \longrightarrow f_*(\mathcal{O}_X).$$

It is easily verified that the ringed spaces form a category, which we denote by $\mathcal{R}s$.

Note that whenever $(X, \mathcal{O}_X)$ is a ringed space, and $f : X \longrightarrow Y$ is a continuous mapping, then $Y = (Y, f_*(\mathcal{O}_X))$ is a ringed space and the pair (f, id) is a morphism from X to Y.

The most common ringed spaces are topological spaces X with various kinds of function sheaves, which usually take their values in a field K. Frequently the field is either $\mathbb{R}$ or $\mathbb{C}$. The sheaf $\mathcal{O}_X$ may be the sheaf of all continuous functions on the respective open subsets, or when X looks locally like an open subset of $\mathbb{R}^n$ or $\mathbb{C}^n$ we may consider functions which are n times differentiable, or algebraic functions when X is an algebraic variety over the field K, and so on.

A. Holme, *A Royal Road to Algebraic Geometry*,
DOI 10.1007/978-3-642-19225-8_10, © Springer-Verlag Berlin Heidelberg 2012

If $(X, \mathcal{O}_X)$ and $(Y, \mathcal{O}_Y)$ are the ringed spaces obtained by taking the sheaves of continuous functions (say to $\mathbb{R}$ or to $\mathbb{C}$) on the two topological spaces X and Y, and if $f : X \longrightarrow Y$ is a continuous mapping, then we obtained a morphism from the continuous mapping f by composition with the restriction of f by defining

$$\theta : \mathcal{O}_Y \longrightarrow f_*(\mathcal{O}_X),$$

as follows: For all open $U \subset Y$ there are homomorphisms

$$\theta_U : \mathcal{O}_Y(U) \longrightarrow f_*(\mathcal{O}_X)(U) = \mathcal{O}_X(f^{-1}(U))$$

$$(U \xrightarrow{\varphi} K) \mapsto (\theta_U(\varphi) : f^{-1}(U) \xrightarrow{f_{|f^{-1}(U)}} U \xrightarrow{\varphi_{|U}} K),$$

where K is $\mathbb{R}$, $\mathbb{C}$ or for that matter, any ring.

Similarly, if the topological spaces have more structure, like being differentiable manifolds, algebraic varieties etc., then this also works if we use morphisms in the category to which X and Y belong, instead of just continuous mappings. The details of these considerations are left to the reader.

Another type of ringed spaces is obtained by taking a topological space X and letting $\mathcal{O}_X$ be the sheaf associated to the presheaf defined by

$$\mathcal{O}'(U) = A,$$

where A is a fixed ring. The sheaf $\mathcal{O}_X$ so defined is referred to as the *constant sheaf of A on X.*

Clearly $\mathrm{Spec}(A)$ which we have defined above is a ringed space. Moreover, if $\varphi : A \longrightarrow B$ is a homomorphism of Comm, then we obtain a morphism of ringed spaces

$$\mathrm{Spec}(\varphi) : \mathrm{Spec}(B) \longrightarrow \mathrm{Spec}(A)$$

as follows: The mapping of topological spaces $f : \mathrm{Spec}(B) \longrightarrow \mathrm{Spec}(A)$ is given by

$$\mathfrak{q} \mapsto \varphi^{-1}(\mathfrak{q}).$$

As is easily seen, we then have

$$f^{-1}(D(\mathfrak{a})) = D((\varphi(\mathfrak{a})B),$$

hence f is a continuous mapping.

Recall the notation of Example 6.11. We then have the

Proposition 10.1 *There is an isomorphism, functorial in M:*

$$\rho_M : f_*(\widetilde{M}) \longrightarrow \widetilde{M_{[\varphi]}}.$$

Proof The assertion of the proposition is immediate from the following general and useful lemma, when applied to the basis consisting of the open subsets of the form $D(a)$:

Lemma 10.2 *Let X be a topological space, and let $\mathcal{B}$ be a basis for the topology on X. Let $\mathcal{F}$ and $\mathcal{G}$ be two sheaves of $\mathcal{A}b$ on X, such that for all $W \in \mathcal{B}$ there is an isomorphism*

$$\varphi_W : \mathcal{F}(W) \xrightarrow{\cong} \mathcal{G}(W),$$

which is compatible with the restriction homomorphisms in the sense that all the diagrams

$$
\begin{array}{ccc}
\mathcal{F}(V) & \xrightarrow{\rho^{\mathcal{F}}_{V,W}} & \mathcal{F}(W) \\
\varphi_V \downarrow & & \downarrow \varphi_W \\
\mathcal{G}(V) & \xrightarrow{\rho^{\mathcal{G}}_{V,W}} & \mathcal{G}(W)
\end{array}
$$

are commutative. Then $\mathcal{F}$ and $\mathcal{G}$ are isomorphic.

Proof of the lemma We have to define isomorphisms φ_U for all open subsets $U \subset X$, not just the basis open subsets. This is a simple application of the definition of sheaves: Let U be any open subset, and let $f \in \mathcal{F}(U)$. We have $U = \bigcup_{i \in I} W_i$, a covering by open subsets from $\mathcal{B}$. Let $g_i = \varphi_{W_i}(f|_{W_i})$, the image by φ_{W_i} of the restriction of f to W_i. For a basis open set $W \subset W_i \cap W_j$ we then have $g_i|_W = g_j|_W$, since the two diagrams

$$
\begin{array}{ccc}
\mathcal{F}(W_i) & \xrightarrow{\rho^{\mathcal{F}}_{W_i,W}} & \mathcal{F}(W) \\
\varphi_{W_i} \downarrow & & \downarrow \varphi_W \\
\mathcal{G}(W_i) & \xrightarrow{\rho^{\mathcal{G}}_{W_i,W}} & \mathcal{G}(W)
\end{array}
\qquad
\begin{array}{ccc}
\mathcal{F}(W_j) & \xrightarrow{\rho^{\mathcal{F}}_{W_j,W}} & \mathcal{F}(W) \\
\varphi_{W_j} \downarrow & & \downarrow \varphi_W \\
\mathcal{G}(W_j) & \xrightarrow{\rho^{\mathcal{G}}_{W_j,W}} & \mathcal{G}(W)
\end{array}
$$

commute. Thus the g_i's glue to a unique $g \in \mathcal{G}(U)$, we put $\varphi_U(f) = g$. We now have to show that φ_U so defined is in fact an isomorphism of Abelian groups, and that it is compatible with restriction. This is straightforward and is left to the reader.   □

To complete the proof of the proposition, we only need to apply the lemma to the basis for the topology on $\mathrm{Spec}(A)$ consisting of the open subsets $D(a)$.   □

To proceed with the definition of $\mathrm{Spec}(\varphi)$, we note that the homomorphism φ may be seen as a homomorphism of A-modules $\varphi : A \longrightarrow B_{[\varphi]}$, hence yields a morphism of $\mathcal{O}_{\mathrm{Spec}(A)}$-modules

$$\theta = \widetilde{\varphi} : \widetilde{A} = \mathcal{O}_{\mathrm{Spec}(A)} \longrightarrow \widetilde{B_{[\varphi]}} = f_*(\mathcal{O}_{\mathrm{Spec}(B)}).$$

Remark 10.3 We follow [15] and identify $\widetilde{B_{[\varphi]}}$ with $f_*(\mathcal{O}_{\mathrm{Spec}(B)})$ via the canonical isomorphism ρ_B from Proposition 10.1.

We define Spec of a morphism by

$$\mathrm{Spec}(\varphi) = (f, \theta) = (\varphi^{-1}(\ ), \widetilde{\varphi}).$$

From now on we adopt the notation of [15] and write $^a\varphi$ for the mapping $\varphi^{-1}(\ )$.

It is easily seen that Spec of a composition is the composition of the Spec's (in reverse order), and that the Spec of the identity on A is the identity on $\mathrm{Spec}(A)$. We may sum our findings up as follows:

Proposition 10.4 Spec *is a contravariant functor*

$$\mathrm{Spec} : \mathfrak{Comm} \longrightarrow \mathfrak{Rs}.$$

10.2 Second Approximation: Local Ringed Spaces

Some of the ringed spaces X we have seen so far have the important property that for all points $x \in X$ the fiber $\mathcal{O}_{X,x}$ of the structure sheaf $\mathcal{O}_X$ at x is a *local ring*. This is certainly so for $\mathrm{Spec}(A)$, and also for the *function spaces* where the functions take their values in a field. Thus for instance, let $(X, \mathcal{O}_X)$ be the topological space X together with the sheaf $\mathcal{O}_X$ of continuous real valued functions on the open subsets. Then the ring $\mathcal{O}_{X,x}$ is the ring of *germs of continuous functions* at x: It is the ring of equivalence classes of function elements (f, U) where U is an open subset containing x and f is a continuous real valued function defined on U. Recall that two function elements at x are equivalent, $(U, f) \sim (V, g)$ if there is an open subset W contained in $U \cap V$ and containing x such that f and g have the same restriction to W. The equivalence class of the function element (f, U) is denoted by $[(f, U)]$.

We have the evaluation homomorphism

$$\varphi_x : \mathcal{O}_{X,x} \longrightarrow \mathbb{R},$$

$$[(f, U)] \mapsto f(x).$$

Clearly this is well defined, it is a ring-homomorphism and it is surjective as the constant functions are continuous.

Let $\mathfrak{m}_{X,x} = \ker(\varphi_x)$. This is a maximal ideal since $\mathcal{O}_{X,x}/\mathfrak{m}_{X,x} \cong \mathbb{R}$. We show that $\mathfrak{m}_{X,x}$ is the only maximal ideal in $\mathcal{O}_{X,x}$. It suffices to show that if f is a continuous function on the open subset U containing x such that $f(x) \neq 0$, then $[(f, U)]$ is invertible in $\mathcal{O}_{X,x}$. Indeed, as f is continuous and $\{0\}$ is a closed subset of $\mathbb{R}$, $f^{-1}(0)$ is a closed subset of U, not containing x. Thus if $V = U - f^{-1}(0)$, then $f_{|V}$ is invertible on V, so $[(f_{|V}, V)]$ is invertible. Since this element is equal to $[(f, U)]$, we are done.

Definition 10.1 A ringed space $(X, \mathcal{O}_X)$ is called a local ringed space provided that all the fibers $\mathcal{O}_{X,x}$ of the structure sheaf are local rings. A morphism of ringed spaces between two local ringed spaces

$$f = (f, \theta) : (X, \mathcal{O}_X) \longrightarrow (Y, \mathcal{O}_Y)$$

is said to be a morphism of local ringed spaces provided that the morphism of sheaves

$$\theta : \mathcal{O}_Y \longrightarrow f_*(\mathcal{O}_X)$$

has the following property:

Whenever $f(x) = y$, the homomorphism $\theta_x^\sharp$ which is the composition

$$\theta_x^\sharp : f^*(\mathcal{O}_Y)_x = \mathcal{O}_{Y,y} \xrightarrow{\theta_y} f_*(\mathcal{O}_X)_y \xrightarrow{\text{canonical}} \mathcal{O}_{X,x}$$

is a local homomorphism in the sense that the maximal ideal of $\mathcal{O}_{Y,y}$ is mapped into the maximal ideal of $\mathcal{O}_{X,x}$.

We note that this property is equivalent to the assertion that the inverse image of the maximal ideal of the target local ring $\mathcal{O}_{X,x}$ be the equal to the maximal ideal of the source local ring $\mathcal{O}_{Y,y}$.

The category thus obtained is denoted by $\mathcal{L}$rs. We note that $\mathrm{Spec}(\varphi)$ is a morphism of local ringed spaces, and also that the morphism between two function spaces obtained from a continuous mapping by composition is a morphism of local ringed spaces.

For all points x of a local ringed space $(X, \mathcal{O}_X)$ we have a field $k(x) = \mathcal{O}_{X,x}/\mathfrak{m}_{X,x}$, referred to as the local field at the point x. These fields play a key role in the theory. For $X = \mathrm{Spec}(A)$, $k(\mathfrak{p})$ is the quotient field of the integral domain $A/\mathfrak{p}$. Thus this field varies from point to point in general. However, for local ringed spaces where the structure sheaf is a sheaf of functions with values in a fixed field, the local fields $k(x)$ are all equal to this fixed field.

If $U \subset X$ is an open subset and $f \in \mathcal{O}_X(U)$, $f(x)$ denotes the image of f under the composition

$$\mathcal{O}_X(U) \longrightarrow \mathcal{O}_{X,x} \longrightarrow k(x).$$

If $f \in \mathcal{O}_X(X)$, then we put

$$X_f = \{x \in X \mid f(x) \neq 0\}.$$

Then

Lemma 10.5 X_f *is an open subset of* X.

Proof The assertion $f(x) \neq 0$ is equivalent to the assertion that the image of f in $\mathcal{O}_{X,x}$ be a unit. Thus if $x \in X_f$, then there exists an open subset U containing x and an element $g \in \mathcal{O}_X(U)$ such that $f_{|U} g = 1$. Hence $U \subset X_f$. $\square$

We have the following important result, which shows that the definition of morphisms between local ringed spaces made above is exactly right for our purposes:

Proposition 10.6 *Let A and B be two commutative rings, and let*

$$(f, \theta) : \mathrm{Spec}(B) = X \longrightarrow \mathrm{Spec}(A) = S$$

be a morphism of ringed spaces. Then $(f, \theta) = \mathrm{Spec}(\varphi)$ for

$$\varphi : A \xrightarrow{\tau_A} \mathcal{O}_S(S) \xrightarrow{\theta_S} \mathcal{O}_X(X) \xrightarrow{\tau_B^{-1}} B$$

if and only if it is a morphism of local ringed spaces.

Remark 10.7 We shall use the convention that $\tau_A : A \longrightarrow \mathcal{O}_S(S)$ denotes the canonical isomorphism τ_1 for A, similar for τ_B. To avoid unwieldy notation, we adhere from now on to the convention of [15] of identifying the rings A and $\mathcal{O}_S(S)$ via the canonical isomorphism τ_A, when there is no danger of misunderstandings.

Striking as this result may be, it is only the starting point of several generalizations. We present the ultimate version, relying on a remark due to *John Tate*. See [15] II, Errata et addenda on p. 217.
We start with the following

Definition 10.2 The local ringed space $(S, \mathcal{O}_S)$ is called an affine scheme if it is isomorphic as a local ringed space to $\mathrm{Spec}(A)$ for some A.

Theorem 10.8 *Let $(S, \mathcal{O}_S)$ be an affine scheme, and let $(X, \mathcal{O}_X)$ be a local ringed space. Then the mapping*

$$\rho = \rho_{X,S} : \mathrm{Hom}_{\mathcal{L}\mathrm{rs}}((X, \mathcal{O}_X), (S, \mathcal{O}_S)) \longrightarrow \mathrm{Hom}_{\mathcal{C}\mathrm{omm}}(\mathcal{O}_S(S), \mathcal{O}_X(X))$$
$$(f, \theta) \mapsto \theta_S$$

is bijective.

We note that the theorem implies the proposition. Indeed, the "only if" part is trivial as $\mathrm{Spec}(\varphi)$ is a morphism of $\mathcal{L}$rs. The "if" part follows since $\rho((f,\theta)) = \rho(\mathrm{Spec}(\varphi)) = \theta_S$ by the theorem.

Proof of the theorem We may assume that $S = \mathrm{Spec}(A)$. We prove bijectivity of ρ by constructing an inverse. We make the canonical identification of A_f with $\mathcal{O}_S(D(f))$ for all $f \in A$. For any homomorphism $\varphi : \mathcal{O}_S(S) = A \longrightarrow \mathcal{O}_X(X)$, we define a mapping of topological spaces

$$^a\varphi : X \longrightarrow S$$

by letting $^a\varphi(x) = \mathfrak{p}_x$ where

$$\mathfrak{p}_x = \{f \in A \,|\, \varphi(f)(x) = 0\}.$$

$\mathfrak{p}_x$ is a prime ideal, being the kernel of a homomorphism into a field. Note that this definition generalizes the previous definition of $^a\varphi$, made in the case when X is affine.

As is easily checked $^a\varphi^{-1}(D(f)) = X_{\varphi(f)}$, and hence $^a\varphi$ is a continuous mapping. We next define a morphism of $\mathcal{O}_X$-modules on S

$$\tilde{\varphi} : \mathcal{O}_S \longrightarrow {}^a\varphi_*(\mathcal{O}_X)$$

by first defining

$$\tilde{\varphi}_{D(f)} : A_f \longrightarrow \mathcal{O}_X(X_{\varphi(f)})$$
$$\frac{s}{f^n} \mapsto (\varphi(s)_{|X_{\varphi(f)}})((\varphi(f)_{|X_{\varphi(f)}})^{-1})^n.$$

It is easily seen that the following diagram commutes,

$$
\begin{array}{ccc}
A_f & \xrightarrow{\ \tilde{\varphi}_{D(f)}\ } & \mathcal{O}_X(X_{\varphi(f)}) \\
{\scriptstyle \rho^{\mathcal{O}_S}_{D(f),D(fg)}}\Big\downarrow & & \Big\downarrow{\scriptstyle \rho^{\mathcal{O}_S}_{X_{\varphi(f)},X_{\varphi(fg)}}} \\
A_{fg} & \xrightarrow[\ \tilde{\varphi}_{D(fg)}\ ]{} & \mathcal{O}_X(X_{\varphi(fg)})
\end{array}
$$

the two vertical arrows being restrictions to a smaller open set. Hence we may complete the set of homomorphisms $\tilde{\varphi}_{D(f)}$ to a morphism of $\mathcal{O}_S$-modules on S, $\tilde{\varphi} : \mathcal{O}_S \longrightarrow {}^a\varphi_*(\mathcal{O}_X)$, as asserted above.

We thus have defined a morphism of $\mathcal{R}$s:

$$\sigma(\varphi) : (X, \mathcal{O}_X) \longrightarrow (S, \mathcal{O}_S).$$

This is actually a morphism of $\mathcal{L}$rs. Indeed, the homomorphism

$$\mathcal{O}_{S,{}^a\varphi(x)} = A_{\mathfrak{p}_x} \longrightarrow \mathcal{O}_{X,x}$$

maps the element $\frac{s}{f}$, where $f \notin \mathfrak{p}_x$, to the element $(\varphi(s)_{X_{\varphi(f)}})(\varphi(f)|_{X_{\varphi(f)}})^{-1}$.
If $s \in \mathfrak{p}_x$ then $(\varphi(s)_{X_{\varphi(f)}})(\varphi(f)|_{X_{\varphi(f)}})^{-1} \in \mathfrak{m}_{X,x}$ by the definition of $^a\varphi(x)$,
and thus $\widetilde{\varphi}_x^\sharp$ is a local homomorphism.

It remains to show that ρ and σ are inverse to one another.

First of all, with the identifications we have made,

$$\widetilde{\varphi}_S = \varphi.$$

Hence $\rho \circ \sigma$ is the identity on $\mathrm{Hom}_{\mathrm{comm}}(\mathcal{O}_S(S), \mathcal{O}_X(X))$. To show that $\sigma \circ \rho$
is the identity, start with a morphism of local ringed spaces

$$(\psi, \theta) : (X, \mathcal{O}_X) \longrightarrow (S, \mathcal{O}_S)$$

and let $\varphi = \theta_S$. Since

$$\theta_x^\sharp : \mathcal{O}_{S,\psi(x)} \longrightarrow \mathcal{O}_{X,x}$$

is a local homomorphism it induces an embedding of fields

$$\theta^x : k(\psi(x)) \hookrightarrow k(x)$$

such that for all $f \in A$ we have $\theta^x(f(\psi(x))) = \varphi(f)(x)$. Then

$$f(\psi(x)) = 0 \quad \Longleftrightarrow \quad \varphi(f)(x) = 0,$$

thus $\psi = {}^a\varphi$. It remains to show that

$$\widetilde{\varphi} = \theta : \mathcal{O}_S \longrightarrow \psi_*(\mathcal{O}_X)(= {}^a\varphi_*(\mathcal{O}_X)).$$

To prove this we note first that the following two diagrams are commutative:

$$
\begin{array}{ccc}
A & \xrightarrow{\varphi} & \mathcal{O}_X(X) \\
\downarrow & & \downarrow \\
A_{\mathfrak{p}_{\psi(x)}} & \xrightarrow{\widetilde{\varphi}_x^\sharp} & \mathcal{O}_{X,x}
\end{array}
\qquad
\begin{array}{ccc}
A & \xrightarrow{\varphi} & \mathcal{O}_X(X) \\
\downarrow & & \downarrow \\
A_{\mathfrak{p}_{\psi(x)}} & \xrightarrow{\theta_x{}^\sharp} & \mathcal{O}_{X,x}
\end{array}
$$

The diagonal mapping $\alpha : A \longrightarrow \mathcal{O}_{X,x}$ is a homomorphism which maps the
multiplicatively close subset $\Delta = A - \mathfrak{p}_{\psi(x)}$ into the group of units of the local
ring $\mathcal{O}_{X,x}$, since the inverse image of its maximal ideal is $\mathfrak{p}_{\psi(x)}$. Thus by the
universal property of localization α factors uniquely through $A_{\mathfrak{p}_{\psi(x)}}$, and so
the two bottom homomorphisms are equal.

This implies that $\theta^\sharp = \widetilde{\varphi}^\sharp$ and hence that $\theta = \widetilde{\varphi}$. Thus the proof is complete. $\square$

One of several useful consequences of Theorem 10.8 is the following, *loc.
cit.* p. 219:

Corollary 10.9 *A local ringed space Y is an affine scheme if and only if $\rho_{X,Y}$ is bijective for all local ringed spaces X.*

Proof The "if" part is the theorem. Assume that all $\rho_{X,Y}$ are bijective, and put $A = \mathcal{O}_Y(Y)$. We then have isomorphisms of functors

$$\mathrm{Hom}_{\mathcal{L}_{rs}}(\ ,Y) \xrightarrow{\cong} \mathrm{Hom}_{\mathcal{C}_{omm}}(A, \mathcal{O}_{(\ )}(\ )) \xleftarrow{\cong} \mathrm{Hom}_{\mathcal{L}_{rs}}(\ ,\mathrm{Spec}(A))$$

by hypothesis and the theorem. Thus the functors h_Y and $h_{\mathrm{Spec}(A)}$ are isomorphic, thus $Y \cong \mathrm{Spec}(A)$ by Proposition 6.6. $\qquad\square$

For a general local ringed space Z we put

$$S(Z) = \mathrm{Spec}(\mathcal{O}_Z(Z)).$$

We then have functorial mappings

$$\mathrm{Hom}_{\mathcal{L}_{rs}}(X,Z) \xrightarrow{\rho_{X,Z}} \mathrm{Hom}_{\mathcal{C}_{omm}}(\mathcal{O}_Z(Z), \mathcal{O}_X(X)) \xrightarrow{\rho_{X,S(Z)^{-1}}} \mathrm{Hom}_{\mathcal{L}_{rs}}(X,S(Z))$$

which yield a morphism of contravariant functors

$$h_Z \longrightarrow h_{S(Z)}$$

thus a morphism of local ringed spaces

$$\epsilon_Z : Z \longrightarrow S(Z).$$

We obtain the further

Corollary 10.10 *Z is an affine scheme if and only if ϵ_Z is an isomorphism.*

Proof By the previous corollary Z is an affine scheme if and only if all $\rho_{X,Z}$ are bijective. The claim follows from this. $\qquad\square$

Remark In the literature, textbooks and other, we frequently encounter assertions of the following type: *"Let X be an affine scheme. Then $X = \mathrm{Spec}(A)$. . . "* A statement like this is justified when we identify X with $S(X)$ by ϵ_X, and this identification will be made throughout this book without further comments.

We note a final, important corollary:

Corollary 10.11 *The category $\mathcal{C}omm^*$ is equivalent to the category of affine schemes, $\mathcal{A}ff\,\mathcal{S}ch$. More generally, if $S \cong \mathrm{Spec}(A)$ then $\mathcal{A}ff\,\mathcal{S}ch_S$ is equivalent to the dual of the category of commutative A-algebras.*

Proof This is immediate by Proposition 6.4 and the last corollary. A proof using only the definition of equivalent categories runs as follows: Let

$$F : \mathfrak{Comm} \longrightarrow \mathcal{A}\mathrm{ff}\,\mathfrak{Sch}$$

be the functor Spec, and let

$$G : \mathcal{A}\mathrm{ff}\,\mathfrak{Sch}^* \longrightarrow \mathfrak{Comm}$$

be the functor $X \mapsto \mathcal{O}_X(X)$. Then the canonical isomorphism

$$\tau_A : A \longrightarrow \mathcal{O}_{\mathrm{Spec}(A)}(\mathrm{Spec}(A))$$

yields an isomorphism

$$\mathrm{id}_{\mathfrak{Comm}} \longrightarrow G \circ F,$$

and the isomorphism

$$\epsilon_X : X \longrightarrow \mathrm{Spec}(\mathcal{O}_X(X))$$

yields an isomorphism

$$\mathrm{id}_{\mathcal{A}\mathrm{ff}\,\mathfrak{Sch}} \longrightarrow F \circ G.$$

This completes the proof.   □

10.3 Definition of the Category of Schemes

The most important object under study in modern algebraic geometry is that of *a scheme*. A scheme is a geometric object which also embodies a vast generalization of the concept of *a commutative ring*:

Definition 10.3 A scheme is a local ringed space X with the following property:

> For all points $x \in X$ there exists an open subset U in X containing x, such that $(U, \mathcal{O}_X|U)$ is an affine scheme, i.e., the morphism $\epsilon_U : (U, \mathcal{O}_X|U) \longrightarrow \mathrm{Spec}(\mathcal{O}_X(U))$ is an isomorphism.

Morphisms of schemes $f : X \longrightarrow Y$ are defined by setting

$$\mathrm{Hom}_{\mathfrak{Sch}}(X, Y) = \mathrm{Hom}_{\mathcal{L}\mathrm{rs}}(X, Y).$$

The category of schemes is denoted by $\mathfrak{Sch}$. Let S be a scheme. The category $\mathfrak{Sch}_S$ is referred to as the category of S-schemes. Recall that an S-scheme is then a pair (X, φ_X), where $\varphi_X : X \longrightarrow S$ is a morphism, which we refer to as the structure morphism of the S-scheme X. A morphism of S-schemes $f : X \longrightarrow Y$ is a morphism of schemes such that $\varphi_Y \circ f = \varphi_X$.

The first important task is to carry out the construction of finite products in the category of S-schemes. We prove the following, given as Théorème (3.2.6) in [15] I, Sect. 3, and essentially following the treatment given by Grothendieck there. Some of the steps in the following proofs are valid in greater generality than stated, e.g. for local ringed spaces rather than just for schemes.

Theorem 10.12 *Finite products exist in the category* Sch_S.

Proof It suffices to construct the product $X_1 \times_S X_2$ for any two S-schemes X_1 and X_2. This is done in several steps. First of all, we know by Corollary 10.11 that if $S = \text{Spec}(A)$, and $X_i = \text{Spec}(B_i)$, where the B_i are A-algebras, then $\text{Spec}(B_i \otimes_A B_2)$ is the product of X_1 and X_2 in the category of affine schemes over S. But by Theorem 10.8 it follows that this is the product in the larger category $\mathcal{L}\text{rs}_S$, in particular in Sch_S: Indeed, for a local ringed space Z we have to show that there is an isomorphism, functorial in Z,

$$\text{Hom}_{\mathcal{L}\text{rs}_S}(Z, \text{Spec}(B_1 \otimes_A B_2))$$
$$\cong \downarrow$$
$$\text{Hom}_{\mathcal{L}\text{rs}_S}(Z, \text{Spec}(B_1)) \times \text{Hom}_{\mathcal{L}\text{rs}_S}(Z, \text{Spec}(B_2)).$$

This follows by the theorem quoted since it provides functorial isomorphisms

$$\text{Hom}_{\mathcal{L}\text{rs}_S}(Z, \text{Spec}(B_1 \otimes_A B_2)) \xrightarrow{\cong} \text{Hom}_A(B_1 \otimes_A B_2, \mathcal{O}_Z(Z))$$

and

$$\text{Hom}_{\mathcal{L}\text{rs}_S}(Z, \text{Spec}(B_i)) \xrightarrow{\cong} \text{Hom}_A(B_i, \mathcal{O}_Z(Z))$$

for $i = 1, 2$ and moreover,

$$\text{Hom}_A(B_1 \otimes_A B_2, \mathcal{O}_Z(Z)) \xrightarrow{\cong} \text{Hom}_A(B_1, \mathcal{O}_Z(Z)) \times \text{Hom}_A(B_2, \mathcal{O}_Z(Z))$$

by the universal property of $\otimes_A$. Summing up, we have shown the

Lemma 10.13 *If* $S = \text{Spec}(A)$ *and* $X_i = \text{Spec}(B_i)$ *for* $i = 1$ *and* 2, B_i *being an* A-*algebra, then* $X_1 \times_S X_2$ *is* $\text{Spec}(B_1 \otimes_A B_2)$ *in the category of local ringed spaces, the category of schemes and in the category of affine schemes.*

To construct the product $X_1 \times_S X_2$ for any two S-schemes, we first reduce to the case when S is an affine scheme. For this we employ the following general lemma, cf. [15] I, (3.2.4):

Lemma 10.14 *Let* $f : S' \longrightarrow S$ *be a morphism of schemes which is a monomorphism. Assume that the* S-*schemes* X_1 *and* X_2 *are such that for*

$i = 1$ and 2 the structure morphisms $\varphi_i : X_i \longrightarrow S$ factor through S', i.e., that
there are morphisms $\psi_i : X_i \longrightarrow S'$ such that the following diagrams commute:

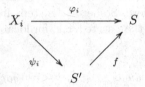

Then

$$X_1 \times_{S'} X_2 = X_1 \times_S X_2,$$

in the sense that if one of the products is defined, then so is the other and
they are canonically isomorphic.

Proof If Z is an S-scheme and $g_i : Z \longrightarrow X_i$ two S-morphisms, then $\varphi_Z = \varphi_1 \circ g_1 = \varphi_2 \circ g_2$, thus

$$f \circ \psi_1 \circ g_1 = f \circ \psi_2 \circ g_2,$$

so as f is a monomorphism,

$$\psi_1 \circ g_1 = \psi_2 \circ g_2.$$

We denote this composition by φ', and may consider Z as an S'-scheme by
φ' and g_1, g_2 as S'-morphisms. This establishes a bijection between pairs of
S-morphisms $g_i : Z \longrightarrow X_i$ and pairs of S'-morphisms $g_i : Z \longrightarrow X_i$, and the
claim follows.   □

Assume that U is an open, non empty subset of the scheme S such that
$(U, \mathcal{O}_S|U)$ is an affine scheme. We then say that U is an open, affine subscheme
(or just subset by abuse of language) of S.

The lemma implies the following (*loc. cit.* Corollaire (3.2.5)):

Proposition 10.15 *Let X_i be two S-schemes with structure morphisms φ_i,
and let U be an open subscheme of S such that $\varphi_i(X_i) \subseteq U$ for $i = 1, 2$. Then*

$$X_1 \times_S X_2 = X_1 \times_U X_2,$$

*in the sense that if one of the products is defined, then so is the other and
they are canonically isomorphic.*

Proof Immediate as the inclusion $U \hookrightarrow S$ obviously is a monomorphism.   □

We need one more general observation, namely that being a product is a
local property. The following proposition is given in [15] I as Lemmas (3.2.6.1)
and (3.2.6.2):

Proposition 10.16 *Let Z be an S-scheme and let $p_i : Z \longrightarrow X_i$ be two S-morphisms.*

1. *Let U and V be open subschemes of X_1 and X_2, respectively. Let*

$$W = p_1^{-1}(U) \cap p_2^{-1}(V).$$

Then if Z is a product of X_1 and X_2, W is a product of U and V.
2. *Assume that*

$$X_1 = \bigcup_{\alpha \in I} X_{1,\alpha} \quad and \quad X_2 = \bigcup_{\beta \in J} X_{2,\beta}$$

are open coverings. For all $(\alpha, \beta) \in I \times J$ put

$$Z_{\alpha,\beta} = p_1^{-1}(X_{1,\alpha}) \cap p_2^{-1}(X_{2,\beta}),$$

and let $p_{1,\alpha,\beta}$ and $p_{2,\alpha,\beta}$ be the restrictions of p_1 and p_2, respectively. Assume that $Z_{\alpha,\beta}$ is the product of $X_{1,\alpha}$ and $X_{2,\beta}$ with these morphisms as the projections. Then Z is the product of X_1 and X_2 with p_1 and p_2 as the projections.

Proof 1. Let

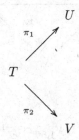

be S-morphisms, i.e., the following diagram commutes:

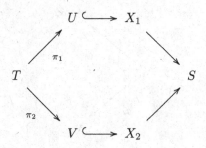

As $Z = X_1 \times_S X_2$ there is a unique $h : T \longrightarrow Z$ such that the diagrams

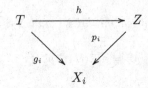

where g_i is the composition of π_i and the inclusion, commute. But this shows that h factors through W, and the claim follows.

2. Let

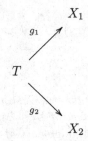

be S-morphisms. To show is that there is a unique S-morphism h such that the diagrams

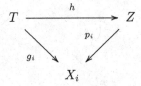

commute.

Uniqueness of h: Put

$$T_{\alpha,\beta} = g_1^{-1}(X_{1,\alpha}) \cap g_2^{-1}(X_{2,\beta}),$$

this yields an open covering of T. We then have the diagram

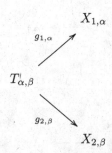

The restriction of h to $T_{\alpha,\beta}$ will then be a morphism

$$T_{\alpha,\beta} \longrightarrow Z_{\alpha,\beta}$$

which corresponds to the universal property of the product $Z_{\alpha,\beta}$ of $X_{1,\alpha}$ and $X_{2,\beta}$. Thus these restrictions are unique, hence so is h itself.

To show existence, define $T_{\alpha,\beta}$, $\pi_{1,\alpha}$ and $\pi_{2,\beta}$ as in the proof of uniqueness above. We get unique morphisms

$$h_{\alpha,\beta} : T_{\alpha,\beta} \longrightarrow Z_{\alpha,\beta}$$

such that the diagrams

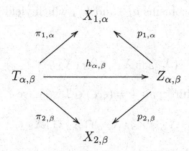

commute. It suffices to show that these $h_{\alpha,\beta}$ may be glued to a morphism $h : T \longrightarrow Z$. Thus we have to show that for all $\alpha, \gamma \in I$ and $\beta, \delta \in J$

$$h_{\alpha,\beta}|T_{\alpha,\beta} \cap T_{\gamma,\delta} = h_{\gamma,\delta}|T_{\alpha,\beta} \cap T_{\gamma,\delta}.$$

But by part 1. we have

$$Z_{\alpha,\beta} \cap Z_{\gamma,\delta} = (X_{1,\alpha} \cap X_{1,\gamma}) \times_S (X_{2,\beta} \cap X_{2,\delta})$$

and moreover

$$T_{\alpha,\beta} \cap T_{\gamma,\delta} = g_1^{-1}(X_{1,\alpha} \cap X_{1,\gamma}) \cap g_2^{-1}(X_{2,\beta} \cap X_{2,\delta})$$

and thus $h_{\alpha,\beta}|T_{\alpha,\beta} \cap T_{\gamma,\delta}$ is the unique morphism coming from the universal property of the product $(X_{1,\alpha} \cap X_{1,\gamma}) \times_S (X_{2,\beta} \cap X_{2,\delta})$, hence it is equal to $h_{\gamma,\delta}|T_{\alpha,\beta} \cap T_{\gamma,\delta}$ as claimed.   □

We are now ready to prove the key result which establishes the existence of finite fibered products in the category $\mathcal{S}ch$, presented as (3.2.6.3) in [15] I on p. 107:

Proposition 10.17 *Let X_1 and X_2 be S-schemes, and let*

$$X_1 = \bigcup_{\alpha \in I} X_{1,\alpha} \quad \text{and} \quad X_2 = \bigcup_{\beta \in J} X_{2,\beta}$$

be open coverings. Assume that all the products $X_{1,\alpha} \times_S X_{2,\beta}$ exist. Then $X_1 \times_S X_2$ also exists.

Proof, essentially following [15] Let $i = (\alpha, \beta) \in I \times J = \mathfrak{I}$, and put

$$Z_i' = X_{1,\alpha} \times_S X_{2,\beta}.$$

Let $j = (\gamma, \delta) \in \mathfrak{I}$, and define the open subscheme $Z_{i,j}'$ of Z_i' by

$$Z_{i,j}' = \mathrm{pr}_{X_{1,\alpha}}^{-1}(X_{1,\alpha} \cap X_{1,\gamma}) \cap \mathrm{pr}_{X_{2,\beta}}^{-1}(X_{2,\beta} \cap X_{2,\delta}).$$

Since $Z_{i,j}'$ is the product of the two intersections (Proposition 10.16 part 1), there are unique isomorphisms $h_{i,j}$ and $h_{j,i}$ which yield isomorphisms $f_{i,j}$ by the compositions

$$f_{i,j} : Z_{i,j}' \xrightarrow{h_{i,j}} (X_{1,\alpha} \cap X_{1,\gamma}) \times_S (X_{2,\beta} \cap X_{2,\delta}) \xrightarrow{h_{j,i}^{-1}} Z_{j,i}'.$$

Moreover, for any third pair $k = (\epsilon, \zeta) \in \mathfrak{I}$ we have

$$(X_{1,\alpha} \cap X_{1,\gamma} \cap X_{1,\epsilon}) \times_S (X_{2,\beta} \cap X_{2,\delta} \cap X_{2,\zeta}) = Z_{k,i}' \cap Z_{k,j}',$$

since the right hand side is easily seen to satisfy the universal property defining the left hand side. It follows that

$$f_{i,k} = f_{i,j} \circ f_{j,k} \quad \text{on } Z_{k,i}' \cap Z_{k,j}',$$

again applying Proposition 10.16 part 1 to the open subschemes $X_{1,\alpha} \cap X_{1,\gamma} \cap X_{1,\epsilon}$ and $X_{2,\beta} \cap X_{2,\delta} \cap X_{2,\zeta}$ of $X_{1,\gamma}$ and $X_{2,\delta}$, respectively.

This condition is some times referred to as the *Cocycle Condition*, and is visualized below. The important condition is that *the inner triangle commutes*.

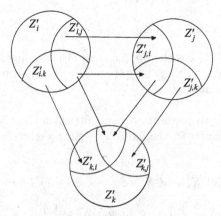

We now need the following general

Lemma 10.18 (Gluing-Lemma) *Given a collection of ringed spaces* $\{Z_i'\}_{i \in \mathcal{I}}$ *with open sub-ringed spaces* $Z_{i,j}'$ *and isomorphisms* $f_{i,j}$ *as above, satisfying the cocycle condition. Then there exists a ringed space* Z, *with an open covering*

$$Z = \bigcup_{i \in \mathcal{I}} Z_i,$$

and isomorphisms $\varphi_i : Z_i' \longrightarrow Z_i$ *such that* $Z_{i,j}'$ *is mapped to* $Z_i \cap Z_j$. *If all the* Z_i' *are local ringed spaces, respectively schemes, then so is* Z.

Proof The last assertion is of course obvious. To perform the gluing, we first put $Z_{i,i}' = Z_i'$, and let $f_{i,i}$ be the identity. We first glue the underlying topological spaces by introducing a relation $\sim$ in the disjoint union of the sets Z_i as follows:

$$x \sim y \iff x \in Z_i' \text{ and } y \in Z_j' \text{ and } f_{i,j}(x) = y.$$

It follows in an obvious manner that this is an equivalence relation, transitivity uses the cocycle condition. As a set we then define Z as the set of equivalence classes of this relation $\sim$. We get injective mappings

$$\varphi_i : Z_i \hookrightarrow Z,$$

and clearly the images $\varphi_i(Z_i') = Z_i$ do have the property that $Z_i \cap Z_j = \varphi_i(Z_{i,j})$. Letting $\mathcal{B}$ be the set of all images under φ_i of the open subsets of Z_i', for all $i \in \mathcal{I}$, we get a basis for a topology on Z, where $Z = \bigcup_{i \in \mathcal{I}} Z_i$ is an open covering. Thus we are done gluing the topological spaces.

We now need to glue the structure sheaves as well. For this we have the following

Lemma 10.19 *Let* Z *be a topological space, with an open covering* $Z = \bigcup_{\lambda \in L} Z_\lambda$. *For all* $\lambda \in L$ *there is given a sheaf* $\mathcal{F}_\lambda$ *of* $\mathcal{A}b$ *on* Z_λ, *such that for all* $\lambda, \mu \in L$ *we have isomorphisms*

$$\varphi_{\lambda,\mu} : \mathcal{F}_\lambda | Z_\lambda \cap X_\mu \xrightarrow{\cong} \mathcal{F}_\mu | Z_\mu \cap X_\lambda$$

satisfying the cocycle condition on $Z_\lambda \cap X_\mu \cap Z_\nu$ *for all* λ, μ *and* ν *in* L. *Then there exists a sheaf* $\mathcal{F}$ *on* Z *with isomorphisms* $\psi_\lambda : \mathcal{F} | Z_\lambda \xrightarrow{\cong} \mathcal{F}_\lambda$ *such that*

$$
\begin{array}{ccc}
\mathcal{F}_\lambda | Z_\lambda \cap Z_\mu & \xrightarrow{\ \varphi_{\lambda,\mu}\ } & \mathcal{F}_\mu | Z_\lambda \cap Z_\mu \\[2mm]
{\scriptstyle \psi_\lambda | Z_\lambda \cap Z_\mu} \searrow & & \swarrow {\scriptstyle \psi_\mu | Z_\lambda \cap Z_\mu} \\[2mm]
& \mathcal{F} | Z_\lambda \cap Z_\mu &
\end{array}
$$

commutes.

Proof Let $\mathcal{B}$ be a basis for the topology on Z consisting of the open subsets contained in Z_λ as λ runs through L. By the remark following Proposition 8.4 it is then enough to define $\mathcal{F}(V)$ for an open subset $V \subset U$ where $U \in \mathcal{B}$, yielding a presheaf on U. We then form the *sheaf* $\mathcal{F}$ on Z by the general procedure outlined in that remark.

This completes the proof of the final lemma, and hence of Proposition 10.17.   □

The proposition has the following

Corollary 10.20 *Let* $\varphi_i : X_i \longrightarrow S$, $i = 1, 2$ *be morphisms of schemes, and let* $S = \bigcup_{j \in J} S_j$ *be an open covering. Let* $X_{i,j} = \varphi_i^{-1}(S_j)$ *for* $i = 1, 2$ *and* $j \in J$. *Then, if all* $X_{1,j} \times_{S_j} X_{2,j}$ *exist,* $X_1 \times_S X_2$ *exists.*

Proof Immediate form the proposition by letting $Z'_i = X_{1,i} \times_S X_{2,i} = X_{1,j} \times_{S_j} X_{2,j}$, for all $i \in J$, and $Z'_{i,j} = p_{X_{1,i}}^{-1}(X_{1,i} \cap X_{1,j}) \cap p_{X_{2,i}}^{-1}(X_{2,i} \cap X_{2,j})$. $Z'_{i,j}$ is isomorphic with $(X_{1,i} \cap X_{1,j}) \times_S (X_{2,i} \cap X_{2,j})$, we get isomorphisms $\varphi_{i,j} : Z_{i,j} \longrightarrow Z'j, i$, and any three of these do satisfy the cocycle condition.   □

We may now complete the proof of Theorem 10.12. It suffices to construct the product $X_1 \times_S X_2$ in the case when $S = \mathrm{Spec}(A)$. For this we take affine open coverings $X_i = \bigcup_{j \in J_i}$ for $i = 1, 2$, with $X_{i,j} = \mathrm{Spec}(B_{i,j})$. For $\alpha \in J_1$, $\beta \in J_2$ we then have $Z_{\alpha,\beta} = X_{1,\alpha} \times_S X_{2,\beta} = \mathrm{Spec}(B_{1,\alpha} \otimes_A B_{2,\beta})$. We are then done by Proposition 10.17. This completes the proof of the theorem.   □

10.4 Formal Properties of Products

Finite products of S-schemes have a collection of formal properties, all of which are easy to prove and actually hold for products in any category: They are consequences of the universal property which defines the product. We give a brief summary below, by abuse of notation canonical isomorphisms are denoted as equalities.

Proposition 10.21 1. *Let* X_i *be* S-*schemes, for* $i = 1, 2$. *Then*

$$X_1 \times_S X_2 = X_2 \times_S X_1.$$

2. *Let* X_i *be* S-*schemes, for* $i = 1, 2, 3$. *Then*

$$(X_1 \times_S X_2) \times_{S'} X_3 = X_1 \times_S (X_2 \times_{S'} X_3),$$

in the sense that if one of the products makes sense, then so does the other and equality holds. Moreover, all the similar relations of associativity hold for any finite number of schemes.

Proof 1. By the universal property.

2. The last assertion is a consequence of the formula given, by repeated application. The formula is immediate from the universal property. □

Remark We say that products are commutative and associative.

We also have following proposition, some times referred to as the triviality rule:

Proposition 10.22 *For any S-scheme X, $X \times_S S = X$.*

We have some basic constructions of morphisms. First of all, if $f_i : Z \longrightarrow X_i$, $i = 1, 2$, are two S-morphisms then the unique S-morphism given by the universal property of the product is denoted by $(f_1, f_2)_S : Z \longrightarrow X_1 \times_S X_2$. When no confusion is possible we write simply (f_1, f_2). When $g_i : Z_i \longrightarrow X_i$, $i = 1, 2$ are two S-morphisms, then composing with the first and the second projection yield two morphisms

$$f_i : Z_1 \times_S Z_2 \xrightarrow{\mathrm{pr}_{Z_i}} Z_i \xrightarrow{g_i} X_i$$

$i = 1, 2$. We then put

$$g_1 \times_S g_2 = (f_1, f_2)_S,$$

in other words,

$$g_1 \times_S g_2 = (g_1 \circ \mathrm{pr}_{Z_1}, g_2 \circ \mathrm{pr}_{Z_2})_S.$$

Whenever we have an S-morphism $f : X \longrightarrow Y$, then we have the *graph* of f, which is defined as the morphism

$$\Gamma_f = (\mathrm{id}_X, f) : X \longrightarrow X \times_S Y.$$

A special case is the diagonal of $X \times_S X$ for an S-scheme X, which is defined as

$$\Delta_{X/S} = \Gamma_{\mathrm{id}_X} : X \longrightarrow X \times_S X.$$

If $\varphi : S' \longrightarrow S$ is a morphism of schemes and $f : X \longrightarrow S$ is a morphism, so X is an S-scheme, then we frequently denote the projection to S' by

$$f_{S'} : X_{S'} = X \times_S S' \longrightarrow S',$$

referring to the morphism and the scheme with the subscript S' as the *extension to S'* of the morphism f or the scheme X, respectively. Bearing in mind that $S \times_S S' = S'$, we have more generally for any morphism $f : X \longrightarrow Y$ the notation $f_{S'} = f \times \mathrm{id}_{S'} : X_{S'} \longrightarrow Y_{S'}$.

This general concept of base extension is transitive in the following sense:

Proposition 10.23 *For two morphisms* $S'' \longrightarrow S' \longrightarrow S$ *we have* $(X_{S'})_{S''} = X_{S''}$, *and the similar relation for morphisms.*

Proof The claim follows by the universal property. Indeed for an S'-scheme Z the mapping

$$\mathrm{Hom}_{S'}(Z, X_{S'}) \longrightarrow \mathrm{Hom}_S(Z, X)$$

$$f \mapsto \mathrm{pr}_X \circ f$$

is bijective, since an S-morphism $g : Z \longrightarrow X$ yields a unique $f = (g, \psi) : Z \longrightarrow X_{S'}$ where $\psi : Z \longrightarrow S'$ is the structure morphism such that $g = \mathrm{pr}_X \circ f$. Repeated application implies, in the situation of the proposition, that

$$\mathrm{Hom}_{S''}(Z, (X_{S'})_{S''}) = \mathrm{Hom}_S(Z, X) = \mathrm{Hom}_{S''}(Z, X_{S''}),$$

and the claim follows.   □

Along the same lines we have the

Proposition 10.24 1. *The following formula holds*

$$X_{S'} \times_{S'} Y_{S'} = (X \times_S Y)_{S'}.$$

2. *Let* Y *be an* S-scheme, $f : X \longrightarrow Y$ *and* $S' \longrightarrow S$ *morphisms. Then*

$$X_{S'} = X \times_Y Y_{S'},$$

and under this identification the second projection corresponds to $f_{S'}$.

Proof 1. As in the proof of Proposition 10.23 we find that

$$\mathrm{Hom}_{S'}(Z, X_{S'}) \times \mathrm{Hom}_{S'}(Z, Y_{S'})$$

$$= \mathrm{Hom}_S(Z, X) \times \mathrm{Hom}_S(Z, Y)$$

$$= \mathrm{Hom}_S(Z, X \times_S Y) = \mathrm{Hom}_{S'}(Z, (X \times_S Y)_{S'}),$$

and the claim follows.

2. This follows by the associativity of products and Proposition 10.22. Indeed, we get

$$X \times_Y Y_{S'} = X \times_Y (Y \times_S S') = (X \times_Y Y) \times_S S') = X \times_S S' = X_{S'}$$

and the claim follows.   □

As an application of these ideas, we prove the following:

Proposition 10.25 *If the S-morphisms $f : X \longrightarrow X'$ and $g : Y \longrightarrow Y'$ are monomorphisms, then so is $f \times_S g : X \times_S Y \longrightarrow X' \times_S Y'$. In particular, the property of being a monomorphism is preserved by base extension.*

Proof The latter assertion follows from the former, the identities being monomorphisms. If $h_i : Z \longrightarrow X \times_S Y$, $i = 1, 2$ are two morphisms such that

$$(f \times_S g) \circ h_1 = (f \times_S g) \circ h_2,$$

then the compositions

$$Z \overset{\mathrm{pr}_X \circ h_i}{\longrightarrow} X \overset{f}{\longrightarrow} X'$$

are the same, and so are

$$Z \overset{\mathrm{pr}_Y \circ h_i}{\longrightarrow} Y \overset{g}{\longrightarrow} Y'.$$

Thus since f and g are monomorphisms,

$$\mathrm{pr}_X \circ h_1 = \mathrm{pr}_X \circ h_2 \quad \text{and} \quad \mathrm{pr}_Y \circ h_1 = \mathrm{pr}_Y \circ h_2.$$

Hence $h_1 = h_2$ by the universal property of the product $X \times_S Y$. $\qquad\square$

An even simpler fact is the following observation:

Proposition 10.26 *For any S-morphism $f : X \longrightarrow Y$ the graph $\Gamma_f : X \longrightarrow X \times_S Y$ is a monomorphism*

Proof Suppose that the two compositions

$$Z \overset{h_1}{\underset{h_2}{\rightrightarrows}} X \overset{\Gamma_f}{\longrightarrow} X \times_S Y$$

are the same. Composing with pr_X we then get $h_1 = h_2$. $\qquad\square$

Chapter 11
Properties of Morphisms of Schemes

This chapter contains such important concepts as sheaves of modules and algebras on schemes, quasi coherence and coherence for these sheaves, Spec of a sheaf of algebras on a scheme, reduced structure and the generalization of the field of functions from the classical theory to the algebra of fractions for a scheme. Also treated here are irreducible components of Noetherian schemes, embeddings, the graph and the diagonal in the category of schemes, and finally, separated morphisms and separated schemes.

11.1 Modules and Algebras on Schemes

Definition 11.1 An $\mathcal{O}_X$-module on the scheme X is a sheaf $\mathcal{F}$ of $\mathcal{A}b$ on X, such that for all open $U \subset X$, $\mathcal{F}(U)$ is an $\mathcal{O}_X(U)$-module and all restrictions $\rho_{U,V}^{\mathcal{F}} : \mathcal{F}(U) \longrightarrow \mathcal{F}(V)$ are $\mathcal{O}_X(U) - \mathcal{O}_X(V)$-homomorphisms.

We have seen one example, namely the sheaf $\widetilde{M}$ on $\mathrm{Spec}(A)$, for an A-module M. Moreover, we make the following

Definition 11.2 A homomorphism of $\mathcal{O}_X$-modules on X is a morphism of sheaves of $\mathcal{A}b$,

$$\varphi : \mathcal{F} \longrightarrow \mathcal{G},$$

such that all φ_U are $\mathcal{O}_X(U)$-homomorphisms.

The kernel, denoted $\ker(\varphi)$ is defined as the *sheaf* $\mathcal{K}$ defined by

$$\mathcal{K}(U) = \ker(\varphi_U),$$

and the cokernel $\mathrm{coker}(\varphi)$ is the *associated sheaf* of the presheaf

$$\mathcal{C}(U) = \mathrm{coker}(\varphi_U).$$

These are all $\mathcal{O}_X$-modules on X, as is easily seen.

A. Holme, *A Royal Road to Algebraic Geometry*,
DOI 10.1007/978-3-642-19225-8_11, © Springer-Verlag Berlin Heidelberg 2012

With these notions available we define *exact sequences* in the standard way, and note that the functor

$$\text{Mod}_A \longrightarrow \mathcal{O}_X\text{-modules} \quad \text{on } X = \text{Spec}(A)$$

$$M \mapsto \widetilde{M}$$

is an exact functor.

An $\mathcal{O}_X$-module $\mathcal{F}$ on s scheme X as defined in Definition 11.1 is too general to be really useful. Specifically, the modules $\mathcal{F}(U)$ may vary from one open U to another V without any sufficiently strong mathematical connection between them, analytic or algebraic. The concept of *coherency* formalizes this kind of connection from an open subscheme U to another open subscheme V of X near U:

Definition 11.3 An $\mathcal{O}_X$-module $\mathcal{F}$ on the scheme X is said to be quasi coherent if for all $x \in X$ there exists an open affine $U = \text{Spec}(A)$ containing x such that $\mathcal{F}|U = \widetilde{M}$ for some A-module M. If the M can be taken as finitely presented, then $\mathcal{F}$ is said to be coherent. In particular this is the case if A is Noetherian and N is finitely generated.

Remark 11.1 It might be appropriate to say a few words on notation, which could otherwise appear confusing. The categories of quasi coherent $\mathcal{O}_X$-modules or algebras, etc. are so central in modern algebraic geometry that it would be natural to single them out in some way. In an early version of this book I therefore followed Grothendieck's practice from [15], and capitalized these names as Module, Algebra etc. But for various reasons I decided to abandon this notation.

It is an important fact that the concept of quasi coherency has an appealing local description, namely:

Proposition 11.2 *An $\mathcal{O}_X$-module $\mathcal{F}$ on the scheme X is quasi coherent if and only if for all open affine subschemes of X $U = \text{Spec}(A)$, we have that $\mathcal{F}|U = \widetilde{\mathcal{F}(U)}$.*

This proposition is an immediate consequence of the following more extensive theorem:[1]

Theorem 11.3 *Let $\mathcal{F}$ be an $\mathcal{O}_X$-module on $X = \text{Spec}(A)$. Then the following are equivalent:*

---

[1] We follow [15] and [35].

(a) $\mathcal{F} = \widetilde{M}$ for some A-module M.
(b) There exists a finite open covering $X = \bigcup_{i=1}^{r} D(f_i)$ where $f_i \in A$ such that $\mathcal{F}|D(f_i) = \widetilde{M_i}$ for an A_{f_i}-module $\widetilde{M_i}$.
(c) For all $x \in X$ there exists an open $U \ni x$, possibly infinite indexing sets I_U and J_U and an exact sequence of $\mathcal{O}_U$-modules

$$\mathcal{O}_U^{I_U} \longrightarrow \mathcal{O}_U^{J_U} \longrightarrow \mathcal{F}|U \longrightarrow 0.$$

(d) $\mathcal{F}$ satisfies the following two conditions:
  (i) For each $g \in A$ and each $s \in \mathcal{F}(D(g))$ the section $g^n s$ can be extended to a section $\sigma \in \mathcal{F}(X)$ for some integer n.
  (ii) For each $g \in A$ and each $t \in \mathcal{F}(X)$ which restricts to 0 over $D(g)$, $g^n t = 0$ for some integer n.

Proof We prove the following implications:

$$(a) \implies (b) \iff (c)$$

$$(d)$$

(a) $\Rightarrow$ (b): Take $r = 1$ and $f_1 = 1$.

(b) $\Rightarrow$ (c): We show that the sets $D(f_i)$ work for (c). Replacing A_{f_i} by A and M_i by M it suffices to show that there exists such an exact sequence on X. Let $\{m_i | i \in J\}$ be a set of generators for M, this yields a surjective A-homomorphism $A^J \twoheadrightarrow M$, and repeating this for its kernel we obtain an exact sequence

$$A^I \longrightarrow A^J \longrightarrow M \longrightarrow 0.$$

We then have the diagram

$$\mathcal{O}^I \longrightarrow \mathcal{O}^J \longrightarrow \mathcal{F} = \widetilde{M} \longrightarrow 0$$

and we are done.

(a) $\Rightarrow$ (d): Let $g \in A$ and $s \in \mathcal{F}(D(g)) = M_g$. Then $s = \frac{m}{g^n}$ for some integers n and $m \in M$. Thus $g^n s = \frac{m}{1} \in M_g$, and (i) follows. For (ii), let $g \in A$ and $t \in M$ be such that $\frac{t}{1}$, the restriction of t to $D(g)$, is zero. Then for some integer n we have $g^n t = 0$, and the claim is proven.

(b) $\Rightarrow$ (d): We assume (b), and first prove (ii) in (d). We are given $g \in A$ and $t \in \mathcal{F}(X)$ such that $t|D(g) = 0$. Then for all i we have $t|D(gf_i) = 0$. Since (ii) holds over $D(f_i)$, there exists an integer n_i such that $(gf_i)^{n_i} t|D(f_i) = 0$.

Increasing some n_i if necessary we may assume that all n_i are equal to some n, so $f_i^n g^n t | D(f_i) = 0$ for all i. But in A_{f_i} the element f_i^n is a unit, thus for all i we have $g^n t | D(f_i) = 0$. Hence $g^n t = 0$ by the sheaf-condition.

We next show (i). Since we have shown that (a) $\Rightarrow$ (d) we know that (i) holds for all $\mathcal{F} | D(f_i)$. We apply (i) to $g_i = f_i g \in A_{f_i}$ and $s | D(f_i g), D(f_i g) \subset D(f_i)$, and get an integer n_i and a section $s_i' \in \mathcal{F}(D(f_i))$ which extends $g^{n_i} s | D(f_i g)$ to all of $D(f_i)$. Increasing some of the n_i if necessary we may assume all $n_i = n$. We now have sections s_i' on each of the $D(f_i)$ which extend $g^n s$ over $D(f_i g)$ to $D(f_i)$. If any two of these, say s_i' and s_j', always had the same restriction to $D(f_i f_j)$, then we would be finished. In that case the sections s_i could be glued to a section σ on all of $X = \bigcup_{i=1}^t D(f_i)$. But unfortunately the difference of the restrictions

$$t_{i,j} = s_i' | D(f_i f_j) - s_j' | D(f_i f_j)$$

need not be zero.

However, by assumption $\mathcal{F} | D(f_i) = \widetilde{M_i}$, hence $\mathcal{F} | D(f_i f_j)$ satisfies (a) and thus (d) by what we have proved above. In particular we may apply (ii) to $\mathcal{F} | D(f_i f_j)$ and the element $g_{i,j} = f_i f_j g$. Since the restriction of $t_{i,j}$ to $D(g_{i,j})$ is zero, there exist integers $m_{i,j}$ such that $g_{i,j}^{m_{i,j}} t_{i,j} \in \mathcal{F}(D(f_i f_j))$ is zero. As before we may assume that all these integers $m_{i,j}$ are equal to some m, so $(f_i f_j g)^m t_{i,j} = 0$. But as $f_i f_j$ is s unit over $D(f_i f_j)$, this yields $g^m t_{i,j} = 0$. Hence replacing the sections s_i' by the modified sections $\sigma_i = g^m s_i'$, we get a family of extensions σ_i of $g^{m+n} s_i \in \mathcal{F}(D(f_i g))$ to the larger open subsets $D(f_i)$, which now do agree on the intersections of these open subsets, and which may therefore be glued to a section $\sigma \in \mathcal{F}(X)$, extending $g^{n+m} s$ to a section over X.

(c) $\Rightarrow$ (b): By refining the open covering in (c) we obtain an open covering of X by open subsets $U_i = D(f_i)$, over which we still have an exact sequence

$$\mathcal{O}_{D(f_i)}^{I_i} \longrightarrow \mathcal{O}_{D(f_i)}^{J_i} \longrightarrow \mathcal{F} | D(f_i) \longrightarrow 0.$$

Moreover, a finite number of these open subsets suffice to cover $X = \mathrm{Spec}(A)$. Indeed, the $D(f_i)$ cover $\mathrm{Spec}(A)$ if and only if the ideal $\mathfrak{a}$ generated by all the f_i is not contained in a prime ideal in A. This means that $A = \mathfrak{a}$, thus

$$1 = a_1 f_{i_1} + a_2 f_{i_2} + \cdots + a_1 r f_{i_r}$$

so

$$X = \bigcup_{j=1}^r D(f_{i_j}).$$

By the short exact sequence above we have a right exact sequence

$$A_{f_i}^{I_i} \longrightarrow A_{f_i}^{J_i} \longrightarrow M_i \longrightarrow 0.$$

M_i being the cokernel. But since $M \mapsto \widetilde{M}$ is an exact functor, this yields

$$
\begin{array}{ccccccc}
\widetilde{A^{I_i}_{f_i}} & \longrightarrow & \widetilde{A^{J_i}_{f_i}} & \longrightarrow & \widetilde{M_i} & \longrightarrow & 0 \\
\| & & \| & & & & \\
\mathcal{O}^{I_i}_{D(f_i)} & \longrightarrow & \mathcal{O}^{J_i}_{D(f_i)} & \longrightarrow & \mathcal{F}|D(f_i) & \longrightarrow & 0
\end{array}
$$

Thus the two modules to the right are identified as cokernels of the same homomorphism up to canonical isomorphisms, and thus they are canonically isomorphic.

(d) $\Rightarrow$ (a): Let $M = \mathcal{F}(X)$, and define $\varphi : \widetilde{\mathcal{F}(X)} \longrightarrow \mathcal{F}$ as follows: On a basic open set $D(f) \subset X = \mathrm{Spec}(A)$, let

$$
\varphi_f\left(\frac{m}{f^n}\right) = ((f|D(f))^{-1})^n(m|D(f)).
$$

Clearly φ_f so defined is compatible with restriction from $D(f_1)$ to $D(f_1 f_2)$. Thus there is a unique homomorphism of $\mathcal{O}_X$-modules on X extending the φ_f to all open subsets. To show that φ so defined is an isomorphism, it suffices to show that all φ_f are isomorphisms, i.e. that φ_f is surjective and injective. φ_f is surjective because if $s \in \mathcal{F}(D(f))$ then there exists by (d) (i) an integer n and a section $\sigma \in \mathcal{F}(X)$ which lifts $f^n s$. Then $\varphi_f(\frac{\sigma}{f^n}) = s$. φ_f is injective because if $t \in \mathcal{F}(D(f))$ is such that $\varphi_f(\frac{t}{f^n}) = 0$, then $\varphi_f(\frac{t}{1}) = 0$ as f is a unit over $D(f)$. Then by (d) (ii) there exists an integer n such that $f^n t = 0$, thus $\frac{t}{f^n} = 0$. $\square$

Kernels and cokernels of homomorphisms of quasi coherent $\mathcal{O}_X$-modules on X are again quasi coherent, as one immediately verifies from the local structure as an $\widetilde{A} = \mathcal{O}_X$-module. The following notation is used:

$$
\Gamma(U, \mathcal{F}) = \mathcal{F}(U).
$$

The additive *functor of global sections*

$$
\mathcal{F} \mapsto \Gamma(X, \mathcal{F})
$$

from the category of $\mathcal{O}_X$-modules to the category of $\Gamma(X, \mathcal{O}_X)$-modules plays an important role in algebraic geometry.

The theorem has an immediate corollary:

Corollary 11.4 *On an affine scheme the functor* $\Gamma(X, -)$ *is exact on quasi coherent* $\mathcal{O}_X$*-modules.*

An *ideal* on X is defined as a quasi coherent sub-$\mathcal{O}_X$-module $\mathcal{I}$ of $\mathcal{O}_X$. A quasi coherent $\mathcal{O}_X$-algebra on X, $\mathcal{A}$, is an $\mathcal{O}_X$-module such that all $\mathcal{A}(U)$ are $\mathcal{O}_X(U)$-algebras, and the restriction homomorphisms are homomorphisms of algebras as well. We define ideals in a quasi coherent algebra on X as we did for ideals on X: Quasi-coherent submodules with the usual multiplicative ideal-property over all open subsets.

An important example of an ideal on X is $\mathcal{N}_X$, the ideal of nilpotent elements. This is a special case of the following general proposition, given in [15] II on p. 127:

Proposition 11.5 *Let X be a scheme, and let $\mathcal{B}$ be a quasi coherent $\mathcal{O}_X$-algebra on X. Then there exists a unique quasi coherent sub-$\mathcal{O}_X$-module $\mathcal{N}$ of $\mathcal{B}$ (actually it is an ideal in $\mathcal{B}$) such that for all $x \in X$, $\mathcal{N}_x$ is the nilpotent radical of $\mathcal{B}$. If $X = \mathrm{Spec}(A)$ and consequently $\mathcal{B} = \widetilde{B}$, where B is an A-algebra, then $\mathcal{N} = \widetilde{N}$ where N is the nilpotent radical of the A-algebra B.*

Proof Let $\mathcal{N}$ be the sheaf associated to the presheaf $\mathcal{N}'$ where

$$\mathcal{N}'(U) = \{ f_U \in \mathcal{B}(U) \mid \text{There exists } n \text{ with } f_U^n = 0 \}.$$

For all $x \in X$ we then have

$$\mathcal{N}'_x = \{ f \in \mathcal{B}_x \mid \text{There exists } n \text{ with } f^n = 0 \}$$

(and consequently the same holds for $\mathcal{N}$). Indeed, if $f \in \mathcal{B}_x$ satisfies $f^n = 0$ for some n, then the element comes from an $f_U \in \mathcal{B}(U)$ for some open U with the same property. And the property $f_U^n = 0$ is preserved passing to the inductive limit.

To show that $\mathcal{N}$ is quasi coherent it suffices to prove the last part of the proposition. So we assume that $X = \mathrm{Spec}(A)$, $\mathcal{B} = \widetilde{B}$ and let

$$N = \mathcal{N}'(X) = \{ f \in B \mid \text{There exists } n \text{ with } f^n = 0 \}.$$

For all $a \in A$ we have the restriction mapping

$$\varphi : N \longrightarrow \mathcal{N}'(D(a))$$

and thus

$$\varphi_a : N_a = \widetilde{N}(D(a)) \longrightarrow \mathcal{N}'(D(a)) \quad \text{given by} \quad \frac{f}{a^r} \mapsto (a^{-1})^n (f|D(a)).$$

This yields a morphism of presheaves

$$\Psi' : \widetilde{N} \longrightarrow \mathcal{N}'$$

and hence, composing with the canonical morphism from a presheaf to the associated sheaf, a homomorphism of $\mathcal{O}_X$-modules

$$\Psi : \widetilde{N} \longrightarrow \mathcal{N}.$$

Since $\widetilde{N}_x = (A - \mathfrak{p}(x))^{-1} N$ where $\mathfrak{p}$ is the prime ideal corresponding to the point $x \in \operatorname{Spec}(A)$, it follows easily that Ψ_x is an isomorphism for all $x \in X$, and the proposition is proven. $\qquad\square$

For all open subsets U in X we let $\mathcal{N}_X(U)$ be the ideal of nilpotent elements in $\mathcal{O}_X(U)$, the nilpotent radical of that ring. We then obtain a quasi coherent ideal on X.

The quotient of a quasi coherent $\mathcal{O}_X$-algebra on X by a quasi coherent ideal, is again a quasi coherent $\mathcal{O}_X$-algebra on X. The usual algebraic operations of sum, intersection, radical etc. also carry over to this general situation.

11.2 Spec of an $\mathcal{O}_X$-Algebra on a Scheme X

An extensive treatment of the material presented in this section may be found in [15] II, starting on p. 6.

Let $\mathcal{A}$ be a quasi coherent $\mathcal{O}_X$-algebra on a scheme X. For all open affine subschemes U of X we then have $\mathcal{A}|U = \widetilde{\mathcal{A}(U)}$. Let $Z_{(U)} = \operatorname{Spec}(\mathcal{A}(U))$. We then have morphisms $\pi_U : Z_U \longrightarrow U$, and if $U \supset V$ are two open affine subschemes, then we have the obvious commutative diagram.

Proposition 11.6 *The $\pi_U : Z_U \longrightarrow U$ may be glued to $\pi : Z \longrightarrow X$, in such a way that $Z_{(U)}$ is identified with the open subset $\pi^{-1}(U) \subset Z$, and the open subsets*

$$\pi_U^{-1}(U \cap V) \quad and \quad \pi_V^{-1}(U \cap V)$$

are identified.

Proof The proof makes essential use of the quasi coherent property, and proceeds along similar lines to the construction of the product of S-schemes. The details are left for the reader to look up in [15] II. $\qquad\square$

Definition 11.4 The scheme Z of Proposition 11.6 is denoted by $\operatorname{Spec}(\mathcal{A})$.

Remark 11.7 The procedure of Proposition 11.6 may be applied, with the obvious necessary adjustments of notation, to any quasi coherent graded $\mathcal{O}_X$-algebra $\mathcal{S}$ on X to yield a morphism $\operatorname{Proj}(\mathcal{S}) \longrightarrow X$. The details are left for the reader to look up in the above given reference.

We note the following general fact:

Proposition 11.8 *Let $f : X \longrightarrow Y$ be a morphism and let $\mathcal{A}$ be an algebra on X. Then $f_*(\mathcal{A})$ is an algebra on Y via $\theta : \mathcal{O}_Y \longrightarrow f_*(\mathcal{O}_X)$, and*

$$\operatorname{Spec}(f_*(\mathcal{A})) \longrightarrow Y$$

is the composition

$$\mathrm{Spec}(\mathcal{A}) \longrightarrow X \longrightarrow Y.$$

Proof It suffices to check this locally on Y, so we may assume that $Y = \mathrm{Spec}(A)$. Then we may assume that $X = \mathrm{Spec}(B)$, since the equality of two given morphisms is a local question on the source scheme. But in the affine case the claim is obvious. $\square$

If $\mathcal{I}$ is an ideal on X, then the morphism

$$i = \pi : \mathrm{Spec}(\mathcal{O}_X / \mathcal{I}) \longrightarrow \mathrm{Spec}(\mathcal{O}_X) = X$$

is called *a canonical closed embedding*. A composition of an isomorphism and a canonical closed embedding is referred to as *a closed embedding*. An *open embedding* is just the inclusion of an open subscheme. For all practical purposes any closed embedding may be regarded as a canonical one.

We immediately note that as a mapping of topological spaces, a closed embedding $i : Z \longrightarrow X$ identifies the source space with a closed subset of the target space. The corresponding $\theta : \mathcal{O}_X \longrightarrow i_*(\mathcal{O}_Z)$ is surjective as a morphism of sheaves.

We may define the *polynomial algebra in* $X_1, \ldots, X_N$ over a scheme X, denoted by

$$\mathcal{A} = \mathcal{O}_X[X_1, \ldots, X_N]$$

by putting $\mathcal{A}(U) = \mathcal{O}_X(U)[X_1, \ldots, X_N]$ for all open subschemes $U \subset X$. This is an $\mathcal{O}_X$-algebra on X, as is immediately verified. We put

$$\mathbb{A}_X^N = \mathrm{Spec}(\mathcal{O}_X[X_1, \ldots, X_N]),$$

referring to this scheme as the affine N-space over X. When $N = 1$ we speak of the affine line over X, etc.

11.3 Reduced Schemes and the Reduced Subscheme X_{red} of X

An important example of a closed embedding is the case when $\mathcal{I} = \mathcal{N}_X$. In that case the source scheme is denoted by X_{red}, and the closed embedding is a homeomorphism as a mapping of topological spaces.

Since forming the nilpotent radical is compatible with localization, it follows that X_{red} is *reduced* in the following sense:

Definition 11.5 A scheme X is said to be reduced if all its local rings are without nilpotent elements.

We have the following:

Proposition 11.9 *The assignment*

$$X \mapsto X_{\text{red}}$$

is a covariant functor from the category of schemes to itself.

Proof We verify that a morphism $f : X \longrightarrow Y$ gives rise to a morphism f_{red} which makes the following diagram commutative:

$$
\begin{array}{ccc}
X_{\text{red}} & \xrightarrow{\ f_{\text{red}}\ } & Y_{\text{red}} \\
{\scriptstyle i}\downarrow & & \downarrow{\scriptstyle j} \\
X & \xrightarrow{\ f\ } & Y
\end{array}
$$

where i and j are the closed embeddings. This is easily reduced to the fact that whenever $\varphi : A \longrightarrow B$ is a homomorphism of commutative rings, then the nilpotent radical $\mathfrak{N}_A$ of A is mapped into the nilpotent radical $\mathfrak{N}_B$ of B, and thus there is a ring homomorphism φ_{red} which makes the diagram below commutative:

$$
\begin{array}{ccc}
A/\mathfrak{N}_A & \xrightarrow{\ \varphi_{\text{red}}\ } & B/\mathfrak{N}_B \\
{\scriptstyle \tau_A}\uparrow & & \uparrow{\scriptstyle \tau_B} \\
A & \xrightarrow{\ \varphi\ } & B
\end{array}
$$

Instead of piecing this together to obtain the globally defined morphism f_{red}, perfectly feasible as this may be, we now proceed by observing that the diagram above holds with $\mathcal{A}$ and $\mathcal{B}$ instead of A and B, i.e. for quasi coherent algebras on X, and Spec on such algebras is a contravariant functor.   □

11.4 Reduced and Irreducible Schemes and the Field of Functions

A scheme X is said to be *irreducible* if it is not the union of two proper closed subsets:

Definition 11.6 The scheme X is said to be irreducible if

$$X = X_1 \cup X_2 \quad \text{where}$$
$$X_1 \text{ and } X_2 \text{ are closed in } X \implies X_1 = X \text{ or } X_2 = X.$$

The concept of an irreducible scheme is particularly powerful when the scheme is also reduced and has the property of being *locally Noetherian*. We make the following

Definition 11.7 A scheme which is a union of open affine subschemes whose coordinate rings are Noetherian is called a *locally Noetherian scheme*. If the union can be taken to be finite, then the scheme is said to be Noetherian.

We have the following:

Proposition 11.10 *Let X be a reduced and irreducible, locally Noetherian, scheme. Then there exists a unique point $x_0 \in X$ such that $\overline{[x_0]} = X$. Moreover, the local ring $\mathcal{O}_{X,x_0}$, which we denote by $K(X)$, is a field, and as $x_0 \in U$ for all non empty open subsets of X there are canonical homomorphisms*

$$\rho_U : \mathcal{O}_X(U) \longrightarrow K(X),$$

which identify these rings as well as the local rings at all points of X with subrings of $K(X)$, in such a way that the restriction homomorphisms from the ring of an open subset to the ring of a smaller open subset are identified with the inclusion mappings. $K(X)$ is the quotient field of all the rings $\mathcal{O}_X(U)$.

Proof Let $U = \mathrm{Spec}(A)$ be an open affine subscheme, where A is Noetherian, and let x_0 be the point which corresponds to the prime ideal $(0) \subset A$ of the integral domain A. Indeed, A is necessarily an integral domain as the existence of more than one minimal prime ideals would yield a decomposition of X as a union of a finite number of proper closed subsets, namely the complement of U and the closures of the points corresponding to the minimal primes of A. Since the local ring at x_0 is without nilpotent elements, and has only one prime ideal, it is a field. The rest of the assertion of the proposition is immediate.   □

Definition 11.8 The field $K(X)$ will be referred to as the field of functions of the scheme X.

11.5 Irreducible Components of Noetherian Schemes

Let X be a topological space. If X may be written as the union of two proper non empty closed subsets, then X is said to be *reducible*. Otherwise X is called *irreducible*.

It now follows easily, by imitating the corresponding fact for the ideals in a Noetherian ring, that

Proposition 11.11 *For a topological space the following are equivalent*:

1. *The set of closed subsets of X satisfy the descending chain condition.*
2. *A collection of closed subsets of X has a minimal element.*
3. *All closed subsets of X may be written as the union of closed irreducible subsets.*

A topological space which satisfies one and hence all of these conditions is referred to as a *Noetherian topological space*. In particular the Noetherian space X itself may be written as a union of irreducible subsets

$$X = X_1 \cup X_2 \cup \cdots \cup X_r,$$

and if we assume that for $i \neq j$, $X_i \nsubseteq X_j$, then the X_i are unique up to rearrangement. These subsets are then referred to as the *irreducible components* of X.

A Noetherian scheme has a Noetherian underlying topological space, as is easily verified. But the latter property is much weaker than being a Noetherian scheme.

11.6 Embeddings, Graphs and the Diagonal

We now know open and closed embeddings. We have the

Definition 11.9 A composition

$$Z \xrightarrow{\ i\ } U \xrightarrow{\ j\ } X$$

where j is an open embedding and i is a closed embedding is referred to as an embedding.

We shall derive several properties of embeddings. We start out with the following:

Proposition 11.12 *Let*

$$f : X \longrightarrow Y \quad and \quad f' : X' \longrightarrow Y'$$

be two S-morphisms which are embeddings. Then so is

$$f \times_S f' : X \times_S X' \longrightarrow Y \times_S Y',$$

and if the two embeddings are open, respectively closed, then so is the product.

Proof Whenever we have S-morphisms

$$X \xrightarrow{f_1} Y_1 \xrightarrow{f_2} Y$$
$$X' \xrightarrow{f'_1} Y'_1 \xrightarrow{f'_2} Y'$$

then

$$(f_2 \circ f_1) \times_S (f'_2 \circ f'_1) = (f_2 \times_S f'_2) \circ (f_1 \times_S f'_1),$$

since they both solve the same universal problem. Hence it suffices to prove the assertions for open and closed embeddings. For open embeddings the claim is obvious, as $U \subset Y$ and $U' \subset Y'$ being two open subschemes yield the open subscheme $U \times_S U' \subset X \times X'$.

For closed embeddings, we may assume that $S = \mathrm{Spec}(A)$, essentially by the same argument used to reduce the existence of $X \times_S X'$ to the case of S being affine. Since the question of being a closed embedding is local on the target space, we may assume that $Y = \mathrm{Spec}(B)$ and $Y' = \mathrm{Spec}(B')$, B and B' being A-algebras. Then we must have $X = \mathrm{Spec}(B/\mathfrak{b})$ and $X' = \mathrm{Spec}(B'/\mathfrak{b}')$, and hence $X \times_S X' = \mathrm{Spec}((B \otimes_A B')/(\mathfrak{b}, \mathfrak{b}'))$. Thus the claim follows. $\square$

In particular it follows from the proposition that being an embedding, open or closed, is preserved by a base extension.

We next deduce several elementary properties of embeddings and monomorphisms.

Proposition 11.13 *All embeddings are monomorphisms.*

Proof This is immediate for open embeddings. For closed embeddings we may assume that the target scheme is affine. Then so is the source scheme. In the situation

$$X \underset{\longrightarrow}{\overset{\longrightarrow}{}} \mathrm{Spec}(B) \hookrightarrow \mathrm{Spec}(A)$$

the two morphisms to the left coincide if they do so on some open covering of X, hence we may assume that $X = \mathrm{Spec}(C)$. We then have the situation

$$C \underset{\longleftarrow}{\overset{\longleftarrow}{}} B \xleftarrow{\varphi} A$$

where φ is surjective, hence an epimorphism in Comm. Thus the claim follows. $\square$

We have seen that the diagonal of an S-scheme, and more generally the graph of any morphism, is a monomorphism. We have a stronger result:

Proposition 11.14 *The diagonal of an S-scheme*

$$\Delta_{X/S} : X \longrightarrow X \times_S X$$

is an embedding of X into $X \times_S X$.[2]

Proof We may assume that $S = \mathrm{Spec}(A)$. Now cover X by open affine subsets,

$$X = \bigcup_{i \in I} U_i, \quad \text{where } U_i = \mathrm{Spec}(B_i).$$

Then if $V = \bigcup_{i \in I} U_i \times_S U_i$, the diagonal factors

$$\Delta_{X/S} = \Gamma_{\mathrm{id}_X} : X \longrightarrow V \hookrightarrow X \times_S X,$$

as is easily seen. We show that the leftmost morphism is a closed embedding. It suffices to show that $U_i \longrightarrow U_i \times_{\mathrm{Spec}(A)} U_i$ is a closed embedding for all $i \in I$. But this is clear, as the morphism

$$\mathrm{Spec}(B) \longrightarrow \mathrm{Spec}(B) \times_{\mathrm{Spec}(A)} \mathrm{Spec}(B) = \mathrm{Spec}(B \otimes_A B)$$

corresponds to the multiplication map

$$B \otimes_A B \longrightarrow B,$$

which is surjective.   □

Let $f : Z \longrightarrow X$ and $g : Z \longrightarrow Y$ be S-morphisms. Then

Proposition 11.15 *The morphism $(f,g)_S$ is the composition*

$$Z \xrightarrow{\Delta_{Z/S}} Z \times_S Z \xrightarrow{f \times_S g} X \times_S Y.$$

Proof The composition solves the same universal problem as does $(f,g)_S$. □

The proposition has the following immediate consequence

Corollary 11.16 *If f and g are embeddings, then so is $(f,g)_S$. If they, as well as the diagonal $\Delta_{Z/S}$ are closed embeddings, then so is $(f,g)_S$.*

Now let X and Y be S-schemes, with structure morphisms $f : X \longrightarrow S$ and $g : Y \longrightarrow S$, and let $\varphi : S \longrightarrow T$ be a morphism, by means of which X

---

[2]Compare with [15] III No. 17, **Err**$_{\mathrm{III}}$ on p. 88. A subscheme is by definition locally closed, a closed subscheme of an open subscheme.

and Y may also be viewed as T-schemes. Denote by p and q the projection morphisms from $X \times_S Y$ to X and Y, respectively. The structure morphism of the S-scheme $X \times_S Y$ is then $\pi = f \circ p = g \circ q$. We now have the canonical morphism

$$(p,q)_T : X \times_S Y \longrightarrow X \times_T Y.$$

We claim the following, cf. [15], I Proposition (5.3.5) on p. 132:

Proposition 11.17 *The following diagram is commutative, and is a product diagram over* $S \times_T S$:

$$
\begin{array}{ccc}
X \times_S Y & \xrightarrow{(p,q)_T} & X \times_T Y \\
{\scriptstyle \pi}\downarrow & & \downarrow{\scriptstyle f \times_T g} \\
S & \xrightarrow[\Delta_{S/T}]{} & S \times_T S
\end{array}
$$

Proof Suppose that we have morphisms h_1 and h_2 making the following diagram commutative:

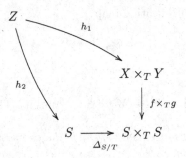

Z is then an S-scheme via h_2 and a T-scheme via h_1 and the latter structure is derived from the former by φ. We need to show that there is a unique h making the following commute:

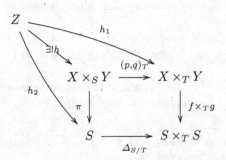

Indeed, by the universal property of $X \times_T Y$ we have $h_1 = (h_3, h_4)_T$, where h_3 and h_4 are T-morphisms from Z to X and Y, respectively. If we can show that these are actually S-morphisms, then we get h as $h = (h_3, h_4)_S$, and the rest will be obvious. We do this for h_3 only, as h_4 is analogous. As $\mathrm{pr}_1 \circ \Delta_{S/T} = \mathrm{id}_S$, we have the commutative diagram

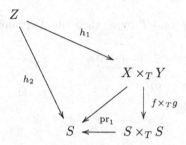

Hence we have the following commutative diagram:

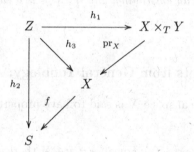

In particular it follows that h_3 is not only a T-morphism, but in fact also an S-morphism.   □

Remark 11.18 See *loc.cit*: Alternatively, we may work as follows, and use a technique which is valid in a number of situations for any category: The object $X \times_S Y$ represents the functor from the category $\mathcal{S}\mathrm{ch}_S$ to $\mathcal{S}\mathrm{et}$

$$P(T) = X(T)_S \times Y(T)_S$$

where

$$X(T)_S = \mathrm{Hom}_S(X, T)$$

denotes the T-valued S-points of X and $\times$ is usual product for sets. Thus one is reduced to showing the corresponding assertion for the category of sets, replacing X, Y, S by $X(Z), Y(Z), S(Z)$, for an arbitrary T-scheme Z. This amounts to showing that the analogous diagram to the one in the proposition is a product diagram in the category $\mathcal{S}\mathrm{et}$. This simple verification is left to the reader.

We note the

Corollary 11.19 *The morphism* $(p, q)_T$ *is an embedding, and if* $\Delta_{S/T}$ *is a closed embedding, then* $(p, q)_T$ *is a closed embedding.*

Proof The claim follows by Proposition 11.14 and Corollary 11.16. □

If we replace S by Y and T by S, then the diagram of Proposition 11.17 becomes

$$
\begin{array}{ccc}
X & \xrightarrow{\ \Gamma_f = (\mathrm{id}_X, f)_S\ } & X \times_S Y \\
f \downarrow & & \downarrow f \times_S \mathrm{id}_Y \\
Y & \xrightarrow[\ \Delta_{Y/S}\ ]{} & Y \times_S Y
\end{array}
$$

We therefore have the following result:

Corollary 11.20 Γ_f *is an embedding, and if* $\Delta_{Y/S}$ *is a closed embedding, then so is* Γ_f.

11.7 Some Concepts from General Topology: A Reminder

Recall that a topological space X is said to have property T_0 if the following holds:

Definition 11.10 (Property T_0) For all $x \neq y \in X$ there either exists an open subset $U \ni x, U \not\ni y$, or there exists an open subset $V \not\ni x, V \ni y$, or both.

The stronger condition of being T_1 is the following:

Definition 11.11 (Property T_1) For all $x \neq y \in X$ there exists an open subset $U \ni x, U \not\ni y$.

Remark Of course it follows that there also exists an open subset $V \not\ni x$, $V \ni y$.

We have the following observation:

Proposition 11.21 *The underlying topological space of a scheme is* T_0, *but in general not* T_1. *However, the subspace consisting of all the closed points of* X *is* T_1.

Proof We may assume that $X = \mathrm{Spec}(A)$, since if the two points x, y are not contained in the same open affine subset, then the condition T_0 is trivially true for them. So let x, y correspond to the primes $\mathfrak{p}, \mathfrak{q} \subset A$. Since they are different, we either have some $a \in \mathfrak{p}, a \notin \mathfrak{q}$, or some $b \notin \mathfrak{p}, b \in \mathfrak{q}$, or both. Then take $V = D(a)$ and $U = D(b)$. The rest of the claim is obvious. $\qquad\square$

The strongest concept of point wise separation in the Axiom of Hausdorff:

Definition 11.12 (Property T_2: The Hausdorff Axiom) The topological space X is said to be Hausdorff if for all $x \neq y \in X$ there exists an open subset $U \ni x$ and an open subset $V \ni y$ such that $U \cap V = \emptyset$.

A topological space which satisfies the Hausdorff Axiom is also called *a separated* topological space. We have the

Proposition 11.22 *A topological space X is separated if and only if the diagonal $\Delta \subset X \times X$ is closed in the product topology.*

Proof Recall that the product topology is the topology given by the base $\mathcal{B}$ consisting of all sets $U \times V$ where U and V are open in X. For the diagonal to be closed, it is necessary and sufficient that $X \times X - \Delta$ be open, thus all points (x, y) in this complement must have an open neighborhood not meeting Δ, or equivalently: Be contained in a set from $\mathcal{B}$ not meeting Δ. If $U \times V$ is this basis open subset, then U and V satisfy the assertion of the Hausdorff Axiom. $\qquad\square$

We finally have the important concept of compactness:

Definition 11.13 (Quasi-Compact and Compact Spaces) A topological space is said to be quasi compact if any open covering of it has a finite sub-covering. If in addition the space is Hausdorff, then it is said to be compact.

We note the following simple but fundamental fact:

Proposition 11.23 *The underlying topological space of* $\mathrm{Spec}(A)$ *is quasi compact.*

Proof Let $\mathrm{Spec}(A) = \bigcup_{i \in I} U_i$. We wish to show that there is a finite subset $\{i_1, i_2, \ldots, i_r\}$ of I such that $\mathrm{Spec}(A) = \bigcup_{\ell=1}^{r} U_{\ell_i}$. Covering all the U_i's by basis open sets $D(a)$, we get a covering of $\mathrm{Spec}(A)$ by such open sets, and to find a finite sub-covering of the former, we need only find one for the latter. Thus we may assume that $U_i = D(a_i)$. Then $\bigcap_{i \in I} V(a_i) = \emptyset$, as the complement of this intersection is the union of all the $D(a_i)$'s. Now $\bigcap_{i \in I} V(a_i) = V(\mathfrak{a})$, where $\mathfrak{a}$ is generated by all the a_is. But as $V(\mathfrak{a}) = \emptyset$, we must have $1 \in \mathfrak{a}$. So there are elements $a_{i_1}, a_{i_2}, \ldots, a_{i_r}$ such that $1 = a_{i_1} b_1 + a_{i_2} b_2 + \cdots + a_{i_r} b_r$. Then the elements $a_{i_1}, a_{i_2}, \ldots, a_{i_r}$ generate the unit ideal A, so $D(a_{i_1}) \cup D(a_{i_2}) \cup \cdots \cup D(a_{i_r}) = \mathrm{Spec}(A)$. $\qquad\square$

The following observation is some times useful:

Proposition 11.24 *Let*

$$f : X \longrightarrow Y$$

be a surjective mapping of T_0 topological spaces. Assume that X and Y have bases $\mathcal{B}_X$ and $\mathcal{B}_Y$ for their topologies such that the mapping f induces a surjective mapping

$$V \mapsto f^{-1}(V)$$

$$\mathcal{B}_Y \longrightarrow \mathcal{B}_X.$$

Then f is a homeomorphism (i.e., is bijective and bi-continuous).

Proof It suffices to show that f is injective. Indeed, it then obviously establishes a bijection between the bases $\mathcal{B}_X$ and $\mathcal{B}_Y$ as well, whence is bicontinuous.

So assume that $x_1 \neq x_2$ are mapped to the same point $y \in Y$. By T_0 we get, if necessary after renumbering the x's, an open subset U in X such that $x_1 \in U$, and $x_2 \notin U$. We may assume $U \in \mathcal{B}_X$, thus there is a $V \in \mathcal{B}_Y$ such that $U = f^{-1}(V)$. But then we also have $x_2 \in U$, a contradiction. $\square$

11.8 Separated Morphisms and Separated Schemes

In analogy with Proposition 11.22 we make the following

Definition 11.14 An S-scheme X is said to be separated if the diagonal

$$\Delta_{X/S} : X \longrightarrow X \times_S X$$

is a closed embedding.

. In this case we also refer to the structure-morphism $\varphi : X \longrightarrow S$ as being a separated morphism. Thus a morphism $f : X \longrightarrow Y$ is called separated if it makes X into a separated Y-scheme.

In the proof of Proposition 11.14, that the diagonal is always an embedding, it was noted that for $S = \mathrm{Spec}(A)$ and $X = \mathrm{Spec}(B)$ where B is an A-algebra, the diagonal $\Delta_{X/S} : X \longrightarrow X \times_S X = \mathrm{Spec}(B \otimes_A B)$ corresponds to the multiplication mapping $B \otimes_A B \longrightarrow B$, and is therefore a closed embedding. Hence we have the

Proposition 11.25 *Any morphism of affine schemes is separated.*

We note the following general result:

Proposition 11.26 *A morphism $f : X \longrightarrow Y$ is separated if and only if for all open $U \subset Y$ the restriction $f|f^{-1}(U) : f^{-1}(U) \longrightarrow U$ is separated. For this to be true, it suffices that there is an open covering of Y with this property.*

Proof With U an open subscheme of Y, we have that

$$V = f^{-1}(U) \times_U f^{-1}(U)$$

is an open subscheme of $X \times_Y X$, and the inverse image of V by the diagonal morphism is U. The claim follows from this.   □

Definition 11.15 (Local Property of a Morphism) Whenever a property of a morphism satisfies the criterion in the proposition above, we say that the property is local on the target scheme.

We collect some observations on separated morphisms:

Proposition 11.27 1. *If $\varphi : S \longrightarrow T$ is a separated morphism and X, Y are S-schemes, then the canonical embedding $X \times_S Y \longrightarrow X \times_T Y$ is a closed embedding.*

2. *If $f : X \longrightarrow Y$ is an S-morphism and Y is separated over S, then the graph Γ_f is a closed embedding.*

3. *Let*

$$h : X \overset{f}{\longrightarrow} Y \overset{g}{\longrightarrow} Z$$

be a closed embedding where g is separated. Then f is a closed embedding.

4. *Let Z be a separated S-scheme and let $g : X \longrightarrow Z$ and $j : X \longrightarrow Y$ be S-morphisms, the latter a closed embedding. Then $(j, g)_S$ is a closed embedding.*

5. *If $\varphi : X \longrightarrow S$ is a separated morphism and $\sigma : S \longrightarrow X$ is a section of φ, i.e. $\varphi \circ \sigma = \mathrm{id}_S$, then σ is a closed embedding.*

Proof 1. Follows by Corollary 11.19. 2. follows by Corollary 11.20. For 3. we have the commutative diagram

Since g is separated, Γ_f is a closed embedding by 2. h is a closed embedding, thus so is $h_Y = h \times_Z \mathrm{id}_Y$. Finally the right pr_Y is an isomorphism. Thus 3. follows. 4. is shown by applying 3. to the situation

$$j : X \overset{(j,g)_S}{\longrightarrow} Y \times_S Z \overset{\mathrm{pr}_Y}{\longrightarrow} Y,$$

and 5. follows by applying 3. to

$$S \xrightarrow{\sigma} X \xrightarrow{\varphi} S.$$

This completes the proof.   □

Remark The proof of 3. above also proves the

Corollary 11.28 (of proof) *If g is a morphism and h is an embedding, then so is f.*

The property of being separated fits into the following general setup, which holds for a variety of other important properties of morphism. It is generally referred to as *la Sorite*. The proposition is given in [15], I, No 4 on p. 136. See also Sect. 13.4.

Proposition 11.29 (i) *Every monomorphism, in particular every embedding, is separated.*

(ii) *The composition of two separated morphisms is again separated.*

(iii) *The product $f \times_S g$ of two separated S-morphisms $f : X \longrightarrow Y$ and $g : X' \longrightarrow Y'$ is again separated.*

(iv) *The property of being separated is preserved by base extensions: If $f : X \longrightarrow Y$ is a separated S-morphism, then so is $f_{S'} : X_{S'} \longrightarrow Y_{S'}$, for any $S' \longrightarrow S$.*

(v) *If the composition $g \circ f$ of two morphisms is separated, then so is f.*

(vi) *$f : X \longrightarrow Y$ is separated if and only if $f_{\mathrm{red}} : X_{\mathrm{red}} \longrightarrow Y_{\mathrm{red}}$ is separated.*

Proof (i) follows since $f : X \longrightarrow Y$ is a monomorphism if and only if $\Delta_{X/Y}$ is an isomorphism. For (ii), let $f : X \longrightarrow Y$ and $g : Y \longrightarrow Z$ be two morphisms. We have the commutative diagram

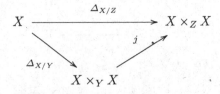

Here the down-right arrow is a closed embedding since f is separated, and the up-right arrow is a closed embedding since g is separated. Thus the composition is a closed embedding, hence $g \circ f$ is separated. Having (i) and (ii), (iii) and (iv) are equivalent. (iv) follows since the diagonal of $X_{S'}$ is the extension to S' of the diagonal of X. (v) was shown above. Finally, (vi) follows by first observing that $X_{\mathrm{red}} \times_Y X_{\mathrm{red}}$ is canonically isomorphic with $X_{\mathrm{red}} \times_{Y_{\mathrm{red}}} X_{\mathrm{red}}$,

$Y_{\mathrm{red}} \hookrightarrow Y$ being a monomorphism. Further, we have the commutative diagram

$$\begin{array}{ccc}
X_{\mathrm{red}} & \xrightarrow{\Delta_{X_{\mathrm{red}}}} & X_{\mathrm{red}} \times_Z X_{\mathrm{red}} \\
j \downarrow & & \downarrow j \times_Y j \\
X & \xrightarrow[\Delta_X]{} & X \times_Y X
\end{array}$$

Since the down-arrows are homeomorphisms on underlying topological spaces, the claim follows.   □

We now have the following important criterion for separatedness, which is useful in general since the property is local on the target scheme. The result is given as Proposition (5.5.6), *loc.cit.* page 138:

Proposition 11.30 (Affine Criterion for Separated) *A morphism* $f : X \longrightarrow S = \mathrm{Spec}(A)$ *is separated if and only if for any two open affine* $U = \mathrm{Spec}(B_1)$ *and* $V = \mathrm{Spec}(B_2)$ *of* X *for which* $U \cap V \neq \emptyset$ *we have* $U \cap V = \mathrm{Spec}(C)$ *where the ring homomorphisms corresponding to the inclusions* ρ_1 *and* ρ_2,

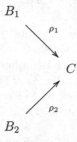

are such that C *is generated as an* A-algebra by $\rho_1(B_1)$ *and* $\rho_2(B_2)$. *It is sufficient that this holds for an open affine covering of* X.

Proof Assume first that X is a separated S-scheme. Then $\Delta_{X/S} : X \longrightarrow X \times_S X$ is a closed embedding. Now $U \times_S V = \mathrm{Spec}(B_1 \otimes_A B_2)$ is an open affine subscheme of $X \times_S X$, hence $\Delta_{X/S}^{-1}(U \times V) = \mathrm{Spec}(C)$, where $C = (B_1 \otimes_A B_2)/\mathfrak{c}$. But we easily see that $\Delta_{X/S}^{-1}(U \times_S V) = U \cap V$, so the statement in the criterion holds. Conversely, assume that there exists an open covering by affine open subschemes so the assertion in the criterion holds for any two members. To show that $\Delta_{X/Y}$ is a closed embedding, we need only check locally on $X \times_S X$: It suffices to show that $\Delta_{X/S}^{-1}(U \times_S V) \longrightarrow U \times_S V$ is a closed embedding for U and V members of the covering given above. But this is clear from the assertion in the criterion.   □

Example 11.1 (Affine Line with the Origin Doubled) Let A be a commutative ring, and put $X_1 = \text{Spec}(B_1)$, where $B_1 = A[t]$ and $X_2 = \text{Spec}(B_2)$, where $B_2 = A[u]$. Of course this is two copies of the affine line over $\text{Spec}(A)$. Further, let $X_{1,2} = D(t)$ and $X_{2,1} = D(u)$. We shall now glue the two affine lines over $\text{Spec}(A)$ in two radically different ways, one way yielding what is known as the *projective line* over $\text{Spec}(A)$, which is a separated scheme over $\text{Spec}(A)$, and the other way of gluing giving us a relatively exotic, non-separated scheme over $\text{Spec}(A)$, which is referred to as the *affine line with the origin doubled*. This is the simplest case of a non-separated scheme over $\text{Spec}(A)$. The first gluing is given by the isomorphisms

$$f_{1,2} : X_{1,2} \longrightarrow X_{2,1}$$

which corresponds to

$$\varphi_{1,2} : A\left[u, \frac{1}{u}\right] \longrightarrow A\left[t, \frac{1}{t}\right], \quad u \mapsto \frac{1}{t},$$

and

$$f_{2,1} : X_{2,1} \longrightarrow X_{1,2}$$

which corresponds to

$$\varphi_{2,1} : A\left[t, \frac{1}{t}\right] \longrightarrow A\left[u, \frac{1}{u}\right], \quad t \mapsto \frac{1}{u}.$$

$X_{1,1} = X_1$, $X_{2,2} = X_2$, moreover $f_{1,1}$ and $f_{2,2}$ are the identities. As the cocycle-condition here is trivially satisfied, we obtain a gluing by these data, temporarily denoted by Z. We have the affine covering $Z = X_1 \cup X_2$, where $U = X_1 \cap X_2 = \text{Spec}(C)$ and $C = A[x, \frac{1}{x}]$. The inclusion morphisms from U to X_1 and from U to X_2 are given by $t \mapsto x$ and $u \mapsto \frac{1}{x}$, respectively. Thus the images of B_1 and B_2 generate C as an A-algebra, and Z is separated over $\text{Spec}(A)$.

On the other hand we may glue by defining the isomorphism $f_{1,2}$ as Spec of

$$\psi_{1,2} : A\left[u, \frac{1}{u}\right] \longrightarrow A\left[t, \frac{1}{t}\right], \quad u \mapsto t,$$

and $f_{2,1}$ as Spec of

$$\psi_{2,1} : A\left[t, \frac{1}{t}\right] \longrightarrow A\left[u, \frac{1}{u}\right], \quad t \mapsto u.$$

Now the resulting scheme Z' still is the union of two open subschemes (isomorphic to) X_1 and X_2, and their intersection is still the open affine subscheme $U = \text{Spec}(A[z, \frac{1}{z}]) \cong X_{1,2}$. But this time the images of B_1 and B_2 only generate the subring $A[z]$ of $A[z, \frac{1}{z}]$, hence Z' is not a separated $\text{Spec}(A)$-scheme.

Chapter 12
Modules, Algebras and Bundles on a Scheme

Here we continue the study of sheaves of modules, algebras and bundles on a scheme. In Sect. 11.1 we defined the categories of $\mathcal{O}_X$-modules and algebras on a scheme X. We now examine their properties more closely.

12.1 The Category of $\mathcal{O}_X$-Modules on a Scheme X

For a commutative ring A we have the categories of A-modules and A-algebras, as well as their graded counterparts. Morphisms, kernels and cokernels, images, direct and inverse images etc. are defined, as well as constructions like tensor products $M_1 \otimes_A M_2 \otimes_A \cdots \otimes_A M_p$, and tensor powers $M^{\otimes n} = T_A^n(M)$, symmetric powers $S_A^n(M)$ which is $T_A^n(M)$ modulo the submodule generated by all elements like $m_1 \otimes m_2 - m_2 \otimes m_1$, etc., exterior powers $\Lambda_A^n(E)$ which is $T_A^n(M)$ modulo the submodule generated by all elements like $m_1 \otimes m_1$, etc., tensor algebras $S_A(M) = \bigoplus_{n \geq 0} T_A^n(M)$, symmetric algebras $\bigoplus_{n \geq 0} S_A^n(M)$ as well as exterior algebras.

Most of these constructions may be carried out in an analogous manner for $\mathcal{O}_X$-modules on a local ringed space, as given in Definition 10.1. However, we shall limit ourselves to the category of quasi coherent sheaves of modules on a scheme X.

For the category of A-modules we also have the internal Hom-construction, $\mathrm{Hom}_A(M, N)$, which is itself an A-module, the tensor product $M \otimes_A N$, etc.

Corresponding to this we have the *sheaf Hom* of two $\mathcal{O}_X$-modules on the scheme X, $\mathcal{H}om_{\mathcal{O}_X}(\mathcal{F}, \mathcal{G})$, tensor product $\mathcal{M} \otimes_X \mathcal{N}$, the dual sheaf $\mathcal{F}^* = \mathcal{H}om_{\mathcal{O}_X}(\mathcal{F}, \mathcal{O}_X)$ and so on.

For locally free sheaves on X all the above constructions yield locally free sheaves.

We have the following:

Proposition 12.1 *All concepts and constructions as well as their basic properties (morphisms, isomorphisms, etc.) over a commutative ring A which are*

A. Holme, *A Royal Road to Algebraic Geometry*,
DOI 10.1007/978-3-642-19225-8_12, © Springer-Verlag Berlin Heidelberg 2012

formulated in the notation of the category of commutative rings and modules over them carry over with the necessary adjustments of notation and language, to the corresponding quasi coherent concepts on a scheme X.

Proof The constructions are first carried out at the level of presheaves: For example, let $\mathcal{F}$ and $\mathcal{G}$ be two (sheaves of) $\mathcal{O}_X$-modules on the scheme X. We form the a presheaf $\mathcal{T}$ by letting

$$\mathcal{T}(U) = \mathcal{F}(U) \otimes_{\mathcal{O}_X(U)} \mathcal{G}(U)$$

for all open U subsets U of X. The associated sheaf is denoted by $\mathcal{F} \otimes \mathcal{G}$. With the usual $\mathcal{O}_X$-homomorphisms this $\mathcal{F} \otimes_{\mathcal{O}_X} \mathcal{G}$ has the usual universal property of a tensor product. A similar construction will yield the sheaf of modules $\mathcal{F} \oplus_{\mathcal{O}_X} \mathcal{G}$ with the usual universal property of a coproduct, except that the presheaf in this case is already a sheaf. By the same procedure we get the sheaf Hom

$$\mathcal{H}om_X(\mathcal{F}, \mathcal{G}).$$

The verifications of these and all the other constructions are simple and will be left to the reader to be verified as the need arises.   □

In particular, a subsheaf of $\mathcal{O}_X$ is a (quasi coherent) *sheaf of ideals* on X, we refer to this as an ideal on X.

If $f : X \longrightarrow Y$ is a morphism, then f_* and f^* are adjoint functors between the categories of $\mathcal{O}_X$-modules and the category of $\mathcal{O}_Y$-modules, via the canonical isomorphism of Abelian groups

$$\mathrm{Hom}_{\mathcal{O}_X}(f^*(\mathcal{G}), \mathcal{F}) \cong \mathrm{Hom}_{\mathcal{O}_Y}(\mathcal{G}, f_*(\mathcal{F})).$$

12.2 Linear Algebra on a Scheme

As we have seen in the previous paragraph, many standard constructions and theory for linear algebra over a commutative ring carry over to the analogies on any scheme, the ring A, or rather, its affine spectrum $\mathrm{Spec}(A)$, being generalized to any scheme X. One approach for carrying out this program is to construct presheaves of modules first, and then take the associated sheaves. This was done for tensor product in the previous paragraph.

Also, in the previous paragraph we used the procedure to introduce the sheaf-Hom, $\mathcal{H}om_{\mathcal{O}_X}(\mathcal{F}, \mathcal{G})$, as well as the dual sheaf $\mathcal{F}^{\vee}$, which is only of real interest if the sheaf $\mathcal{F}$ is locally free.

Along the same lines we introduce the *tensor algebra* of a sheaf of $\mathcal{O}_X$-modules $\mathcal{F}$,

$$\mathcal{T}_{\mathcal{O}_X}(\mathcal{F}) = \bigoplus_{i \geq 0} \mathcal{T}_{\mathcal{O}_X}^i(\mathcal{F})$$

which is the sheaf-theoretic version of the tensor algebra of the A-module M over the commutative ring A,

$$T_A(M) = \bigoplus_{i \geq 0} T_A^i(M)$$

namely

$$\mathcal{T}_{\mathcal{O}_X}(\mathcal{F}) = \bigoplus_{i \geq 0} \mathcal{T}_{\mathcal{O}_X}^i(\mathcal{F}).$$

For a module M over A the symmetric algebra is defined as

$$S_A(M) = T_A(M)/\mathfrak{A}$$

where $\mathfrak{A}$ is the two-sided ideal generated by all elements of the form $x \otimes y - y \otimes x$. The sheaf-theoretic version is defined analogously to the tensor-algebra version.

Moreover, we also have the *exterior algebra*

$$\bigwedge_{\mathcal{O}_X}(\mathcal{F}) = \bigoplus_{i \geq 0} \overset{i}{\bigwedge_{\mathcal{O}_X}}(\mathcal{F})$$

which is the sheaf-theoretic version of the exterior algebra associated to an A-module M over the commutative ring A, defined as the quotient of the tensor algebra $S_A(M)$ by the two sided ideal generated by all elements of the form $x \otimes y - y \otimes x$.

Most, if not all, of the usual properties carry over from linear algebra over a commutative ring to the present more general situation. See [18], Exercise II 5.16 for more details.

12.3 Locally Free Modules on a Scheme

Let X be a locally Noetherian scheme, and let $\mathcal{F}$ be a quasi coherent $\mathcal{O}_X$-module on X.

Definition 12.1 $\mathcal{F}$ is said to free of rank r if $\mathcal{F} \cong \mathcal{O}_X^r$ for some integer r. If this holds over some open subset $U \ni x$ at all points $x \in X$, then $\mathcal{F}$ is said to be locally free of rank r.

We have the following simple result:

Proposition 12.2 $\mathcal{F}$ *is locally free of rank r at x if and only if $\mathcal{F}_x \cong \mathcal{O}_{X,x}^r$.*

Proof One way is obvious. So assume that $\mathcal{F}_x$ is free on the generators $\varphi_1, \ldots, \varphi_r$. They may be extended to sections $f_1, \ldots, f_r \in \mathcal{F}(U)$ for some open affine subset U,

$$x \in U = \operatorname{Spec}(A) \subset X.$$

Let $F = \mathcal{F}(U)$. We then have $\mathcal{F}_U = \widetilde{F}$, and obtain the A- homomorphism $\psi : A^r \longrightarrow F$ which maps $(a_1, \ldots, a_r)$ to $a_1 f_1 + \cdots + a_r f_r$.

Now $\ker(\psi)$ and $\operatorname{coker}(\psi)$ are A-modules, and their supports are closed subsets of $\operatorname{Spec}(A)$ which do not contain the point x. Thus by making U smaller, if necessary, we obtain $\mathcal{F}_U \cong \widetilde{A^r} = \mathcal{O}_U^r$.　　　□

We have the following three results on locally free sheaves on a scheme X, valid even more generally than stated here. Our first proposition is the following, see [17] (5.4.2) and (5.4.3):

Proposition 12.3 *Let $\mathcal{L}$ and $\mathcal{F}$ be two $\mathcal{O}_X$-modules on X. Then there is a functorial canonical homomorphism*

$$\psi : \mathcal{H}om_{\mathcal{O}_X}(\mathcal{L}, \mathcal{O}_X) \otimes_{\mathcal{O}_X} \mathcal{F} \longrightarrow \mathcal{H}om_{\mathcal{O}_X}(\mathcal{L}, \mathcal{F}).$$

If $\mathcal{L}$ or $\mathcal{F}$ is locally free of finite rank, then this is an isomorphism. In particular, if $\mathcal{L}$ is locally free of rank 1, then

$$\mathcal{H}om_{\mathcal{O}_X}(\mathcal{L}, \mathcal{L}) \cong \mathcal{O}_X,$$

so

$$\mathcal{L}^{\vee} \otimes_{\mathcal{O}_X} \mathcal{L} \cong \mathcal{O}_X.$$

Proof We define ψ on the level of presheaves, then pass to the associated sheaves. This process being functorial, the homomorphism of presheaves gives rise to a homomorphism of sheaves. So let $U \subset X$ be open, put $A = \mathcal{O}_X(U)$ and define

$$\psi_U : \operatorname{Hom}_A(\mathcal{L}(U), A) \otimes_A \mathcal{F}(U) \longrightarrow \operatorname{Hom}_A(\mathcal{L}(U), \mathcal{F}(U))$$

as being induced from

$$u \otimes f \mapsto \varphi \in \operatorname{Hom}_A(\mathcal{L}(U), \mathcal{F}(U)) \quad \text{where } \varphi(\ell) = u(\ell)f.$$

If $\mathcal{L}$ is locally free this is an isomorphism, indeed we may check this locally and thus assume $\mathcal{L}(U) = A^r$, in which case we have both sides equal to $\mathcal{F}(U)^r$ and readily verify that ψ_U is an isomorphism. This also proves the last assertion by taking $\mathcal{F} = \mathcal{L}$, as $r = 1$ in that case. $\square$

Secondly, we note that

Proposition 12.4 *Let $\mathcal{E}$, $\mathcal{F}$ and $\mathcal{G}$ be coherent $\mathcal{O}_X$-modules on X. Then there is a functorial canonical homomorphism of $\mathcal{O}_X(X)$-modules*

$$\psi : \mathrm{Hom}_{\mathcal{O}_X}(\mathcal{E} \otimes_{\mathcal{O}_X} \mathcal{F}, \mathcal{G}) \longrightarrow \mathrm{Hom}_{\mathcal{O}_X}(\mathcal{F}, \mathcal{Hom}_{\mathcal{O}_X}(\mathcal{E}, \mathcal{G})).$$

If $\mathcal{E}$ is locally free of finite rank r then this is an isomorphism.

Proof Let

$$h \in \mathrm{Hom}_{\mathcal{O}_X}(\mathcal{E} \otimes_{\mathcal{O}_X} \mathcal{F}, \mathcal{G}).$$

We define the morphism

$$\psi(h) : \mathcal{F} \longrightarrow \mathcal{Hom}_{\mathcal{O}_X}(\mathcal{E}, \mathcal{G})$$

as follows. Let $U = \mathrm{Spec}(A)$ be an open affine subset of X. Define

$$\psi(h)_U : \mathcal{F}(U) \longrightarrow \mathrm{Hom}_A(\mathcal{E}(U), \mathcal{G}(U))$$

by letting $\psi(h)_U(f) : \mathcal{E}(U) \to \mathcal{G}(U)$, where $f \in \mathcal{F}(U)$, be the A-homomorphism which maps $e \in \mathcal{E}(U)$ to $h_U(e \otimes_A f) \in \mathcal{G}(U)$. Since the affine subsets form a base $\mathcal{B}$ for the topology on X, and this construction is compatible with restrictions to smaller subsets of $\mathcal{B}_X$, we can use this to define $\psi(h)_V$ over all open subsets V of X by piecing together sections over an affine open covering of V.

If $\mathcal{E}$ is locally free of finite rank r then this is shown to be an isomorphism as in the proof of Proposition 12.3. $\square$

The third property is known as the *Projection Formula*, cf. [18], II Exercise 5.1(d):

Proposition 12.5 *Let $\mathcal{F}$ be a coherent $\mathcal{O}_X$-module on X and $\mathcal{E}$ be a coherent $\mathcal{O}_Y$-module on Y, and let $f : X \longrightarrow Y$ be a morphism of schemes. Then there is a functorial canonical isomorphism of $\mathcal{O}_Y$-modules*

$$\psi : f_*(\mathcal{F} \otimes_{\mathcal{O}_X} f^*(\mathcal{E})) \longrightarrow f_*(\mathcal{F}) \otimes_{\mathcal{O}_Y} \mathcal{E}.$$

Proof Assume first that $Y = \mathrm{Spec}(A)$ and $X = \mathrm{Spec}(B)$, let f be given by $\varphi : A \longrightarrow B$, and let $\mathcal{E} = \widetilde{E}$ and $\mathcal{F} = \widetilde{F}$ where E is an A-module and F is a B-module. Then the claim amounts to the canonical isomorphism of A-modules

$$(F \otimes_B (E \otimes_A B))_{[\varphi]} \cong F_{[\varphi]} \otimes_A E$$

by noting that $f_*(\mathcal{F})$ corresponds to $F_{[\varphi]}$ and $f^*(\mathcal{G})$ corresponds to $G \otimes_A B$. Since we have canonical isomorphisms $(F \otimes_B (E \otimes_A B)) \cong F \otimes_B (B \otimes_A E) \cong (F \otimes_B B) \otimes_A E \cong F \otimes_A E$, the claim in the proposition follows immediately by Proposition 12.1.   □

12.4 Vector Bundles on a Scheme X

Another term for a locally free sheaf $\mathcal{E}$ of rank r on the scheme X, is a *vector bundle of rank r on X*. The reason for this name is largely historical, but from our point of view we may understand it better by the following considerations. Before we can explain this, we need some new concepts.

Definition 12.2 Let $f : Z \longrightarrow X$ be a morphism of schemes. Then a morphism $\sigma : X \longrightarrow Z$ is called a section of f if $f \circ \sigma = \mathrm{id}_X$, for an open subset $U \subset X$ a section over U is a section of $p_U : p^{-1}(U) \longrightarrow U$.

We also need the term of a *vector bundle* (or an *affine bundle*), the projective counterpart is treated in Sect. 14.5.

Definition 12.3 Let $\mathcal{E}$ be an $\mathcal{O}_X$-module on X. The vector bundle (or affine bundle) associated to $\mathcal{E}$ is the morphism

$$p_{\mathcal{E}} : \mathbb{A}(\mathcal{E}) = \mathrm{Spec}(S_X(\mathcal{E})) \longrightarrow X.$$

The term "bundle" is misleading unless $\mathcal{E}$ is locally free of finite rank, say r. But with this assumption, the term agrees with our intuition.

Moreover, we may define a sheaf $\mathcal{F}$ corresponding to the morphism $p_{\mathcal{E}} : \mathbb{A}(\mathcal{E}) \longrightarrow X$ by letting

$$\mathcal{F}(U) = \{\sigma \,|\, \sigma \text{ is a section of } p_{\mathcal{E}} \text{ over } U\}.$$

Thus we have for all open affine subsets $U = \mathrm{Spec}(A) \subset X$, writing $\mathcal{E}|U = \widetilde{E}$,

$$\sigma : U = \mathrm{Spec}(A) \longrightarrow \mathrm{Spec}(S_A(E))$$

$$\Updownarrow$$

$$A \longleftarrow S_A(E)$$

$$\Updownarrow$$

$$A \longleftarrow E$$

$$\Updownarrow$$

$$\sigma \in \mathrm{Hom}_A(E, A) = \mathcal{E}^*(U).$$

Thus the sheaf $\mathcal{F}$ of local sections of vector bundle $p_{\mathcal{E}} : \mathbb{A}(\mathcal{E}) \longrightarrow X$ is canonically isomorphic to the *dual* of the sheaf $\mathcal{E}$.

This situation has caused an endemic confusion of notation in the fields of algebraic geometry and topology. But it reflects a fact we certainly have to live with.

Chapter 13
More Properties of Morphisms, Scheme Theoretic Image and the "Sorite"

This chapter gives more of the important properties of morphisms, and explains scheme theoretic images, finiteness conditions and the *"Sorite"*.

13.1 Dominant Morphisms and Scheme Theoretic Image of a Subscheme

Definition 13.1 A morphism

$$f : X \longrightarrow Y$$

is said to be dominant if the point set image $f(X)$ is a dense subset of the topological space of X.

We now consider a morphism $f : Z_1 \longrightarrow Z_2$, and consider an embedding $i : X \hookrightarrow Z_1$. Corresponding to f and i we have the homomorphisms

$$\varphi : \mathcal{O}_{Z_2} \longrightarrow f_*(\mathcal{O}_{Z_1})$$
$$\psi : \mathcal{O}_{Z_1} \longrightarrow i_*(\mathcal{O}_X)$$

which yields

$$\mathcal{O}_{Z_2} \xrightarrow{\ \varphi\ } f_*(\mathcal{O}_{Z_1}) \xrightarrow{\ f_*(\psi)\ } f_*(i_*(\mathcal{O}_X))$$

Definition 13.2 (Scheme Theoretic Image) The kernel of this composition is an ideal on Z_2, which defines a closed subscheme of Z_2 referred to as the (closed) scheme theoretic image of X under f. It is denoted by $f(X)$.

A. Holme, *A Royal Road to Algebraic Geometry*,
DOI 10.1007/978-3-642-19225-8_13, © Springer-Verlag Berlin Heidelberg 2012

We must distinguish between the scheme theoretic image and the point set image, however f induces a dominant morphism

$$\overline{f} : X \longrightarrow f(X).$$

If i is a closed embedding and f is proper, see Sect. 15.1, then $\overline{f}$ is even surjective.

13.2 Quasi Compact Morphisms

The following concept is important for the study of certain morphisms:

Definition 13.3 A morphism

$$f : X \longrightarrow Y$$

of schemes is said to be quasi compact if for all open quasi compact subsets V of Y, $f^{-1}(V) \subset X$ is quasi compact.

We immediately verify that if $\mathfrak{B}$ is a base for the topology of X, then f is quasi compact if and only if $f^{-1}(B)$ is quasi compact for all $B \in \mathfrak{B}$. Furthermore, the property is local on the target scheme Y in the sense that

Proposition 13.1 *If $f : X \longrightarrow Y$ is a quasi compact morphism, then for all open subschemes $V \subset Y$ the induced $f_V : f^{-1}(V) \longrightarrow V$ is quasi compact.*

Moreover, if the above holds for all members of an open covering of Y, then f is quasi compact.

The following concept is useful:

Definition 13.4 Whenever a subset $Z \subset Y$ has the property

$$z_0 \in Z \quad \text{and} \quad z \in \overline{[z_0]} \quad \Longrightarrow \quad z \in Z$$

then we say that Z is closed under specialization.

Now note the following property of quasi compact morphisms:

Proposition 13.2 *Let*

$$f : X \longrightarrow Y$$

be a quasi compact morphism of Noetherian schemes.[1] Then $f(X)$ is closed if and only if for all $x_0 \in X$ and $y \in \overline{\{f(x_0)\}}$ there exists $x \in X$ such that $f(x) = y$.

---

[1] The assumption of being Noetherian is not needed, although it simplifies proofs. See [15] I or [18], p. 98.

Proof If $f(X)$ is closed, it is closed under specialization:

Namely, take $f(x_0) \in f(X)$, then $\overline{\{f(x_0)\}} \subset f(X)$, thus for all $y \in \overline{\{f(x_0)\}}$ there is $x \in X$ such that $y = f(x)$.

For the converse, assume that f is quasi compact, and that $f(X)$ is closed under specialization. We wish to show that $f(X)$ is a closed subset of Y. We may replace Y with the closure $\overline{f(X)}$ with its reduced scheme structure, in other words we replace Y by $f(X)_{\mathrm{red}}$ where now $f(X)$ denotes the scheme theoretic image, see Sect. 13.1. We also may replace X by X_{red}, and thus may assume that f is a dominant morphism of reduced schemes. To show is that it is *surjective*.

Let $y \in Y$ be a point, we wish to find $x \in X$ such that $f(x) = y$. We may replace Y by an open affine neighborhood V of y, X by $f^{-1}(V)$ and f by $f_V : f^{-1}(V) \longrightarrow V$, thus we may assume that Y is affine.

Now we conclude as follows: Since f is dominant, all generic points of Y are in $f(X)$. The point y lies in some irreducible component of Y, say $Y_0 = \overline{[y_0]}$ Thus y is a specialization of $y_0 \in f(X)$, hence as $f(Y)$ is stable under specialization, $y \in f(X)$. □

13.3 Finiteness Conditions

We have previously defined *affine spectra of finite type over a field*. This concept is merely a very special case of an extensive set of conditions:

Definition 13.5 A morphism $f : X \longrightarrow Y$ is said to be:

(QC) Quasi compact if the inverse image of all quasi compact open subsets of Y are quasi compact.

(LFT) Locally of finite type if there exists an open affine covering of Y,

$$Y = \bigcup_{i \in I} U_i \quad \text{where } U_i = \mathrm{Spec}(A_i),$$

such that for all $i \in I$,

$$f^{-1}(U_i) = \bigcup_{j \in J_i} V_{i,j} \quad \text{where } V_{i,j} = \mathrm{Spec}(B_{i,j}),$$

and for all i and $j \in J_i$ the restriction of f, $f_{i,j} : V_{i,j} \longrightarrow U_i$ is Spec of a homomorphism $\varphi_{i,j} : A_i \longrightarrow B_{i,j}$ which makes $B_{i,j}$ an A_i-algebra of finite type, i.e., a quotient of a polynomial ring in finitely many variables over A_i.

(FT) Of finite type if all the indexing sets J_i in (LFT) may be taken to be finite sets. (However, I may be an infinite set.)

(A) Affine if $f^{-1}(U_i) = \mathrm{Spec}(B_i)$.

(F) Finite if (A) holds and in addition B_i is finite as an A_i-module.

The situations become simpler if the schemes involved are Noetherian, or even just locally Noetherian. The details are left to the reader.

The picture emerging from the above definition is completed by

Proposition 13.3 *If one of the above conditions holds, then the relevant condition on U_i holds for any open affine subscheme of Y.*

Proof We refer to [15]: I Sect. (6.6.1) for (QC), I Sect. (6.6.2) for (LFT), II, Sect. 1.6 for (A).   □

The proof of the following proposition is a simple exercise:

Proposition 13.4 *A morphism $f : X \longrightarrow Y$ is an affine morphism if and only if there exists a quasi coherent $\mathcal{O}_X$-algebra $\mathcal{A}$ on X such that X and $\mathrm{Spec}(\mathcal{A})$ are isomorphic over Y.*

13.4 The Sorite for Properties of Morphisms

In order to study the different properties of morphisms of schemes, we need a systematic framework. The following simple but clarifying devise is due to Grothendieck, [15], No. 4, Chap. I, 5.5.12:

Proposition 13.5 *Let $\mathbf{P}$ be a property of morphisms of schemes. We consider the following statements about $\mathbf{P}$:*

(i) *Every closed embedding has property $\mathbf{P}$.*

(ii) *The composition of two morphisms which have property $\mathbf{P}$ again has property $\mathbf{P}$.*

(iii) *The product $f \times_S g$ of two S-morphisms $f : X \longrightarrow Y$ and $g : X' \longrightarrow Y'$ which have property $\mathbf{P}$ again has property $\mathbf{P}$.*

(iv) *If $f : X \longrightarrow Y$ is an S-morphism which has property $\mathbf{P}$, and $S' \longrightarrow S$ is a morphism, then the base-extension of f to S', $f_{S'} : X_{S'} \longrightarrow Y_{S'}$ has property $\mathbf{P}$.*

(v) *If the composition $h = g \circ f$ of two morphisms f and g*

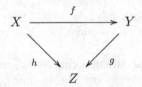

has property $\mathbf{P}$, and if g is a separated morphism, then f has property $\mathbf{P}$.

(vi) $f : X \longrightarrow Y$ *has property* **P** *if and only if* $f_{\text{red}} : X_{\text{red}} \longrightarrow Y_{\text{red}}$ *has property* **P**.

Then we have the following: If (i) *and* (ii) *holds,* (iii) *and* (iv) *are equivalent. Moreover,* (i), (ii) *and* (iii) *together imply* (v) *and* (vi).

indexseparated scheme

Proof The proof follows the same lines as the proof of Proposition 11.29, and may be found in *loc. cit.* on p. 140.   □

As an application of this proposition we have the

Proposition 13.6 *The properties for morphisms listed in Definition* 13.5 *satisfy* (i), (ii) *and* (iii) *and hence* (i)–(vi) *in Proposition* 13.5.

Proof (i) is immediate in each case. (ii) is also clear from the definitions. For (iii) we may assume that S and S' are affine, in which case the verification of (iii) is straightforward.   □

Chapter 14
Projective Schemes and Bundles

The chapter starts by defining Proj(S) for a graded ring $k[X_0, \ldots, X_N]$, the scheme-theoretic version of projective N-space. We then immediately move on to Proj(S) for a graded A-algebra S. It is first defined as a Spec(A)-scheme when A is a commutative ring, then we proceed to extended the definition to the important global case of a graded $\mathcal{A}$-algebra $\mathcal{S}$ on a scheme X. Likewise the module corresponding to a graded $\mathcal{S}$-module $\mathcal{M}$ is defined, as well as the global counterpart for a graded $\mathcal{S}$-module on a scheme X.

14.1 Algebraic Schemes over k and k-Varieties

We consider schemes over the base $S = \text{Spec}(k)$ where k is a (not necessarily algebraically closed) field, and make the

Definition 14.1 A scheme X over $\text{Spec}(k)$ which is separated and of finite type as a $\text{Spec}(k)$-scheme is called an algebraic scheme. If in addition the scheme $X_{\overline{k}}$ is reduced and irreducible, where $\overline{k}$ denotes the algebraic closure of k, then X is called a k-variety.

We have the following fact:

Proposition 14.1 *Let X be a scheme, algebraic over a field k. Then a point $x \in X$ is a closed point if and only if the field $\kappa_X(x) = \mathcal{O}_{X,x}/\mathfrak{m}_{X,x}$ is an algebraic extension of k.*

Proof Given a finite open covering of X, then x is closed if and only if it is closed in each of the open subsets.

Thus we may assume that $X = \text{Spec}(A)$ where $A = k[X_1, \ldots, X_N]/\mathfrak{a}$. The point x is closed if and only if the corresponding prime ideal in A is a maximal ideal, and the claim follows by the theorem below. It is proven as a lemma on p. 165 in vol. 2 of [42], as a key step in proving the *Hilbert Nullstellensatz*:

A. Holme, *A Royal Road to Algebraic Geometry*,
DOI 10.1007/978-3-642-19225-8_14, © Springer-Verlag Berlin Heidelberg 2012

Theorem 14.2 *If a finitely generated integral domain $R = k[x_1, \ldots, x_n]$ over a field k is a field, then the x_i are algebraic over k.*

Assuming this theorem, if x is a closed point, then $R = A/\mathfrak{p}_x$ is a field, hence an algebraic extension of k. If conversely the point x is such that the quotient field of $A/\mathfrak{p}_x$ is an algebraic extension of k, then in particular all elements of $A/\mathfrak{p}_x$ must be algebraic over k. But then the Hilbert Nullstellensatz applied to $R = A/\mathfrak{p}_x$ shows that this ring must be a field, hence that $\mathfrak{p}_x$ is a maximal ideal, and the claim follows. $\square$

Proposition 14.1 has the following

Corollary 14.3 *The point x is closed in the algebraic k-scheme X if and only if it is closed in any open k-subscheme $U \subset X$ containing it.*

Proof The property of being closed is expressed by a property of $\kappa_X(x)$, invariant by passing to an open subscheme containing x. $\square$

Remark The assertion of the corollary is false without the assumption of X being algebraic over a field, and with "subset" instead of "subscheme". As a counterexample, consider the Spec of a local ring of Krull dimension greater than 0, for example the ring $R = k[x]_{(x)}$. $X = \mathrm{Spec}(R)$ has two points, the closed point m corresponding to the maximal ideal $\mathfrak{m} = (x)R$ and the generic point x_0 corresponding to the zero ideal. The subset $U = \{x_0\}$ is open in X, $x_0 \in U$ is closed in U as it is the only point of the subset, but x_0 is not closed in X.

Moreover, we have the

Proposition 14.4 *Let X be an algebraic scheme over the field k. Then the set of closed points in X is dense in X.*

Proof Assume the converse, and let $Y \subset X$ be the closure of the set of all the closed points of X. Then $X - Y = U$ is a non empty open subscheme, thus contains an open affine subscheme $V = \mathrm{Spec}(A)$, where A is a finitely generated algebra over k. But then V has closed points, and by Corollary 14.3 above these are also closed points of X, a contradiction. $\square$

We finally note the

Proposition 14.5 *A k-morphism between schemes algebraic over k maps closed points to closed points.*

Proof Let $f : X \longrightarrow Y$ be the morphism and let $x \in X$ be a closed point. Then the injective k-homomorphism $\kappa(f(x)) \hookrightarrow \kappa(x)$ shows that $\kappa(f(x))$ is an algebraic extension of k. $\square$

14.2 Definition of Proj(S) as a Topological Space

Let S be a graded A-algebra, where as usual A is a commutative ring with 1. We assume that S is positively graded, that is to say that

$$S = S_0 \oplus S_1 \oplus S_2 \oplus \cdots \oplus S_s \oplus \cdots,$$

where all the S_d's are A-modules and the multiplication in S satisfies $S_i S_j \subseteq S_{i+j}$. An element $f \in S$ may be written uniquely as

$$f = f_{\nu_1} + \cdots + f_{\nu_r},$$

where $\nu_1 < \cdots < \nu_r$ and $f_{\nu_i} \in S_{\nu_i}$. The elements f_{ν_i} are referred to as the homogeneous components of f. Recall also that an ideal $\mathfrak{a} \subset S$ is called a homogeneous ideal if, equivalently,

1. If $f \in \mathfrak{a}$ then all $f_{\nu_i} \in \mathfrak{a}$.
2. $\mathfrak{a}$ has a homogeneous set of generators.

Note that the subset $S_+ = S_1 + S_2 + \cdots \subset S$ is a homogeneous ideal. It is referred to as the irrelevant ideal of S.

Example 14.1 1. Let $S = A[X_0, X_1, \ldots, X_N]$. Then $S_0 = A$, and S_d is generated as an $A = S_0$-module by the monomials of degree d.

2. If $\mathfrak{I}$ is a homogeneous ideal in S above then $T = S/\mathfrak{I}$ is another example of a graded A-algebra.

We define the topological space $\mathrm{Proj}(S)$ as the set of all homogeneous prime ideals in $\mathrm{Spec}(S)$ which do not contain S_+, with the induced topology from $\mathrm{Spec}(S)$.

Let $f \in S_d$ be a homogeneous element. Define

$$D_+(f) = D(f) \cap \mathrm{Proj}(S) \quad \text{and} \quad V_+(f) = V(f) \cap \mathrm{Proj}(S).$$

We have the following

Proposition 14.6 *As h runs through the set L of all homogeneous elements in S the subsets $D_+(h)$ constitutes a basis for the topology on $\mathrm{Proj}(S)$.*

Proof If $\mathfrak{p}$ is a homogeneous ideal then the element f is in $\mathfrak{p}$ if and only if all the homogeneous components of f are in $\mathfrak{p}$.   □

As usual we let S_f denote the localization of S in the multiplicatively closed set

$$\Delta(f) = \{1, f, f^2, \ldots, f^r, \ldots\}$$

when f is not a nilpotent element. If f is a homogeneous element of S, say $f \in S_d$, then S_f is a graded A-algebra, but in this case graded by $\mathbb{Z}$, in the sense that

$$S_f = \left\{ \frac{g}{f^n} \;\middle|\; g \in S_m, m = 0, 1, 2, \ldots, n = 0, 1, 2, \ldots \right\}$$

$$= \cdots \oplus (S_f)_{-2} \oplus (S_f)_{-1} \oplus (S_f)_0 \oplus (S_f)_1 \oplus (S_f)_2 \oplus \cdots.$$

The homogeneous piece of degree zero is of particular interest, we put

$$S_{(f)} = (S_f)_0 = \left\{ \frac{g}{f^n} \;\middle|\; g \in S_{nd} \right\}.$$

We now define a mapping of topological spaces

$$\psi_f : D_+(f) \longrightarrow \mathrm{Spec}(S_{(f)})$$

by

$$\mathfrak{p} \mapsto \mathfrak{q} = \left\{ \frac{g}{f^n} \;\middle|\; g \in \mathfrak{p}_{nd} \right\}.$$

We have to show that $\mathfrak{q}$ is a prime ideal in $S_{(f)}$. It clearly is a subset of $S_{(f)}$, to show it's an additive subgroup it suffices to show it's closed under subtraction. Let $\frac{g_1}{f^n}, \frac{g_2}{f^m} \in \mathfrak{q}$. Then $g_1 \in \mathfrak{p}_{nd}$ and $g_2 \in \mathfrak{p}_{md}$, thus $f^m g_1 - f^n g_2 \in \mathfrak{p}_{dm+dn}$ hence

$$\frac{f^m g_1 - f^n g_2}{f^{m+n}} = \frac{g_1}{f^n} - \frac{g_2}{f^m} \in \mathfrak{q}.$$

The multiplicative property is also immediate, thus $\mathfrak{q}$ is at least an ideal in $S_{(f)}$. To show that $\mathfrak{q}$ is prime, assume that

$$\left(\frac{g_1}{f^n} \right) \left(\frac{g_2}{f^m} \right) = \frac{g_1 g_2}{f^{m+n}} = \frac{G}{f^N} \quad \text{where } G \in \mathfrak{p}_{Nd}.$$

Then there exists r such that

$$f^r (f^N g_1 g_2 - f^{m+n} G) = 0,$$

thus $f^{r+N} g_1 g_2 = f^{m+n} G \in \mathfrak{p}$. Since $f \notin \mathfrak{p}$, we get $g_1 g_2 \in \mathfrak{p}$, thus g_1 or $g_2 \in \mathfrak{p}$, and the claim follows.

We now have the

Proposition 14.7 ψ_f *is a homeomorphism of topological spaces.*

Proof By Proposition 11.24 it suffices to show that ψ_f is a surjective mapping, and that it establishes a surjection from a basis for the topology of $\mathrm{Spec}(S_{(f)})$ to a basis for the topology on $D_+(f)$. We first show surjectivity of ψ_f.

So let $\mathfrak{q}$ be a prime ideal in $S_{(f)}$. For all $n \geq 0$ let

$$\mathfrak{p}_n = \left\{ g \in S_n \;\middle|\; \frac{g^d}{f^n} \in \mathfrak{q} \right\}.$$

To show is that

$$\mathfrak{p} = \mathfrak{p}_0 \oplus \mathfrak{p}_1 \oplus \cdots \oplus \mathfrak{p}_d \oplus \cdots$$

is a homogeneous prime such that $\psi_f(\mathfrak{p}) = \mathfrak{q}$.

We first show that $\mathfrak{p}$ is, equivalently that for all n $\mathfrak{p}_n$ is, an additive subgroup of S_+, and do so by showing that it is closed under subtraction. It may come as a slight surprise that this argument needs $\mathfrak{q}$ to be a radical ideal, which is OK as it is actually a prime. Let $g_1, g_2 \in \mathfrak{p}_n$, i.e., $\frac{g_1^d}{f^n}$ and $\frac{g_2^d}{f^n} \in \mathfrak{q}$. Expanding by the binomial formula we then find that $\frac{(g_1-g_2)^{2d}}{f^{2n}} \in \mathfrak{q}$, thus $\frac{(g_1-g_2)^d}{f^n} \in \mathfrak{q}$, as $\mathfrak{q}$ is prime and hence radical. Thus $g_1 - g_2 \in \mathfrak{p}_n$.

For the multiplicative property, it evidently suffices to show that $\mathfrak{p}_n S_m \subset \mathfrak{p}_{n+m}$. This is completely straightforward.

We now have that $\mathfrak{p}$ is a homogeneous ideal in S, to show that it is prime we need to show that the graded A-algebra $T = S/\mathfrak{p}$ is without zero-divisors. Clearly, it suffices to show that there are no homogeneous ones.

For this, assume g_1 and g_2 to be elements of S_m and S_n, respectively, such that $g_1 g_2 \in \mathfrak{p}_{m+n}$ while $g_1 \notin \mathfrak{p}_m$ and $g_2 \notin \mathfrak{p}_n$. But then $\frac{g_1^d}{f^m}$ and $\frac{g_2^d}{f^n}$ are not in $\mathfrak{q}$, while their product is, a contradiction.

We next show that $\psi_f(\mathfrak{p}) = \mathfrak{q}$. We have

$$\psi_f(\mathfrak{p}) = \left\{ \frac{g}{f^n} \;\middle|\; g \in \mathfrak{p}_{nd} \right\}$$

and by definition

$$g \in \mathfrak{p}_{nd} \quad \Longleftrightarrow \quad \frac{g^d}{f^{nd}} \in \mathfrak{q}$$

which as $\mathfrak{q}$ is radical is equivalent to $\frac{g}{f^n} \in \mathfrak{q}$, and the claim follows.

Finally we show that the mapping

$$V \mapsto \psi_f^{-1}(V)$$

maps the basis for $\mathrm{Spec}(S_{(f)})$

$$\mathcal{B}_1 = \left\{ D\left(\frac{g}{f^n} \right) \;\middle|\; g \in S_{dn}, n = 0, 1, 2, \ldots \right\}$$

onto the basis for the topology on $D_+(f)$,

$$\mathcal{B}_2 = \{ D_+(gf) \,|\, g \in S_{dn}, n = 0, 1, 2, \ldots \}.$$

As evidently $\psi_f^{-1}(D(\frac{g}{f^n})) = D_+(gf)$, we need only show that $\mathcal{B}_2$ is a basis for the topology on $D_+(f)$. Now all the sets $D_+(h)$, as h runs through the homogeneous elements on S, form a base for the topology on $\mathrm{Proj}(S)$. Thus the sets $D_+(hf)$ constitute a base for the topology on $D_+(f)$.

This completes the proof of the proposition. $\qquad\qquad\qquad\qquad\qquad\square$

14.3 The Spec(A)-Scheme Proj(S) and Its Sheaves of Modules

Let $M = \bigoplus_{n \in \mathbb{Z}} M_n$ be a graded module over the graded A-algebra S. In most cases one encounters at the present level of sophistication the following *Finiteness Condition* holds:

$$S_0 = A, \quad \text{and} \quad S \text{ is generated as an } A\text{-algebra by } S_1. \qquad (14.1)$$

We shall assume this simplifying property in the following. The general theory is available by straightforward modification of the present text, best carried out consulting Chap. II of [15].

Define a sheaf of Comm on the topological space $\mathrm{Proj}(S)$ introduced in the previous paragraph, proceeding as follows:

Let $f \in S_d$, on $\mathrm{Spec}(S_{(f)})$ we put $\mathcal{M}_f = \widetilde{M_{(f)}}$, where $M_{(f)}$ is the homogeneous part of degree zero in M_f, which evidently is an $S_{(d)}$-module. By the homeomorphisms ψ_f this sheaf is transported to $D_+(f)$, denoted by $\widetilde{M}_{(f)}$. The canonical isomorphisms for $f \in S_d$ and $g \in S_e$,

$$S_{(fg)} \cong (S_{(f)})_{\frac{g^d}{f^e}} \quad \text{and} \quad M_{(fg)} \cong (M_{(f)})_{\frac{g^d}{f^e}}$$

identifies $\widetilde{M}_{fg}$ with the restriction of $\widetilde{M}_f$ to $D(\frac{g^d}{f^e})$. Thus we may glue the $\widetilde{M}_f$ to a sheaf on all of $\mathrm{Proj}(S)$, which we denote by $\widetilde{M}$, using Lemma 10.18 after checking the cocycle condition. Checking the cocycle condition in this case amounts to a simple computation with fractions.

For the graded S-module M we define another graded module $M(m)$ by shifting the grading: We put

$$M(m)_n = M_{m+n}.$$

Obviously this is again a graded S-module, and we define

$$\widetilde{M}(m) = \widetilde{M(m)}.$$

We now define $\mathcal{O}_{\mathrm{Proj}(S)} = \widetilde{S}$. We obtain a scheme in this way, also denoted by $\mathrm{Proj}(S)$. The open subset $D_+(f)$ is then identified with the affine open subscheme $\mathrm{Spec}(S_{(f)})$ of $\mathrm{Proj}(S)$. Also, $\widetilde{M}$ is an $\mathcal{O}_{\mathrm{Proj}(S)}$-module on $\mathrm{Proj}(S)$.

In particular, if $X = \mathrm{Proj}(S)$ then

$$\mathcal{O}_X(m) = \widetilde{S(m)}.$$

Moreover, as is easily verified using Lemma 10.18 as above, the morphisms

$$\pi_f : \mathrm{Spec}(S_{(f)}) \longrightarrow \mathrm{Spec}(A)$$

glue to a morphism

$$\pi : \mathrm{Proj}(S) \longrightarrow \mathrm{Spec}(A),$$

and thus Proj(S) becomes a Spec(A)-scheme. This Spec(A) scheme is indeed a *separated scheme*, and thus a Spec(A)-*scheme* in Grothendiecks original notation from [15], and not just what he called a *prescheme* there: We have the

Proposition 14.8 $\pi : \mathrm{Proj}(S) \longrightarrow \mathrm{Spec}(A)$ *is a separated morphism.*

Proof By the affine criterion for separated morphisms, Proposition 11.30. □

We also note the

Proposition 14.9 *The* $\mathcal{O}_X$-*module* $\widetilde{M}$ *defined above on* $X = \mathrm{Proj}(S)$ *is quasi-coherent, and if M is finitely presented it is coherent. In particular this holds if M is finitely generated, A is Noetherian and S is a quotient of some polynomial ring over A by a homogeneous ideal,* $S = k[X_0, \dots, X_N]/I$.

Proof Checking this is a simple exercise. □

Now let S be as above, and let r be a positive integer. We define a new graded ring $S^{(r)}$ as follows:

$$S_m^{(r)} = S_{rm} \quad \text{and} \quad S^{(r)} = \bigoplus_{m \in \mathbb{Z}} S_{rm}.$$

This is a graded ring, and we find, in view of the Finiteness Condition 14.1, that $\mathrm{Proj}(S) \cong \mathrm{Proj}(S^{(r)})$. The details of this easy verification is left to the reader. The result is important for understanding the *Segre embeddings*, which we come to in Sect. 15.5.

Now let $X = \mathrm{Proj}(S)$, where S is a an A-algebra, finitely generated by S_1 as A-algebra. Let $X = \mathrm{Proj}(S)$. We define a functor Γ_* from the category of quasi coherent sheaves of $\mathcal{O}_X$-modules $\mathcal{F}$ to graded S-modules as follows:

$$\Gamma_*(\mathcal{F}) = \bigoplus_{n \in \mathbb{Z}} \Gamma(X, \mathcal{F}(n)).$$

This is a graded module over S, indeed it is a graded module over $\Gamma_*(\mathcal{O}_X)$ and thus over S via the canonical $S \longrightarrow \Gamma_*(\mathcal{O}_X)$. The following proposition is easily verified, and left as an exercise:[1]

Proposition 14.10 *There is a canonical and functorial isomorphism*

$$\widetilde{\Gamma_*(\mathcal{F})} \overset{\cong}{\longrightarrow} \mathcal{F}.$$

14.4 Proj of a Graded $\mathcal{O}_X$-Algebra on X

From now on we assume that all $\mathcal{O}_X$-modules and algebras on a scheme X be quasi coherent. Strictly speaking, we do not need to assume that X be separated, but as this frequently simplifies arguments, there is no harm in making this assumption as well.

We have defined the morphism $Y = \mathrm{Spec}(\mathcal{A}) \longrightarrow X$, where $\mathcal{A}$ is an $\mathcal{O}_X$-algebra on the scheme X. We did so by gluing the $\mathrm{Spec}(\mathcal{A}(U)) = Y_U$ for all $U \subset X$ which are open and affine.

Moreover, if $\mathcal{M}$ is an $\mathcal{O}_X$-*module* on X, we have the $\mathcal{O}_Y$-*module* $\widetilde{\mathcal{M}}$ on Y, defined by gluing all $\widetilde{\mathcal{M}(U)}$ on $\mathrm{Spec}(\mathcal{A}(U)) = Y_U$ for all $U \subset X$ which are open and affine.

We similarly define $Z = \mathrm{Proj}(\mathcal{S}) \longrightarrow X$, where $\mathcal{S}$ is a graded O_X-algebra on the scheme X. We do so by gluing the $\mathrm{Proj}(\mathcal{S}(U)) = Z_U$ for all $U \subset X$ which are open and affine.

We extend this to the graded case and obtain the following important notion:

Let $\mathcal{M}$ be a graded quasi coherent $\mathcal{O}_X$-module on X, we define the $\mathcal{O}_Z$-module $\widetilde{\mathcal{M}}$ on Z by gluing all $\widetilde{\mathcal{M}(U)}$ on $\mathrm{Proj}(\mathcal{S}(U)) = Z_U$ for all $U \subset X$ which are open and affine.

All general properties deduced for $\pi : \mathrm{Proj}(S) \longrightarrow \mathrm{Spec}(A)$ and for $\widetilde{M}$ in this and the previous section hold for the corresponding global constructions $\pi : \mathrm{Proj}(\mathcal{S}) \longrightarrow X$ and $\widetilde{\mathcal{M}}$.

14.5 Projective Bundles and the Projective N-space over a Scheme

Let $\mathcal{F}$ be a quasi coherent $\mathcal{O}_X$- module on the locally Noetherian scheme X. Form the symmetric algebra $\mathrm{Sym}_{\mathcal{O}_X}(\mathcal{F})$, it will be denoted by $S_X(\mathcal{F})$ from now on.

---

[1]See, e.g. [18], p. 119.

For a scheme X and $\mathcal{O}_X$-algebra $\mathcal{A}$, we use Sect. 14.4 and first define the $\mathcal{O}_X$-algebra

$$\mathcal{B} = \mathcal{A}[X_1, X_2, \ldots, X_N]$$

by first for each open affine $U = \mathrm{Spec}(A)$ letting $\mathcal{B}(U) = A[X_1, \ldots, X_N]$, and then extending to all open subsets of X in the standard way.

Moreover, we make the

Definition 14.2 The scheme $\mathrm{Proj}(S_X(\mathcal{F}))$ is denoted by $\mathbb{P}(\mathcal{F})$, and referred to as the projective bundle of $\mathcal{F}$ on X.[2] We also write

$$\mathbb{A}_X^N = \mathrm{Spec}(\mathcal{O}_X[X_1, X_2, \ldots, X_N]),$$

and

$$\mathbb{P}_X^N = \mathrm{Proj}(\mathcal{O}_X[X_0, X_1, X_2, \ldots, X_N]).$$

A closed subscheme Z of $\mathbb{P}_X^N$ is referred to as a *projective scheme over* X, if $X = \mathrm{Spec}(A)$ then such a scheme is $\mathrm{Proj}(S)$ for some finitely generated graded A-algebra S, and X is referred to as a projective A-scheme. Open subschemes of projective schemes over X are referred to as quasi-projective over X.

Whenever we have a surjective homomorphism

$$\varphi : \mathcal{E} \twoheadrightarrow \mathcal{F}$$

of coherent $\mathcal{O}_X$-modules on X, we get a closed embedding

$$\mathbb{P}(\mathcal{F}) \hookrightarrow \mathbb{P}(\mathcal{E}).$$

---

[2] Some references, such as [12], use this designation for the scheme $\mathrm{Proj}(S_X(\mathcal{F}^\vee))$.

Chapter 15
Further Properties of Morphisms

The chapter starts by treating affine and projective morphisms, then proceeds to proper morphisms. The important valuational criteria for separated and for proper are then given. Very ample sheaves and the Segre embedding are treated, in the context of tensor products of graded S-modules.

15.1 Affine and Projective Morphisms

We now introduce the notion of a *projective morphism*:

Definition 15.1 A morphism $f : Z \longrightarrow X$ is said to be projective if there is a coherent $\mathcal{O}_X$-module on X such that f factors through a closed embedding i

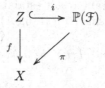

Evidently $f : Z \longrightarrow X$ is projective if and only if $Z \cong \mathrm{Proj}(\mathcal{S})$ where $\mathcal{S}$ is a graded $\mathcal{O}_X$-module on X such that $\mathcal{S}$ is generated by $\mathcal{S}_1$.

If moreover $X = \mathrm{Spec}(A)$, then $f : Z \longrightarrow X = \mathrm{Spec}(A)$ is projective if and only if it factors through a closed embedding $Z \hookrightarrow \mathbb{P}^N_A$.

Projective morphisms are conceptually simple, but difficult to work with in general. One problem is that the property of being *projective* for a morphism $X \longrightarrow Y$ is not local on the target space Y. This follows from examples given by M. Nagata and H. Hironaka in the article "Existence theorems for non projective complete algebraic varieties", *Ill. J. Math.*, t. II (1958). The concept of a *proper morphism* is simpler to work with but conceptually more complicated. Recall that a mapping $f : X \longrightarrow Y$ of topological spaces is called *proper* if the base change mapping $f \times \mathrm{id}_Z : X \times Z \longrightarrow Y \times Z$ sending the point (x, z) to $(f(x), z)$ is always closed. If the target space Y is locally compact, then one may show that a continuous mapping f is proper if and only if the inverse image of a compact subset of Y is compact in X.

Definition 15.2 (Proper Morphism) A morphism of schemes

$$f : X \longrightarrow Y$$

is called proper if the following three conditions hold:

(i) f is separated.
(ii) f is of finite type, Definition 13.5, 2.
(iii) f is universally closed, i.e., for all morphisms

$$Z \longrightarrow Y$$

the base extension of f,

$$f_Z : X_Z \longrightarrow Y_Z,$$

maps closed subsets of the topological space of the source scheme to closed subsets of the topological space of the target scheme.

The property of being proper is a local property on the target scheme, and "proper" as property **P** in Proposition 13.5 will satisfy all of (i)–(vi) there. This is not fully the case for projective morphisms, as (ii) does not hold in general. The following important result establishes the strong link between proper and projective morphisms ([15] II, 5.6.1):

Theorem 15.1 (Chow's Lemma) *Let Y be a Noetherian scheme and*

$$f : X \longrightarrow Y$$

a proper morphism. Then there exists a scheme X' and a commutative diagram

$$
\begin{array}{ccc}
X & \xleftarrow{\;\;g\;\;} & X' \\
{\scriptstyle f}\downarrow & \swarrow{\scriptstyle h=f\circ g} & \\
Y & &
\end{array}
$$

such that X' is projective over Y, and there is a dense open subset U of X such that g induces an isomorphism $g^{-1}(U) \xrightarrow{\;\cong\;} U$.

Proof of this theorem is given as Exercise 4.10 in Hartshorne's book [18], where the proof is outlined in four steps. This program is carried out in detail in a paper by *Hansheng Diao*, available on the web as [8]. □

Corresponding to the relation between the concepts of complete and projective morphisms, we have the two notions of a projective and a complete variety over a field k. We have the following definition:

Definition 15.3 A k-scheme X is said to be complete if the structure morphism $X \longrightarrow \mathrm{Spec}(k)$ is proper.

15.2 Valuational Criteria for Separated and for Proper

For a more extensive and complete treatment, as well as for more detailed proofs of Theorems 15.2 and 15.3, see [15] II, (7.2.3) and (7.3.8), pp. 141–145.

There is a criterion for when a morphism is separated, in terms of discrete valuation rings:

Theorem 15.2 (Valuational Criterion for Separated) *Let $f : X \longrightarrow Y$ be a morphism locally of finite type, with Y locally Noetherian. Then f is separated if and only if the following criterion holds:*

For all discrete valuation rings V with quotient field K and all morphisms $g : Y' = \mathrm{Spec}(V) \longrightarrow Y$ the canonical mapping

$$\mathrm{Hom}_Y(Y', X) \longrightarrow \mathrm{Hom}_Y(\mathrm{Spec}(K), X)$$

is injective.

Proof We note first that the statement about the discrete valuation ring V amounts to asserting that if two Y-morphisms

$$\mathrm{Spec}(V) \underset{h}{\overset{h'}{\rightrightarrows}} X$$

coincide at the generic point of $\mathrm{Spec}(V)$, then they are equal.

In fact, consider the diagram

Here i is the morphism induced by the inclusions of V into its quotient field K. Composition of i with h and h' amounts to note where h and h' send the generic point of V, and to assert that it be sent to the same point by both, is equivalent to the assertion that the two compositions are the same.

To prove the criterion in this last form, assume first that f be separated, and that h and h' are two morphisms coinciding at the generic point, to show is that $h = h'$. Let

$$H = (h, h')_Y : \mathrm{Spec}(V) \longrightarrow X \times_Y X.$$

Clearly H maps the generic point of $v_0 \in \mathrm{Spec}(V)$ to a point in the diagonal of $X \times_Y X$. Since the closed point $v \in \mathrm{Spec}(V)$, corresponding to the maximal ideal, lies in the closure of $\{v_0\}$, $H(v)$ lies in the closure of $\{H(v_0)\}$, and in particular, as the diagonal is closed since f is assumed separated, $H(v)$ lies on the diagonal. It follows that in addition to $h(v_0) = h'(v_0)$, which we now denote by x_0, we also have $h(v) = h'(v)$ which we denote by x. Then $x \in Z = \overline{\{x_0\}}$. We endow this scheme with its reduced subscheme structure and denote the corresponding closed embedding by $i_Z : Z \hookrightarrow X$. We then find that h and h' factor through ℓ and ℓ', respectively, as follows:

$$\mathrm{Spec}(V) \underset{\ell}{\overset{\ell'}{\rightrightarrows}} Z \overset{}{\underset{i_Z}{\hookrightarrow}} X.$$

Let $\varphi_i : \mathcal{O}_{Z,x} \longrightarrow V = \mathcal{O}_{\mathrm{Spec}(V),v}$ be the local homomorphisms of the local rings which corresponds to h_i at v and x, for $i = 1, 2$. At x_0 the local homomorphism is the injective φ:

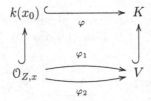

Hence $\varphi_1 = \varphi_2$ and thus $\ell = \ell'$, so $h = h'$. Thus the first part of the proof is complete.

Now assume the criterion. We first show that the diagonal embedding $X \longrightarrow X \times_Y X$ is quasi compact, see Sect. 13.2. Indeed, since Y is locally Noetherian and the morphism is locally of finite type, Proposition 13.1 implies that we may assume Y to be Noetherian and X to be of finite type over Y, hence Noetherian. Now the claim follows by Proposition 13.2.   □

There is also a criterion for when a morphism is proper, in terms of discrete valuation rings:

Theorem 15.3 (Valuational Criterion for Properness) *Let* $f : X \longrightarrow Y$ *be a morphism of finite type. Then f is proper if and only if the following criterion holds:*

For all a discrete valuation rings V with quotient field K and all morphisms $g : Y' = \operatorname{Spec}(V) \longrightarrow Y$ the canonical mapping

$$\operatorname{Hom}_Y(Y', X) \longrightarrow \operatorname{Hom}_Y(\operatorname{Spec}(K), X)$$

is surjective.

15.3 Basic Properties of Proj(S)

Let $Z = \operatorname{Proj}(S)$. Assume that $S_0 = \mathcal{O}_X$ and that S is generated as algebra by S_1. We shall construct a large class of quasi coherent $\mathcal{O}_Z$-modules on Z. In fact, these are all possible quasi coherent modules, but we will neither use nor prove this fact here.

We start out over an affine base. So let $X = \operatorname{Proj}(S)$ where S is a graded A-algebra. Let M be a graded S-module, then the coherent $\mathcal{O}_X$-module $\widetilde{M}$ is defined by gluing all the modules $\widetilde{M_{(g)}}$ on all the $\operatorname{Spec}(S_{(g)})$.

We now define a new graded S-module N by regrading M:

$$N_n = M_{i+n}.$$

We then introduce the following important notation:

$$\widetilde{M}(n) = \widetilde{N}.$$

This construction is carried out in the analogous manner for $Z = \operatorname{Proj}(S)$ and graded modules $\mathcal{M}$ over S, the latter being assumed to have the property prescribed for S above. We get $\widetilde{\mathcal{M}}$ and $\widetilde{\mathcal{M}}(n)$ in this case as well, the verification of the details are left to the reader.

When $\mathcal{M} = S$, we have $\widetilde{S}(n) = \mathcal{O}_Z(n)$. This gives an important class of invertible sheaves on Z.

15.4 Very Ample Sheaves

The following is a special case of a more general situation: We consider a projective scheme $X = \operatorname{Proj}(S)$ over $\operatorname{Spec}(A)$, so assuming S finitely generated over A there is a closed embedding $i : X \hookrightarrow \mathbb{P}_A^N$. We have

$$i^*(\mathcal{O}_{\mathbb{P}_A^N}(n)) = \mathcal{O}_X(n).$$

Letting $S = A[X_0, X_1, \ldots, X_N]$, we have a surjective homomorphism which preserves the grading

$$S^{\oplus(N+1)} \longrightarrow S(1)$$

$$(s_0, s_1, \ldots, s_N) \longrightarrow \sum_{i=0}^{N} X_i s_i.$$

By taking $\widetilde{(\ )}$ we get a surjective homomorphism of $\mathcal{O}_X$-modules on X

$$\mathcal{O}_X^{\oplus(N+1)} \longrightarrow\!\!\!\!\!\rightarrow \mathcal{O}_X(1).$$

Whenever $\mathcal{L}$ is an invertible sheaf on a scheme X, the canonical $\pi : \mathbb{P}(\mathcal{L}) \longrightarrow X$ is an isomorphism, as is easily seen. Thus we get a closed embedding from the surjection above,

$$X = \mathbb{P}(\mathcal{O}_X(1)) \hookrightarrow \mathbb{P}(\mathcal{O}_X^{\oplus(N+1)}) = \mathbb{P}_A^N \times_A X.$$

This holds more generally for an invertible sheaf $\mathcal{L}$ which is globally generated. But the strong property in the present case is that the *composition*

$$X = \mathbb{P}(\mathcal{O}_X(1)) \hookrightarrow \mathbb{P}(\mathcal{O}_X^{\oplus(N+1)}) = \mathbb{P}_A^N \times_A X \xrightarrow{\mathrm{pr}_1} \mathbb{P}_A^N$$

is a closed embedding.

In the above considerations we may replace $\mathrm{Spec}(A)$ by an arbitrary scheme Y, and we obtain the following result, essentially by the same argument:

Proposition 15.4 *Let X be Y-scheme and let $\mathcal{L}$ be an invertible sheaf on X which is globally generated by a finite number $s_0, s_1, \ldots, s_N$ of global sections. Then the surjective homomorphism $\mathcal{O}_X^{\oplus(N+1)} \longrightarrow\!\!\!\!\!\rightarrow \mathcal{L}$ yields a closed embedding*

$$X = \mathbb{P}(\mathcal{L}) \hookrightarrow \mathbb{P}(\mathcal{O}_X^{\oplus(N+1)}) = X \times_Y \mathbb{P}(\mathcal{O}_Y^{\oplus(N+1)}) = X \times_Y \mathbb{P}_Y^N$$

which when followed by the second projection yields a Y-morphism

$$\varphi_{\mathcal{L}} : X \longrightarrow \mathbb{P}_Y^N.$$

We are now ready for the important

Definition 15.4 When $\varphi_{\mathcal{L}}$ is a closed embedding, then $\mathcal{L}$ is said to be very ample. If $\varphi_{\mathcal{L}^{\otimes n}}$ is a closed embedding for some n, then $\mathcal{L}$ is said to be ample.

15.5 The Segre Embedding

Another important concept is the *Segre embedding*. We start out with the following simple observation: Let S and T be graded A-algebras. Then we may consider the product of affine schemes

$$\mathrm{Spec}(S) \times_{\mathrm{Spec}(A)} \mathrm{Spec}(T)$$

and the product of projective ones

$$\mathrm{Proj}(S) \times_{\mathrm{Spec}(A)} \mathrm{Proj}(T).$$

Here the former is the product of the "affine cones" over the factors in the latter product, but unfortunately there is no immediate way of deducing the last product from the first one. The problem is interesting however, it is related to taking general *joins*, and is treated among other places in the articles [1] and in [31].

Here we take a simper path, and note that if we have

$$S = A[X_0, X_1, \dots, X_N] \quad \text{and} \quad T = A[Y_0, Y_1, \dots, Y_M]$$

and put

$$U = A[X_0 Y_0, X_0 Y_1, \dots, X_N Y_M] = A[Z_0, Z_1, \dots, Z_R],$$

where the $X_i Y_j$ are ordered as $Z_0, Z_1, \dots, Z_T$ in some order, and given the degree 1, then we get the canonical isomorphism and closed embedding

$$\mathrm{Proj}(S) \times_{\mathrm{Spec}(A)} \mathrm{Proj}(T) \cong \mathrm{Proj}(U) \hookrightarrow \mathbb{P}^R_{\mathrm{Spec}(A)}.$$

This closed embedding is a special case of the *Segre embedding* over $\mathrm{Spec}(A)$.

It is fairly straightforward to show that this procedure applies to any scheme Y as a base, not just an affine scheme $\mathrm{Spec}(A)$. This is carried out in the following section.

15.6 Algebraic Complement

The following considerations, constructions and notation also apply to more general situations, such as with the ring A below replaced by a ringed space X and the module M replaced by an $\mathcal{O}_X$-module $\mathcal{M}$ on X. But we keep the simple context of modules M over a commutative ring A.

For any commutative ring A we have the category of all A-modules M, frequently denoted by Mod_A, the constructions of direct sums, tensor products and the non-commutative graded tensor algebra

$$T_A(M) = \bigoplus_{n=0}^{\infty} M^{\otimes n}.$$

We define the *commutative* symmetric graded algebra $S_A(M)$ and the canonical surjective A-homomorphism $T_A(M) \longrightarrow S_A(M)$, $S_A(M)$ is defined as $T_A(M)/\mathfrak{C}$ where $\mathfrak{C}$ is the two sided ideal generated by the set

$$\{x \otimes y - y \otimes x \mid x, y \in M\}.$$

While the graded module $S_A(M)$ is the "commutative version" of $T_A(M)$, we also have to work with the "anti-commutative version", namely

$$\bigwedge_A(M) = T_A(M)/\mathfrak{A}$$

where $\mathfrak{A}$ is the homogeneous ideal generated by the elements

$$\{x \otimes y + y \otimes x \mid x, y \in M\}.$$

Clearly $\bigwedge_A(M)$ is an anticommutative graded A-algebra. We denote the homogeneous pieces of degree n of the above graded algebras by, respectively, $T_A^n(M)$, $S_A^n(M)$ and $\bigwedge_A^n(M)$.

In Proposition 12.1 we described the equivalence of the properties of A-modules and $\mathcal{O}_X$-modules on $X = \mathrm{Spec}(A)$, and how this leads to analogues over any scheme. In the same spirit all the above constructions are carried out over any scheme X instead of A or rather, $\mathrm{Spec}(A)$, and any $\mathcal{O}_X$-module instead of the A-module M.

We finally sum up some important properties of these three constructions, following [18], Chap. II, Exercise 5.16. First, a computation of the ranks of the homogeneous pieces of the three algebras on X constructed above:

Proposition 15.5 *If $\mathcal{F}$ is locally free of rank n on the scheme X, then the three $\mathcal{O}_X$-modules the rth. tensor power, $T_X^r(\mathcal{F})$, the rth. symmetric power $S_X^r(\mathcal{F})$ and the rth. exterior power $\bigwedge_X^r(\mathcal{F})$ are locally free of ranks rn, $\binom{n+r-1}{n-1}$, and $\binom{n}{r}$, respectively.*

Next, the following observation is handy for computing duals:

Proposition 15.6 *Let $\mathcal{F}$ be locally free of rank n. Then the multiplication map*

$$\bigwedge_X^r(\mathcal{F}) \otimes \bigwedge_X^{n-r}(\mathcal{F}) \longrightarrow \bigwedge_X^n(\mathcal{F})$$

induces an isomorphism

$$\bigwedge_X^r(\mathcal{F}) \xrightarrow{\cong} \left(\bigwedge_X^{n-r}(\mathcal{F})\right)^\vee \otimes \bigwedge_X^n(\mathcal{F}).$$

The homogeneous pieces of the symmetric powers in the symmetric algebra have important filtrations:

Proposition 15.7 *Let*

$$0 \longrightarrow \mathcal{F}' \longrightarrow \mathcal{F} \longrightarrow \mathcal{F}'' \longrightarrow 0$$

be an exact sequence of locally free sheaves. Then for any r there is a filtration of $S_X^r(\mathcal{F})$

$$S_X^r(\mathcal{F}) = \mathcal{F}^0 \supseteq \mathcal{F}^1 \supseteq \cdots \supseteq \mathcal{F}^r \supseteq \mathcal{F}^{r+1} = 0$$

with quotients

$$\mathcal{F}^p/\mathcal{F}^{p+1} \cong S_X^p(\mathcal{F}') \otimes S_X^{r-p}(\mathcal{F}'')$$

for all p.

We also have the corresponding result for exterior powers:

Proposition 15.8 *Let*

$$0 \longrightarrow \mathcal{F}' \longrightarrow \mathcal{F} \longrightarrow \mathcal{F}'' \longrightarrow 0$$

be an exact sequence of locally free sheaves. Then for any r there is a filtration of $\bigwedge_X^r(\mathcal{F})$

$$\bigwedge_X^r(\mathcal{F}) = \mathcal{F}^0 \supseteq \mathcal{F}^1 \supseteq \cdots \supseteq \mathcal{F}^r \supseteq \mathcal{F}^{r+1} = 0$$

with quotients

$$\mathcal{F}^p/\mathcal{F}^{p+1} \cong \bigwedge_X^p(\mathcal{F}') \otimes \bigwedge_X^{r-p}(\mathcal{F}'')$$

for all p.

15.7 The Functorial Behavior of Proj

Now let $\mathcal{M}$ and $\mathcal{N}$ be two $\mathcal{O}_X$ modules on the scheme X. We now move on to describe the Segre-embedding from Sect. 15.5 as an embedding

$$\mathrm{Proj}(S_X(\mathcal{M})) \times_X \mathrm{Proj}(S_X(\mathcal{N})) \hookrightarrow \mathrm{Proj}(S_X(\mathcal{M} \otimes_X \mathcal{N})).$$

As we have seen, Spec is a functor from the category of commutative rings to the category of schemes. In general the functor Proj behaves differently. We now essentially follow [15], Chap. II 2.8:

Let $\varphi : S' \longrightarrow S$ be a graded homomorphism of graded A-algebras. While we do have the morphism

$$\mathrm{Spec}(\varphi) : \mathrm{Spec}(S) \longrightarrow \mathrm{Spec}(S'),$$

we do not in general obtain a morphism

$$\mathrm{Proj}(\varphi) : \mathrm{Proj}(S) \longrightarrow \mathrm{Proj}(S').$$

Indeed, a homogeneous prime ideal $\mathcal{P}$ in S will yield a homogeneous prime ideal $\mathcal{P}' = \varphi^{-1}(\mathcal{P})$ in S', but this homogeneous prime in S' may not be in $\mathrm{Proj}(S')$. We therefore need to introduce the following open subset of $\mathrm{Proj}(S)$:

Definition 15.5

$$G(\varphi) = D_+(((\varphi(S'_+)))).$$

The induced morphism

$$G(\varphi) = \mathrm{Proj}(S) - V_+(((\varphi(S'_+)))) \longrightarrow \mathrm{Proj}(S')$$

is denoted by $\mathrm{Proj}(\varphi)$.

So $\mathrm{Proj}(\varphi)$ is not defined everywhere in $\mathrm{Proj}(S)$.

Example 15.1 Let $S = k[x_0, x_1, x_2]$ and $S' = k[x_0, x_1]$, k being a field. Let $\varphi :$ $k[x_0, x_1] \hookrightarrow k[x_0, x_1, x_2]$ be the inclusion. Then $G(\varphi)$ is $D_+(x_2)$ and $\mathrm{Proj}(\varphi)$ is projection onto the x_0, x_1-plane with $(0:0:1)$ as center. Or, more accurately speaking, that is what the morphism looks like on the affine level:

$$\mathrm{pr}_{(0:0:1)} : \mathbb{A}^3_k \dashrightarrow \mathbb{A}^2_k$$

Remark 15.9 This procedure is followed in many other situations where exceptional loci of morphisms between projective schemes are involved.

15.8 Tensor Products of Graded S-Modules

Let S be a graded A-algebra, and let M and N be two graded S-modules. We define a graded S-module $M \otimes_S N$ as follows: First, consider the $\mathbb{Z}$-module $M \otimes_\mathbb{Z} N$, graded by $(M \otimes_\mathbb{Z} N)_q = \bigoplus_{m+n=q} M_m \otimes_\mathbb{Z} N_n$. Then let P be the sub $\mathbb{Z}$-module of $M \otimes_\mathbb{Z} M$ generated by all elements $(xs) \otimes y - x \otimes (sy)$ for all $s \in S, x \in M$ and $y \in N$. P being a graded submodule, we may define $M \otimes_S N = M \otimes_\mathbb{Z} N/P$. In particular, if S' is a graded S-algebra, then $M \otimes_S S'$ is an S'-module, and we have the following immediate result:

Proposition 15.10 1. *Let M and N be two graded S-modules and let $X = \mathrm{Proj}(S)$. Then*

$$\widetilde{M} \otimes_{\mathcal{O}_X} \widetilde{N} = \widetilde{M \otimes_S N}.$$

2. *Let M be a graded S-module and S' be a graded S-algebra. Let*

$$f : X' = \mathrm{Proj}(S') \longrightarrow X = \mathrm{Proj}(S),$$

which we now assume to be defined in all of X'. Then

$$f^*(\widetilde{M}) = \widetilde{M \otimes_S S'}.$$

15.9 Basic Facts on Global Proj

So far we have defined and studied in some detail the $\mathrm{Spec}(A)$-scheme $\mathrm{Proj}(S)$ where S is a graded A-algebra. In the spirit of Grothendieck's work in [15] we now move on to a more general concept defined over any scheme X. The following constructions will be needed later, in Sect. 16.3.

So let $\mathcal{S}$ be a coherent graded $\mathcal{O}_X$-algebra on the scheme X. In Sect. 14.4 we defined Proj of a graded quasi coherent $\mathcal{O}_X$ algebra on X, denoted by $\mathcal{S}$, yielding an X-scheme $\pi : \mathrm{Proj}(\mathcal{S}) \longrightarrow X$.

Let $f : X \longrightarrow Y$ be a morphism of schemes. Let $\mathcal{L}$ be an invertible $\mathcal{O}_X$-module on X, and $\mathcal{S}$ a quasi coherent $\mathcal{O}_Y$-algebra on Y, for simplicity we assume $\mathcal{S}_0 = \mathcal{O}_Y$ and that $\mathcal{S}$ be generated by $\mathcal{S}_1$. Let $\mathcal{S}' = S_X(\mathcal{L})$, the symmetric algebra. Also, let there be given an $\mathcal{O}_X$-homomorphism of $\mathcal{O}_X$-algebras,

$$\psi : f^*(\mathcal{S}) \longrightarrow \mathcal{S}'$$

which is equivalent to giving an $\mathcal{O}_Y$-homomorphism

$$\psi^\flat : \mathcal{S} \longrightarrow f_*(\mathcal{S}').$$

Since $\mathcal{L}$ is invertible, $\mathrm{Proj}(\mathcal{S}') = X$, and thus by Remark 15.9 ψ yields an open subscheme $G(\psi)$ of X and a morphism

$$r_{\mathcal{L},\psi} : G(\psi) \longrightarrow \mathrm{Proj}(\mathcal{S})$$

obtained as the composition with the first projection

$$\mathrm{Proj}(\psi) : G(\psi) \longrightarrow \mathrm{Proj}(f^*(\mathcal{S})) = \mathrm{Proj}(\mathcal{S}) \times_Y X \longrightarrow \mathrm{Proj}(\mathcal{S}).$$

Following [15] II, Sect. 3.3, let $\mathcal{M}$ be a quasi coherent $\mathcal{S}$-module on Y. Then local sections of $\mathcal{M}_0$ on Y give rise to local sections of $\widetilde{\mathcal{M}}$ on $\mathrm{Proj}(\mathcal{S})$ by the canonical

$$\alpha_{0,U} : \mathcal{M}_0(U) \longrightarrow \widetilde{\mathcal{M}}(\pi^{-1}(U))$$

where $\pi : \mathrm{Proj}(\mathcal{S}) \longrightarrow Y$ is the canonical morphism. Thus we get a canonical homomorphism $\alpha_0 : \mathcal{M}_0 \longrightarrow \pi_*(\widetilde{\mathcal{M}})$ which when applied to $\mathcal{M}(n)$ yields

$$\alpha_n : \mathcal{M}_n \longrightarrow \pi_*(\widetilde{\mathcal{M}}(n))$$

or

$$\alpha_n^{\sharp} : \pi^*(\mathcal{M}_n) \longrightarrow \widetilde{\mathcal{M}}(n).$$

Chapter 16
Conormal Sheaf and Projective Bundles

This chapter starts with introducing the conormal sheaf, then proceeds to the sheaf of Kähler differentials of an S-scheme X. Then comes the construction of the universal 1-quotient on $\mathbb{P}(\mathcal{E})$, leading up to the Segre embedding for projective bundles. The chapter concludes with base extensions of projective morphisms and the concept of regular schemes, also the concept of a *formally étale* morphism is defined.

16.1 The Conormal Sheaf

Let X be a closed subscheme of the S-scheme Z, in other words there is a closed S-embedding $i : X \hookrightarrow Z$ and X is given as a subscheme of Z by a quasi coherent ideal on Z, we have $i_*(\mathcal{O}_X) = \mathcal{O}_Z/\mathfrak{I}_X$ where $\mathfrak{I}_X$ is the ideal defining X as a closed S-subscheme of Z. In this situation it turns out that the sheaf of $\mathcal{O}_X$-modules on X given as $\mathfrak{I}_X/\mathfrak{I}_X^2$ carries much significant information about X and its embedding into Z. We make the

Definition 16.1 We put

$$\mathcal{N}_{X/Z/S} = \mathfrak{I}_X/\mathfrak{I}_X^2$$

and refer to it as the *conormal sheaf* of X in Z over S. Usually S is omitted in the notation when no confusion is possible.

Example 16.1 If $Z = \mathrm{Spec}(A)$, and $X = \mathrm{Spec}(A/\mathfrak{a})$, then $\mathcal{N}_{X/Z} = \widetilde{\mathfrak{a}/\mathfrak{a}^2}$.

16.2 Kähler Differentials of an S-Scheme X

Let X be an S-scheme, separated over S. Then the diagonal morphism

$$\Delta_{X/S} : X \longrightarrow X \times_S X$$

is a closed embedding, thus we have a conormal sheaf in this case.

A. Holme, *A Royal Road to Algebraic Geometry*,
DOI 10.1007/978-3-642-19225-8_16, © Springer-Verlag Berlin Heidelberg 2012

Definition 16.2 The conormal sheaf of the diagonal embedding over S of the separated S-scheme X is denoted by $\Omega^1_{X/S}$ and referred to as the sheaf of Kähler differentials.

We start out by noting that

Proposition 16.1 *The sheaf $\Omega^1_{X/S}$ defined above is quasi coherent, and coherent if the schemes are locally Noetherian.*

Proof If $S = \bigcup_{i \in I} S_i$ is an open affine covering, then we get an open covering of X as $\bigcup X_{U_i} = \bigcup X_i$. Thus covering each X_i with open affine subsets, the local study of $\Omega^1_{X/S}$ is reduced to the case that the schemes are affine and $f : X \longrightarrow S$ is given by the ring homomorphism $\varphi : A \longrightarrow B$. Then the diagonal morphism

$$\mathrm{Spec}(B) = X \longrightarrow X \times_S X = \mathrm{Spec}(B \otimes_A B)$$

is given by the multiplication map

$$m : B \otimes_A B \longrightarrow B \quad \text{where } b_1 \otimes b_2 \mapsto b_1 b_2.$$

Letting I denote the kernel of m, we get a B-module $\Omega^1_{B/A} = I/I^2$, and have that locally

$$\widetilde{\Omega^1_{B/A}} = I/I^2 = \Omega^1_{X/S}. \qquad \square$$

We note that the mapping

$$d_{B/A} : B \longrightarrow I/I^2 \text{ where } b(\mathrm{mod}\,I^2) \mapsto 1 \otimes b - b \otimes 1 (\mathrm{mod}\,I^2)$$

is well defined. This $d = d_{B/A}$ is an A-derivation of B into $\Omega^1_{B/A}$ in the sense that it is A-linear, zero on elements from A and satisfies the usual rule

$$d_{B/A}(b_1 b_2) = b_1 d_{B/A}(b_2) + b_2 d_{B/A}(b_1).$$

This $d_{B/A}$ is referred to as the *Universal Derivation* of B over A, for the following reason: An A-derivation of B into some B-module Ω is defined by the properties specified above for $d_{B/A}$. The set of all such derivations is denoted by $\mathrm{Der}_A(B, \Omega)$. It is easily seen that the functor $\Omega \mapsto \mathrm{Der}_A(B, \Omega)$ is representable by $\Omega^1_{B/A}$, and the universal element is of course $d_{B/A}$.

Using this it is also elementary to check that

Proposition 16.2 *Let A' and B be A-algebras and $B' = B \otimes_A A'$. Then there is a canonical and functorial isomorphism*

$$\Omega^1_{B'/A'} \cong \Omega^1_{B/A} \otimes_B B'.$$

If S is a multiplicatively closed subset in B, then

$$\Omega^1_{S^{-1}B/A} \cong S^{-1}\Omega^1_{B/A}.$$

There are two fundamental exact sequences for modules of differentials:

Proposition 16.3 (First Exact Sequence for Differentials) *Let*

$$A \longrightarrow B \longrightarrow C$$

be homomorphisms of commutative rings. Then there is an exact sequence of C-modules

$$\Omega^1_{B/A} \otimes_B C \xrightarrow{\alpha} \Omega^1_{C/A} \xrightarrow{\beta} \Omega^1_{C/B} \longrightarrow 0.$$

Proof $\Omega^1_{C/A}$ is generated over C by the elements $d_{C/A}(c)$ for all $c \in C$, and $\beta(d_{C/A}(b)) = d_{C/B}(b) = 0$ for all $b \in B$. Thus $\ker(\beta)$ is generated by the set of all elements $d_{C/A}(b)$. But these elements generate the image of α as a C-module, and exactness follows. □

Proposition 16.4 (Second Exact Sequence for Differentials) *Assume in addition that $C = B/I$. Then there is an exact sequence of C-modules*

$$I/I^2 \xrightarrow{\delta} \Omega^1_{B/A} \otimes_B C \longrightarrow \Omega^1_{C/A} \longrightarrow 0.$$

Proof Using Proposition 16.3 we only need to prove exactness to the left. This is seen by noting that $\Omega^1_{B/A} \otimes_B C = \Omega^1_{B/A}/I\Omega^1_{B/A}$, thus if $b \in I$ we may define

$$\delta(b \ (\mathrm{mod}\ I^2)) = d_{B/A}(b) \ (\mathrm{mod}\ I)$$

I/I^2 has a structure as C-module, since $(b' \ (\mathrm{mod}\ I))(b \ (\mathrm{mod}\ I^2)) = (b'b \ (\mathrm{mod}\ I^2))$ is well defined. □

The verification of the following example is immediate:

Example 16.2 If $B = A[X_1, \ldots, X_N]$ is a polynomial ring, then

$$\Omega^1_{B/A} = B\mathrm{d}X_1 \oplus \cdots \oplus B\mathrm{d}X_N.$$

If B is an algebra generated over A by $\xi_1, \ldots, \xi_N$, or is a localization of such an algebra, then

$$\Omega^1_{B/A} = B\mathrm{d}\xi_1 + \cdots + B\mathrm{d}\xi_N.$$

Proposition 16.5 *Let K be a finitely generated field extension of the field k, of transcendence degree r. Then the K-vector space $\Omega^1_{K/k}$ is of dimension $\geq r$, with equality holding if and only if K is separately generated over k.*

Proof We may assume that $K = k(\xi_1, \ldots, \xi_r, \xi_{r+1}, \ldots, \xi_N)$ where $\xi_1, \ldots, \xi_r$ is a separating transcendence basis. Then $d\xi_1, \ldots, d\xi_r, d\xi_{r+1}, \ldots, d\xi_N$ generate $\Omega^1_{K/k}$ over K by Example 16.2. The last part of the claim follows by applying the First Exact Sequence to

$$k \hookrightarrow L = k[\xi_1, \ldots, \xi_r] \hookrightarrow K,$$

since $\Omega_{K/L} = 0$. □

Proposition 16.6 *Let $\mathcal{O}$ be a local ring with maximal ideal $\mathfrak{m}$, which is a localization of a finitely generated algebra over the field k. Assume that $k = \mathcal{O}/\mathfrak{m}$. Then the*

$$\mathfrak{m}/\mathfrak{m}^2 \xrightarrow{\delta} \Omega^1_{\mathcal{O}/k} \otimes_{\mathcal{O}} k$$

in the Second Exact Sequence is an isomorphism.

Proof Since $\Omega^1_{k/k} = 0$ the Second Exact Sequence shows that δ is surjective. To show injectivity we prove that the dual mapping

$$\delta^\vee : \mathrm{Hom}_k(\Omega^1_{\mathcal{O}/k} \otimes_{\mathcal{O}} k, k) = \mathrm{Hom}_k(\Omega^1_{\mathcal{O}/k}, k) = \mathrm{Der}_k(\mathcal{O}, k) \longrightarrow \mathrm{Hom}_k(\mathfrak{m}/\mathfrak{m}^2, k)$$

is surjective. We first recall how $\delta^\vee$ maps a derivation D of $\mathcal{O}$ into the $\mathcal{O}$-module k to a k linear map $\mathfrak{m}/\mathfrak{m}^2 \longrightarrow k$. Indeed, by restricting D to $\mathfrak{m}$ and remembering that by the action of $\mathcal{O}$ on k we have $ab = 0$ for all $a \in k$ and all and all $b \in \mathcal{O}$, we find $D(\mathfrak{m}^2) = 0$. Thus $\delta^\vee(D)$ is a k-linear map $\mathfrak{m}/\mathfrak{m}^2 \longrightarrow k$. Conversely, letting $\varphi : \mathfrak{m}/\mathfrak{m}^2 \longrightarrow k$ be k-linear, define $d(a) = \varphi(a \mod \mathfrak{m}^2)$. This is easily checked to be a derivation as required. □

Proposition 16.7 *Let k be a perfect field and $\mathcal{O}$ be as in Proposition 16.6 but in addition assumed to be the localization of an algebra of finite type over k and $n = \dim(\mathcal{O})$. Then $\Omega^1_{\mathcal{O}/k}$ is free of rank n if and only if $\mathcal{O}$ is a regular local ring.*

Proof If $\Omega^1_{\mathcal{O}/k}$ is free of rank n then $\mathfrak{m}/\mathfrak{m}^2$ is a k-vector space of dimension n by Proposition 16.6. So $\mathcal{O}$ is a regular local ring of dimension n. Conversely, if $\mathcal{O}$ is a regular local ring of dimension n then by the same proposition $\Omega^1_{\mathcal{O}/k} \otimes_{\mathcal{O}} k$ is an n-dimensional k-space. By Proposition 16.2 $\Omega^1_{K/k} = \Omega^1_{\mathcal{O}/k} \otimes_{\mathcal{O}} K$, since k is perfect K is separately generated over k, thus the dimension as K-vector space of $\Omega^1_{K/k}$ is equal to the transcendence degree of K over k, which in turn is equal to the dimension of $\mathcal{O}$ by standard facts on the Krull dimension of polynomial rings. Hence $\dim_K(\Omega^1_{\mathcal{O}/k} \otimes_{\mathcal{O}} K) = \dim_k(\Omega^1_{\mathcal{O}/k} \otimes_{\mathcal{O}} k)$. Now by Lemma 4.9 in Sect. 4.2, we find that $\Omega^1_{\mathcal{O}/k}$ is generated by n elements. We thus have an exact sequence

$$0 \longrightarrow \Delta \longrightarrow \mathcal{O}^{\oplus n} \longrightarrow \Omega^1_{\mathcal{O}/k} \longrightarrow 0$$

where $\Delta \otimes_{\mathcal{O}/k} K = 0$. But Δ is a submodule of a free module, hence torsion free. So $\Delta = 0$. Thus $\Omega^1_{\mathcal{O}/k}$ is free of rank n, and the proof is complete. $\square$

When $S = \mathrm{Spec}(k)$ for a field k, we write $\Omega^1_{X/k}$ instead of $\Omega^1_{X/S}$. The following result is true more generally, but we shall only use this version:

Proposition 16.8 *Assume that X is a non singular projective variety defined over an algebraically closed field k. Then $\Omega^1_{X/k}$ is a locally free sheaf of rank $n = \dim(X)$.*

Proof This follows by Proposition 16.7 applied to the rings $\mathcal{O}_{X,x}$ for all closed points $x \in X$. $\square$

Definition 16.3 When X is as in the previous proposition, we denote the dual sheaf of the sheaf of differentials $(\Omega^1_{X/k})^\vee$ by $\mathcal{T}_{X/k}$, and refer to it as the *tangent sheaf* of X over k. It is locally free of rank n. The morphisms $\mathbb{P}(\Omega^1_{X/k}) \longrightarrow X$ and $\mathbb{P}(\mathcal{T}_{X/k}) \longrightarrow X$ are referred to as the cotangent, respectively the tangent, bundles of X over k. Moreover, in the same situation we define the canonical sheaf on X as $\omega_{X/k} = \bigwedge^n(\Omega^1_{X/k})$. It is an invertible sheaf.

We have the following by Proposition 15.5, as $\binom{n}{n} = 1$:

Proposition 16.9 *The sheaf $\omega_{X/k} = \bigwedge^n(\Omega^1_{X/k})$ is locally free of rank 1.*

16.3 The Universal 1-Quotient on $\mathbb{P}(\mathcal{E})$

Now let $\mathcal{E}$ be a quasi coherent $\mathcal{O}_Y$-module on Y, and form the morphism $\pi_\mathcal{E} : \mathbb{P}(\mathcal{E}) \longrightarrow Y$.

We apply the construction of α_n and $\alpha_n^\sharp$ in Sect. 15.9 to $\mathcal{M} = S_{\mathcal{O}_Y}(\mathcal{E})$ and get for $n = 1$

$$\psi_\mathcal{E} = \alpha_1^\sharp : \pi_\mathcal{E}^*(\mathcal{E}) \longrightarrow \mathcal{O}_{\mathbb{P}(\mathcal{E})}(1)$$

and note the following

Lemma 16.10 *$\psi_\mathcal{E}$ defined above is surjective.*

Proof $\psi_\mathcal{E} = \alpha_1^\sharp$ is the $\widetilde{(\ )}$ of the canonical

$$\mathcal{E} \otimes_{\mathcal{O}_Y} S_{\mathcal{O}_Y}(\mathcal{E}) \longrightarrow S_{\mathcal{O}_Y}(\mathcal{E})(1).$$

But by definition $\mathcal{E}$ generates $S_Y(\mathcal{E})$, in particular $\mathcal{O}_P(1)$ is globally generated and the mapping above is surjective. $\square$

We define 1-quotients over the scheme Y as follows:

Definition 16.4 Given an $\mathcal{O}_Y$-module $\mathcal{E}$ on the scheme Y. A 1-quotient of $\mathcal{E}$ on the Y-scheme $f : X \longrightarrow Y$ is a pair $(\mathcal{L}, \varphi)$ where $\mathcal{L}$ is an invertible sheaf on X and $\varphi : f^*(\mathcal{E}) \longrightarrow \mathcal{L}$ is a surjective homomorphism of $\mathcal{O}_X$-modules on X. Two 1-quotients $(\mathcal{L}_1, \varphi_1)$ and $(\mathcal{L}_2, \varphi_2)$ are said to be equivalent, written as $(\mathcal{L}_1, \varphi_1) \sim (\mathcal{L}_2, \varphi_2)$, if there is an isomorphism $\psi : \mathcal{L}_1 \longrightarrow \mathcal{L}_2$ making the following diagram commutative

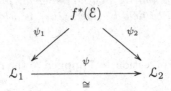

We then define the contravariant functor of 1-quotients over the scheme Y as

$$f : \mathrm{Schemes}_Y \longrightarrow \mathrm{Sets}$$

$$X \to Y \mapsto F(X) = \left\{ (\mathcal{L}, \varphi) \,\middle|\, \begin{array}{l} (\mathcal{L}, \varphi) \\ \text{is a 1 quotient of} \\ \mathcal{E} \text{ on } X \end{array} \right\} / \sim .$$

We have the following important

Theorem 16.11 *The functor F is representable by the object $\mathbb{P}(\mathcal{E})$. The universal element is*

$$(\mathcal{O}_{\mathbb{P}(\mathcal{E})}(1), \pi_{\mathbb{P}(\mathcal{E})}).$$

Proof We have to construct an isomorphism of contravariant functors

$$\Phi : \mathrm{Hom}_Y(\ , \mathbb{P}(\mathcal{E})) \longrightarrow F(\ )$$

i.e., for all Y-schemes X

$$\Phi_X : \mathrm{Hom}_Y(X, \mathbb{P}(\mathcal{E})) \longrightarrow F(X)$$

which is bijective and functorial in X. So let $r \in \mathrm{Hom}_Y(X, \mathbb{P}(\mathcal{E}))$, thus

$$\begin{array}{ccc} \mathbb{P}(\mathcal{E}) & \xleftarrow{\ r\ } & X \\ {\scriptstyle \pi_{\mathbb{P}(\mathcal{E})}} \downarrow & \swarrow{\scriptstyle f} & \\ Y & & \end{array}$$

We then put

$$\Phi_X(r) = (\mathcal{L}_r, \varphi_r) \quad \text{where } \mathcal{L}_r = r^*(\mathcal{O}_{\mathbb{P}(\mathcal{E})}(1)), \varphi_r = r^*(\pi_{\mathbb{P}(\mathcal{E})})).$$

Conversely, we define

$$\Psi_X : F(X) \longrightarrow \mathrm{Hom}_Y(X, \mathbb{P}(\mathcal{E}))$$

by letting

$$(\mathcal{L}, \varphi) \quad \text{where } f^*(\mathcal{E}) \xrightarrow{\varphi} \mathcal{L}$$

first correspond to

$$f^*(S_{\mathcal{O}_Y}(\mathcal{E})) = S_{\mathcal{O}_X}(f^*(\mathcal{E})) \longrightarrow S_{\mathcal{O}_X}(\mathcal{L})),$$

and then, φ being surjective, to the composition

$$\begin{aligned}
X = \mathrm{Proj}(S_{\mathcal{O}_X}(\mathcal{L})) &\longrightarrow \mathrm{Proj}(S_{\mathcal{O}_X}(f^*(\mathcal{E}))) \\
&= \mathrm{Proj}(f^*(S_{\mathcal{O}_Y}(\mathcal{E}))) \\
&= \mathrm{Proj}((S_{\mathcal{O}_Y}(\mathcal{E}))) \times_Y X \\
&= \mathbb{P}(\mathcal{E}) \times_Y X \xrightarrow{\mathrm{pr}_1} \mathbb{P}(\mathcal{E}).
\end{aligned}$$

It is straightforward to verify that Φ_X and Ψ_X are inverse mappings, and the proof is complete. □

We finally note the following important result:

Theorem 16.12 *Assume that $\mathcal{E}$ be locally free on Y, and let*

$$\pi : X = \mathbb{P}(\mathcal{E}) \longrightarrow Y$$

be the structure morphism. As usual denote the canonical invertible sheaf on X by $\mathcal{O}_X(1)$. Then there is a canonical exact sequence on X

$$0 \longrightarrow \Omega^1_{X/Y} \longrightarrow \pi^*(\mathcal{E})(-1) \longrightarrow \mathcal{O}_X \longrightarrow 0.$$

Proof In Sect. 16.3 we have shown that the following homomorphism is surjective

$$\psi_{\mathcal{E}} = \alpha^{\#}_1 : \pi^*_{\mathcal{E}}(\mathcal{E}) \longrightarrow \mathcal{O}_X(1).$$

Twisting by -1 yields an exact sequence of modules on X:

$$\pi^*(\mathcal{E})(-1) \longrightarrow \mathcal{O}_X \longrightarrow 0.$$

It remains to identify the kernel as $\Omega^1_{X/Y}$.

It suffices to check this locally on Y, and we may assume $T = \mathrm{Spec}(A)$ and $\mathcal{E}$ free of rank r, $\mathcal{E} = \widetilde{A^r}$.

This is where we need to do some work, it is carried out in the proof of Theorem 8.13 on pp. 176–177 in [18]. □

16.4 The Segre Embedding for Projective Bundles

We now return to the Segre embedding, a special case of which was treated in Sect. 15.5, when it was defined for free $\mathcal{O}_Y$-modules in Y.

Now let $\mathcal{E}$ and $\mathcal{F}$ be two quasi coherent $\mathcal{O}_Y$-modules on Y, and consider the morphisms

$$\pi_{\mathcal{F}} : \mathbb{P}(\mathcal{F}) \longrightarrow Y \quad \text{and} \quad \pi_{\mathcal{E}} : \mathbb{P}(\mathcal{E}) \longrightarrow Y.$$

Consider the diagram

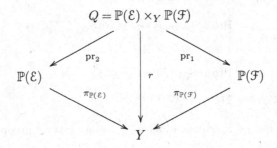

and let

$$\mathcal{L} = \mathrm{pr}_1^*(\mathcal{O}_{\mathbb{P}(\mathcal{E})}(1)) \otimes_{\mathcal{O}_Q} \mathrm{pr}_2^*(\mathcal{O}_{\mathbb{P}(\mathcal{F})}(1)).$$

When the two surjective

$$\pi_{\mathbb{P}(\mathcal{E})}^*(\mathcal{E}) \longrightarrow\!\!\!\!\!\to \mathcal{O}_{\mathbb{P}(\mathcal{E})}(1) \quad \text{and} \quad \pi_{\mathbb{P}(\mathcal{F})}^*(\mathcal{F}) \longrightarrow\!\!\!\!\!\to \mathcal{O}_{\mathbb{P}(\mathcal{F})}(1)$$

are pulled up to Q and tensorized, we get

$$r^*(\mathcal{E} \otimes_{\mathcal{O}_Q} \mathcal{F}) \longrightarrow\!\!\!\!\!\to \mathcal{L}$$

which by Theorem 16.11 corresponds to a morphism

$$\zeta_{\mathcal{E},\mathcal{F}} : \mathbb{P}(\mathcal{E}) \times_Y \mathbb{P}(\mathcal{F}) \longrightarrow \mathbb{P}(\mathcal{E} \otimes_{\mathcal{O}_Y} \mathcal{F})$$

which is referred to as the *Segre morphism*. We then have the

Proposition 16.13 *The Segre morphism $\zeta_{\mathcal{E},\mathcal{F}}$ is a closed embedding for all quasi coherent $\mathcal{E}$ and $\mathcal{F}$.*

Proof By quasi coherency we may assume that $Y = \operatorname{Spec} A$ and $\mathcal{E} = \widetilde{E}, \mathcal{F} = \widetilde{F}$. Then the claim follows for $\mathcal{E}$ and $\mathcal{F}$ locally free by Sect. 15.5. For the general case we refer to [15] II, p. 77. $\square$

16.5 Base Extensions of Projective Morphisms

Let $\pi_X : \mathrm{Proj}(\mathcal{S}) \longrightarrow X$ be the projective morphism given by the graded $\mathcal{O}_X$-algebra $\mathcal{S}$ on X. Let $f : Y \longrightarrow X$ be an affine morphism. Letting $\mathcal{B} = f_*(\mathcal{O}_Y)$, we get a canonical isomorphism $\mathrm{Spec}(\mathcal{B}) \cong Y$, and f may be identified with the canonical $\mathrm{Spec}(\mathcal{B}) \longrightarrow X$. We now have the following

Proposition 16.14 *The following is a product diagram over X:*

$$
\begin{array}{ccc}
\mathrm{Proj}(\mathcal{S} \otimes_{\mathcal{O}_X} \mathcal{B}) & \xrightarrow{\ \mathrm{pr}_1\ } & \mathrm{Proj}(\mathcal{S}) \\[2mm]
{\scriptstyle \mathrm{pr}_2 = \pi_Y} \Big\downarrow & & \Big\downarrow {\scriptstyle \pi_X} \\[2mm]
Y = \mathrm{Spec}(\mathcal{B}) & \xrightarrow[\ f\ ]{} & X
\end{array}
$$

In particular, if f is a closed embedding, then $\mathcal{B}$ is canonically isomorphic to $\mathcal{O}_X/\mathcal{J}$, where $\mathcal{J}$ is the ideal on X defining the closed subscheme Y, and $\mathcal{S} \otimes_{\mathcal{O}_X} \mathcal{B} = \mathcal{S}/\mathcal{J}\mathcal{S} = \overline{\mathcal{S}}$, and pr_1 is a closed embedding. In this case $\mathrm{Proj}(\overline{\mathcal{S}}) = \mathrm{Proj}(\mathcal{S})_Y = \pi_X^{-1}(Y)$.

Proof As the assertions following "In particular" are obvious, we have to show that the diagram is Cartesian.

For this we may assume that $X = \mathrm{Spec}(A)$ and $Y = \mathrm{Spec}(B)$ where B is an A-algebra, and that $\mathcal{S} = \widetilde{S}$, where S is a graded A-algebra.

Then the claim follows by the construction of $\mathrm{Proj}(S)$ and of the fibered product of schemes, as

$$ S_{(g)} \otimes_A B = (S \otimes_A B)_{(g \otimes 1)} \qquad \square $$

16.6 Regular Schemes

Let R be a commutative Noetherian ring. Recall that the *Krull dimension* of R is given as the maximum of all m such that there exists a chain of prime ideals $\mathfrak{p}_0 \subset \mathfrak{p}_1 \subset \cdots \subset \mathfrak{p}_m \subset R$. If I is an ideal in R, then the *dimension of I*, denoted $\dim(I)$, is $\dim(R/I)$, and if I is prime then the *codimension*, denoted $\mathrm{codim}(I)$ is $\dim(R_I)$. If I is not prime, then the codimension is the minimal of the codimensions of all the minimal prime ideals containing I. Finally, recall from commutative algebra that it follows by repeated application of the Principal Ideal Theorem that if $\mathfrak{P} \subset R$ is an ideal and $r_1, r_2, \ldots, r_c \in \mathfrak{P}$, then $\mathrm{codim}(\mathfrak{P}) = \dim(R_{\mathfrak{P}}) \leq c$.

Let A be a local ring with maximal ideal $\mathfrak{m}$, and put $k_A = A/\mathfrak{m}$. Let $n = \dim_{k_A}(\mathfrak{m}/\mathfrak{m}^2)$. By Nakayama's Lemma in the form of Lemma 4.9, $\mathfrak{m}$ is then generated by n elements. Thus $\dim(A) \leq n$ by the facts recalled above, as $A = A_{\mathfrak{m}}$. Recall the

Definition 16.5 A is called a regular local ring if

$$\dim(A) = \dim_{k_A}(\mathfrak{m}/\mathfrak{m}^2).$$

We now introduce the following important concepts, so far treated partially and only for curves in $\mathbb{P}_k^2$.

Definition 16.6 Let X be a scheme, and let $x \in X$ be a point. If $\mathcal{O}_{X,x}$ is a regular local ring, then x is called a regular point of X. Moreover, if x is a closed point in X, then $\dim(\mathcal{O}_{X,x})$ is referred to as the dimension of X at x. A scheme X such that all points $x \in X$ are regular, is called a regular scheme.

Definition 16.7 (Formally Étale) A morphism $f : X \longrightarrow Y$ is said to be formally étale if the following condition holds:

For all affine Y-schemes Z, and every closed subscheme $i : W \hookrightarrow Z$ such that $\mathfrak{J}_W$ is nilpotent, the map given by composition with i,

$$\mathrm{Hom}_Y(i,Y) : \mathrm{Hom}_Y(Z,X) \longrightarrow \mathrm{Hom}_Y(W,X)$$

is bijective.

Chapter 17
Cohomology Theory on Schemes

This chapter opens with a short reminder of some homological algebra, used to treat derived functors and Grothendieck cohomology. This is complemented by the Čech cohomology for Abelian sheaves on a topological space.

17.1 Some Homological Algebra

Let $\mathcal{C}$ be an Abelian category. A complex K of $\mathcal{C}$ consists of

(i) A sequence $K = \{K^n\}_{n \in \mathbb{Z}}$ of objects from $\mathcal{C}$, and
(ii) a sequence $\{d_K^n\}$ of morphisms, called *differentials* or *coboundary maps* of the complex

$$d_K^n : K^n \longrightarrow K^{n+1}$$

such that $d_K^{n+1} \circ d_K^n = 0$.

The complexes of $\mathcal{C}$ themselves form a category with a morphism $f = \{f^n\} : K \longrightarrow L$ defined as a sequence of morphisms $f^n : K^n \longrightarrow L^n$ such that all the diagrams

$$
\begin{array}{ccc}
L^n & \xrightarrow{\ f^n\ } & K^n \\
{\scriptstyle d_L^n}\big\downarrow & & \big\downarrow{\scriptstyle d_K^n} \\
L^{n+1} & \xrightarrow[\ f^{n+1}\ ]{} & K^{n+1}
\end{array}
$$

commute. One verifies that the collection of all complexes with values in the given category $\mathcal{C}$ forms a category, which we shall denote by $\mathcal{K}(\mathcal{C})$.

Proposition 17.1 *If K is a complex, then $\operatorname{im}(d_K^n)$ is a subobject of $\ker(d_K^{n+1})$.*

A. Holme, *A Royal Road to Algebraic Geometry*,
DOI 10.1007/978-3-642-19225-8_17, © Springer-Verlag Berlin Heidelberg 2012

Proof Since the composition

$$K^{n-1} \xrightarrow{d_K^{n-1}} K^n \xrightarrow{d_K^n} K^{n+1}$$

is zero, we have that $\mathrm{im}(d_K^{n-1})$ is a subobject of $\ker(d_K^n)$.   □

· By abuse of language we describe the situation by writing

$$\mathrm{im}(d_K^{n-1}) \subseteq \ker(d_K^n)$$

We introduce the following notation:

Definition 17.1 For all $n \in \mathbb{Z}$ let

$$Z^n(K) = \ker(d_K^n)$$
$$B^n(K) = \mathrm{im}(d_K^{n-1})$$
$$H^n(K) = Z^n(K)/B^n(K)$$

We have the following fact:

Proposition 17.2 *The functors B^n, Z^n and H^n are additive functors.*

Proof Everything follows by simple diagram chasing, using the universal properties of the kernel, image and cokernel objects. In fact, to show functoriality let $f : L \longrightarrow K$ be a morphism of complexes. Then by the universal property of $\ker(d_K^n)$ there exists a unique morphism $\overline{f}_n$ which makes the following diagram commute:

$$
\begin{array}{ccccc}
Z^n(L) = \ker(d_L^n) & \hookrightarrow & L^n & \xrightarrow{d_L^n} & L^{n+1} \\
{\scriptstyle \exists! \overline{f}_n} \downarrow & & {\scriptstyle f^n} \downarrow & & {\scriptstyle f^{n+1}} \downarrow \\
Z^n(K) = \ker(d_K^n) & \hookrightarrow & K^n & \xrightarrow[d_K^n]{} & K^{n+1}
\end{array}
$$

Then define $Z^n(f)$ as this $\overline{f}_n$.

Since $B^n(K) \subset Z^n(K)$ for all n, we may define $H^n(K) = Z^n(K)/B^n(K)$. This yields assignments $H^n : \mathcal{K}(\mathcal{C}) \longrightarrow \mathcal{C}$.

This assignment is functorial since the following diagram commutes

$$
\begin{array}{ccc}
B^n(L) & \hookrightarrow & Z^n(L) \\
{\scriptstyle B^n(f)} \downarrow & & {\scriptstyle Z^n(f)} \downarrow \\
B^n(K) & \hookrightarrow & Z^n(K)
\end{array}
$$

Whenever a general result may be shown in a manner similar to the above, we say simply that the *result follows by diagram chasing*. The detailed implementation being regarded as straightforward and left to the reader.   □

Definition 17.2 A complex K is said do be positive if $K^n = 0$ for all $n < 0$, negative if $K^n = 0$ for all $n > 0$. A positive complex is called a cochain-complex, and $H^n(K)$ is the called the nth cohomology of K, $H^n(\ )$ is called the nth cohomology functor. In the negative case we write $H^{-n}(\ ) = H_n(\ )$ and call it the nth. homology functor.

Two morphisms of complexes of $\mathcal{C}$

$$g, f : L \longrightarrow K$$

are said to be homologous, or *homotopic*, if there exist morphisms in $\mathcal{C}$

$$k^n : L^n \longrightarrow K^{n-1}$$

for all $n \in \mathbb{Z}$ such that

$$g^n - f^n = d_K^{n-1} k^n + k^{n+1} d_L^n,$$

for all n, or written more compactly: There exists a morphism $k : L \longrightarrow K$ of degree -1 such that $f - g = dk + kd$. In this case we write $f \sim g$, as is easily verified this is an equivalence relation. We have the following

Proposition 17.3 *If $f \sim g$ then $H^n(f) = H^n(g)$ for all n.*

Proof See [4], Proposition 7.1: Since H^n is an additive functor, it suffices to show that whenever $f \sim 0$, then $H^n(f) = 0$. This is done as follows by a straightforward diagram chase: Since $f \sim 0$ we have $f = dk + kd$ where $k : L \longrightarrow K$ is a morphism of complexes of degree -1.

In other words, there are morphisms commuting with the $d's$, $k^n : L^n \longrightarrow K^{n-1}$, such that for all n

$$f^n = d_K^{n-1} k^n + k^{n+1} d_L^n$$

as shown in

We are dealing with the situation below, where $\lambda \in \ker(d_L^n)$:

$$
\begin{array}{ccc}
H^n(L) == \ker(d_L^n)/\mathrm{im}(d_L^{n-1}) & \quad & \lambda + \mathrm{im}(d_L^{n-1}) \\
\Big\downarrow H^n(f) & & \Big\downarrow \\
H^n(K) == \ker(d_K^n)/\mathrm{im}(d_K^{n-1}) & & f^n(\lambda) + \mathrm{im}(d_K^{n-1})
\end{array}
$$

By the explicit form of the condition $f \sim 0$ given above, we have

$$f^n(\lambda) = d_K^{n-1} k^n(\lambda) + k^{n+1} d_L^n(\lambda),$$

hence $f^n(\lambda) = d_K^{n-1}(k^n(\lambda)) \in \mathrm{im}(d^{n-1})$. Thus $H^n(f)(\overline{\lambda}) = 0$, and we have shown that $H^n(f) = 0$.   □

Now let $A \in \mathrm{Ob}(\mathcal{C})$. A right resolution of A is a positive complex K and a morphism $\epsilon : A \longrightarrow K^0$ such that the sequence

$$0 \longrightarrow A \longrightarrow K^0 \longrightarrow K^1 \longrightarrow K^2 \cdots$$

is exact.

An object I in $\mathcal{C}$ is called *injective* if the functor from $\mathcal{C}$ to the category of Abelian groups Ab

$$\mathrm{Hom}_{\mathcal{C}}(\ ,I) : \mathcal{C} \longrightarrow \mathrm{Ab}$$

is exact and not merely left exact: In other words, given a morphisms f and a monomorphism i there exists a morphism g which makes the diagram commutative:

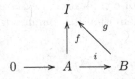

See [4], p. 124.

Let $\mathcal{C}$ be an Abelian category. Recall the definition of the functor of points of an object X

$$h_{\mathcal{C}}^A(X) = \mathrm{Hom}_{\mathcal{C}}(X, A).$$

This functor is contravariant and left exact. It is exact if and only if the object A is injective, this being one of several equivalent definitions of the term. An object is said to be *projective* if it is injective in the dual category.

There is an appealing, more symmetrical form of this definition:

Proposition 17.4 *An object Q is injective if and only if any diagram as below where the leftmost composition is 0, factors through a morphism ψ as shown:*

$$A_1 \xrightarrow{u_1} A_2 \xrightarrow{u_2} A_3$$

with $v: A_2 \to Q$, $0: A_1 \to Q$, $\psi: A_3 \to Q$ into Q.

The *proof* is an easy exercise, and may be found on p. 175 in [4].

The Abelian category $\mathcal{C}$ is said to have *enough injectives* if every object may be embedded into an injective object, that is to say, for all objects A in $\mathcal{C}$ there exists an injective object I and a monomorphism $i: A \longrightarrow I$.

We have the following

Lemma 17.5 *Assume that $\mathcal{C}$ has enough injectives. Then for every object A in $\mathcal{C}$ we may construct an injective resolution, functorial in A. Two such injective resolutions are homotopic in the sense explained below, at the end of the proof of this lemma.*

Proof We follow [4]. By assumption there is a monomorphism $A \xrightarrow{\epsilon} I^0$ for an injective object I^0. Similarly, letting $A^1 = \operatorname{coker}(\epsilon)$ we get a monomorphism $\epsilon_1 : A^1 \longrightarrow I^2$ for some injective object I^2. By composition we obtain the short exact sequence

$$0 \longrightarrow A \xrightarrow{\epsilon} I^0 \xrightarrow{\epsilon^1} I^1.$$

Repetition yields a resolution by injective objects,

$$0 \longrightarrow A \xrightarrow{\epsilon} I^0 \xrightarrow{\epsilon^1} I^1 \xrightarrow{\epsilon^2} I^2 \cdots$$

or for short

$$0 \longrightarrow A \longrightarrow I.$$

It remains to show the assertion about functoriality and uniqueness up to homotopy.

More generally, suppose we have two injective resolutions

$$0 \longrightarrow A \longrightarrow I \quad \text{and} \quad 0 \longrightarrow B \longrightarrow J$$

and a morphism $u : A \longrightarrow B$ as shown below:

$$0 \longrightarrow A \longrightarrow I^0 \longrightarrow I^1 \longrightarrow I^2 \longrightarrow I^3 \longrightarrow \cdots$$
$$\downarrow u$$
$$0 \longrightarrow B \longrightarrow J^0 \longrightarrow J^1 \longrightarrow J^2 \longrightarrow J^3 \longrightarrow \cdots$$

Using injectivity for the J^i we get morphisms f^i making the diagrams commute in

$$0 \longrightarrow A \overset{}{\longrightarrow} I^0 \overset{d_I^0}{\longrightarrow} I^1 \overset{d_I^1}{\longrightarrow} I^2 \overset{d_I^2}{\longrightarrow} I^3 \overset{d_I^3}{\longrightarrow} \cdots$$
$$\downarrow u \quad \downarrow f^0 \quad \downarrow f^1 \quad \downarrow f^2 \quad \downarrow f^3$$
$$0 \longrightarrow B \longrightarrow J^0 \overset{d_J^0}{\longrightarrow} J^1 \overset{d_J^1}{\longrightarrow} J^2 \overset{d_J^2}{\longrightarrow} J^3 \overset{d_J^3}{\longrightarrow} \cdots$$

Indeed, we first get f^0 since ϵ is a monomorphism and J^0 is an injective object. To define the morphism $f^1 : I^1 \longrightarrow J^1$ we consider the diagram

$$A \overset{\epsilon}{\longrightarrow} I^0 \overset{d_I^0}{\longrightarrow} I^1$$
$$\downarrow d_J^0 f^0$$
$$J^1$$

We evidently find ourselves in the situation of Proposition 17.4, hence there is a morphism $f^1 : I^1 \longrightarrow J^1$ which makes the diagram commute, i.e.,

$$d_J^0 f^0 = f^1 d_I^0.$$

Continuing in this way we get by induction morphisms $f^n : I^n \longrightarrow J^n$ such that the diagrams below commute:

$$I^{n-2} \overset{d_I^{n-2}}{\longrightarrow} I^{n-1} \overset{d_I^{n-1}}{\longrightarrow} I^n$$
$$d_J^{n-1} f^{n-1} \downarrow \qquad \swarrow f^n$$
$$J^n$$

or

$$d_J^{n-1} f^{n-1} = f^n d_I^{n-1}.$$

Thus the first part of the assertion is proven, it remains to show uniqueness up to homotopy. We proceed as follows:

Starting with the morphism $u : A \longrightarrow B$ we constructed above a morphisms of complexes

$$f^\bullet : I \longrightarrow J$$

compatible with the morphism u. Let

$$g^\bullet : I \longrightarrow J$$

be another such morphism satisfying the same conditions. To show that $f^\bullet$ and $g^\bullet$ are homotopic, we construct morphisms $k^i : I^i \longrightarrow J^{i-1}$ for all $i \geq 1$ such that

$$f^i - g^i = d_J^{i-1} k^i + k^{i+1} d_I^i.$$

We first construct $k^1 : I^1 \longrightarrow J^0$ from the diagram

$$
\begin{array}{ccccc}
A & \xrightarrow{\ \epsilon\ } & I^0 & \xrightarrow{\ d_I^0\ } & I^1 \\
 & & \Big\downarrow{\scriptstyle f^0 - g^0} & & \\
 & & J^0 & &
\end{array}
$$

Recall the commutative diagrams

$$
\begin{array}{ccc}
A & \xrightarrow{\ \epsilon\ } & I^0 \\
{\scriptstyle u}\Big\downarrow & {\scriptstyle f^0}\Big\downarrow\ {\scriptstyle g^0} & \\
B & \xrightarrow{\ \eta\ } & J^0
\end{array}
$$

which show that $(f^0 - g^0)\epsilon = 0$. Thus by Proposition 17.4 there is a morphism $k^1 : I^1 \longrightarrow J^0$ making the following diagram commutative:

$$
\begin{array}{ccccc}
A & \xrightarrow{\ \epsilon\ } & I^0 & \xrightarrow{\ d_I^0\ } & I^1 \\
 & & \Big\downarrow{\scriptstyle f^0 - g^0} & \swarrow{\scriptstyle k^1} & \\
 & & J^0 & &
\end{array}
$$

We proceed to move up one step, and use Proposition 17.4 once more to find a morphism $k^2 : I^2 \longrightarrow J^1$ such that the diagram below commutes:

$$I^0 \xrightarrow{\; d_I^0 \;} I^1 \xrightarrow{\; d_I^1 \;} I^2$$

with vertical morphism $f^1 - g^1 - d_J^0 k^1 \downarrow$ from I^1 to J^1 and diagonal morphism k^2 from I^2 to J^1.

For this, all we have to do is to show that the composition of the left horizontal morphism with the vertical one, is the zero morphism. This verification presents no problems whatsoever, and is left to the reader.

By this procedure we construct, for $n = 2, 3, 4, \ldots$ the morphisms k^n such that the diagrams below commute:

$$I^{n-1} \xrightarrow{\; d_I^{n-1} \;} I^n \xrightarrow{\; d_I^n \;} I^{n+1}$$

with vertical morphism $f^n - g^n - d_J^{n-1} k^n \downarrow$ from I^n to J^n and diagonal morphism k^{n+1} from I^{n+1} to J^n.

This completes the proof.   □

17.2 Derived Functors

Now let $F : \mathcal{C} \longrightarrow \mathcal{C}'$ be a covariant additive left exact functor. $\mathcal{C}$ is assumed to have enough injectives. For an object A, choose an injective resolution

$$0 \longrightarrow A \xrightarrow{\; \epsilon_A \;} I_A^\bullet.$$

We omit the label ϵ_A and the subscript A in I_A when no confusion is possible.

Form the (positive) complex

$$K = \{0 \longrightarrow F(A) \longrightarrow F(I^\bullet)\}.$$

Taking the cohomology of this complex we obtain objects which are denoted as follows:

$$H^n(K) = R^n F(A).$$

We have the following

Proposition 17.6 *The $R^n F(\ )$ are covariant functors from $\mathcal{C}$ to $\mathcal{C}'$. We have* $R^0 F(\ ) = F(\ ).$

Proof The proof is straightforward, and uses Lemma 17.5. More details and information may be found in Chap. 7 of [4]. □

Among the information of *loc. cit.* referred to above, we single out the following which is particularly important:

Proposition 17.7 (Snake Lemma) *Let*

$$0 \longrightarrow A \xrightarrow{f} B \xrightarrow{g} C \longrightarrow 0$$

be a short exact sequence from $\mathcal{C}$*. Then there are morphisms* ∂^i *for* $i = 1, 2, \ldots$ *such that*

1. *The following long sequence is exact:*

$$
\begin{array}{ccccccc}
0 & \longrightarrow & F(A) & \xrightarrow{F(f)} & F(B) & \xrightarrow{F(g)} & F(C) \\
& & & & & \partial^1 & \\
& & R^1 F(A) & \longrightarrow & R^1 F(B) & \longrightarrow & R^1 F(C), \\
& & & R^1(f) & \partial^2 & R^1(g) & \\
& & R^2 F(A) & \longrightarrow & \cdots & & \cdots \\
& & & R^2(f) & \partial^3 & & \\
& & & & & & \cdots \\
& & & & \partial^n & & \\
& & R^n F(A) & \longrightarrow & R^n F(B) & \longrightarrow & R^n F(C) \\
& & & & \partial^{n+1} & & \\
& & \cdots & & & &
\end{array}
$$

Moreover, we have that

2. *The morphisms* ∂^i *are compatible with morphisms of short exact sequences in the obvious way.*

The functors $R^i F$ are referred to as the right derived functors of the left exact functor F, and the ∂^i's are referred to as the *connecting morphisms*. Passing to the dual category $\mathcal{C}'^{\vee}$ we get the concept of left derived functors of a right exact functor G, with connecting morphisms going the other way.

In general a collection of covariant functors F^i with the properties listed in Proposition 17.7 is referred to as a (covariant) ∂-sequence.

17.3 Grothendieck Cohomology

The purpose of this section is to define *cohomology of sheaves* $H^i(X,\mathcal{F})$ as the right derived functors $R^i\Gamma$ of the left exact functor of global sections

$$\Gamma = \Gamma(X, \ ) : \mathrm{Mod}_X \longrightarrow \mathrm{Mod}_{\mathcal{O}_X(X)}.$$

Theorem 17.8 *The category* Mod_X *of* $\mathcal{O}_X$*-modules on a ringed space* X *has enough injectives.*

Proof We essentially follow Hartshorne's textbook [18], starting on p. 206. We reduce the proof to showing the corresponding result for the category Mod_A of modules over a commutative ring with 1:

Proposition 17.9 *If A is a commutative ring with 1 then every A-module is isomorphic to a submodule of an injective module.*

A proof of this key result is given in [34] Sects. 3A and 3D. See also [33]. Assuming this, we proceed to prove Theorem 17.8: Let $\mathcal{F}$ be a sheaf of $\mathcal{O}_X$-modules on X. For each point $x \in X$ the $\mathcal{O}_{X,x}$-module $\mathcal{F}_x$ can be embedded in an injective $\mathcal{O}_{X,x}$-module I_x, by Proposition 17.9. For each point $x \in X$ we let

$$j_x : \{x\} \hookrightarrow X$$

denote the inclusion mapping of the point x into X, I_x being considered as a sheaf of modules on the one point space $\{x\}$. We define a sheaf of $\mathcal{O}_X$-modules on X by

$$\mathcal{I} = \prod_{x \in X} (j_x)_*(I_x).$$

Now recall Definition 9.2, from which it follows that as $\mathcal{F}$ is a sheaf, it is canonically isomorphic to it's associated sheaf: $[\mathcal{F}] = \mathcal{F}$. It follows that the local homomorphisms $\mathcal{F}_x \longrightarrow I_x$ glue to an $\mathcal{O}_X$-morphism $\mathcal{F} \longrightarrow \mathcal{I}$, which is obviously an injective morphism, as it is so on every stalk.

We show that $\mathcal{I}$ is an injective object in the category Mod_X by verifying that the functor $\mathrm{Hom}_{\mathcal{O}_X}(\ , \mathcal{I})$ is exact. This follows since it is the direct product over all $x \in X$ of exact functors: Namely $\mathcal{G} \mapsto \mathcal{G}_x$ followed by $\mathrm{Hom}_{\mathcal{O}_{X,x}}((\ )_x, I_x)$.   □

From now on the geometric objects are assumed to be *schemes*, not merely ringed spaces, unless specified otherwise. We are going to compute the cohomology and the cohomology sheaves of the $\mathcal{O}_X$-modules on a scheme X introduced in Chap. 12.

Since there are enough injectives in the category Mod_X, we are able to state the following

Proposition-Definition 17.10 *Let $\mathcal{F}$ and $\mathcal{G}$ be objects in Mod_X. As the functors involved have the appropriate exactness properties, we have*

1. $H^i(\mathcal{F}, X) = R^i\Gamma(X,\ )(\mathcal{F})$ *exists and is referred to as the ith cohomology of $\mathcal{F}$ on X.*
2. *If $f : X \longrightarrow Y$ is a morphism of schemes, then f_* is a left exact functor from the category of $\mathcal{O}_X$ modules on X to the category of $\mathcal{O}_Y$-modules on Y. Thus the right derived functors exist, we refer to $(R^i(f_*))(\mathcal{F})$ as the ith higher direct image of $\mathcal{F}$ under f.*
3. $L^i(\mathcal{F} \otimes (\ ))(\mathcal{E}) = \mathrm{Tor}^i_X(\mathcal{F}, \mathcal{E})$. *Moreover, this is canonically isomorphic to $L^i((\ ) \otimes \mathcal{E})(\mathcal{F})$.*
4. *Let $H_{\mathcal{F}}(\ ) = \mathcal{H}om(\mathcal{F},\ )$. Then the functor $H_{\mathcal{F}}$ is right exact, and $L^i H_{\mathcal{F}}(\mathcal{E}) = \mathcal{E}xt^i_X(\mathcal{F}, \mathcal{E})$. Moreover, $V_{\mathcal{E}} = \mathcal{H}om(\ , \mathcal{E})$ is left exact, and $R^i V_{\mathcal{E}}(\mathcal{F})$ is canonically isomorphic to $\mathcal{E}xt^i_X(\mathcal{F}, \mathcal{E})$.*

Proof Immediate by the elementary properties of derived functors.   □

The following theorem sums up important special cases of much more general theorems to be found in [15].

Theorem 17.11 *Assume that X and Y in Proposition-Definition 17.10 are complete schemes over $\mathrm{Spec}(k)$, i.e., the structure morphism is proper, where k is a Noetherian ring (in most applications an algebraically closed field). Assume that f is a proper morphism. Also assume that the $\mathcal{O}_X$-modules $\mathcal{F}$ and $\mathcal{E}$ are coherent. Then all the $\mathcal{O}_X$ and $\mathcal{O}_Y$-modules constructed, in particular all $R^i f_*(\mathcal{F})$, are coherent. Thus in particular all $H^i(\mathcal{F}, X)$ are finite dimensional vector spaces over k. Finally $R^i f_*(\mathcal{F})$ vanish for all $i > \dim(X)$, in particular $H^i(\mathcal{F}, X) = 0$ for all $i > \dim(X)$.*

Proof We first note that the two assertions following "*in particular*", follow by applying the general assertions to the structural morphism $X \longrightarrow \mathrm{Spec}(k)$. But conversely, these special cases imply the general assertions: Indeed, the target scheme Y being complete, is covered by a finite number of affine open subschemes $Y = \bigcup_{i=1}^N U_i$. Coherency being a local property, it suffices to verify the assertions on $R^i f_*(\mathcal{F})$ for $\mathcal{F}_i = \mathcal{F}_{X_i}$ and $f_i = f_{X_i}$. Hence we may assume that $Y = \mathrm{Spec}(A)$. Now, it suffices to prove that $H^i(\mathcal{F}, X)$ is a finite A-module, zero for all $i \gg 0$. In fact, this will imply the assertion on the module $R^i f_*(\mathcal{F})$, since by the universal property the latter is the sheaf associated to the presheaf $\mathcal{H}(V) = H^i(f^{-1}(V), \mathcal{F}_{|f^{-1}(V)})$ on X.

For the same reason we have the

Lemma 17.12 *If* $Y = \mathrm{Spec}(A)$, *then*

$$R^i f_*(\mathcal{F}) = \widetilde{H^i(X, \mathcal{F})}.$$

The proof of the general statements above are completed by this lemma. $\square$

Proposition-Definition 17.13 *In the situation of Proposition-Definition* 17.11 *with* $Y = \mathrm{Spec}(k)$ *for some field* k, *the sum to the right below is finite, it is called the Euler-Poincaré Characteristic of* $\mathcal{E}$ *and denoted by* $\chi(\mathcal{E})$:

$$\chi(\mathcal{E}) = \sum_{i=0}^{\infty} (-1)^i \dim_k(H^i(X, \mathcal{E})).$$

Proof By Theorems 17.11 and Proposition 17.7. $\square$

17.4 Flasque, or Flabby Sheaves

The class of injective sheaves is important for computing cohomology, as we shall see. Another class of sheaves useful in this context, is the class of *flasque* sheaves, in English often called *flabby* sheaves.

A sheaf $\mathcal{F}$ of $\mathcal{A}b$ on a topological space is called flasque if all restriction mappings from an open subset U of X to a smaller open subset $V \subset U$ is surjective.

We have the following:

Proposition 17.14 *Any injective* $\mathcal{O}_X$-*module* $\mathcal{I}$ *on a ringer space* X *is flasque.*

Proof For all open $U \subset X$ we let $j : U \hookrightarrow X$ be the inclusion and $j_!(\mathcal{O}_X|U) = \mathcal{O}_U$ be the extension by zero of $\mathcal{O}_X|U$ from U to X. For open subsets $V \subset U$ we get

$$0 \longrightarrow \mathcal{O}_V \longrightarrow \mathcal{O}_U$$

which yields

$$\mathrm{Hom}(\mathcal{O}_U, \mathcal{I}) \longrightarrow \mathrm{Hom}(\mathcal{O}_V, \mathcal{I}) \longrightarrow 0$$

that is to say,

$$\mathcal{I}(U) \longrightarrow \mathcal{I}(V) \longrightarrow 0.$$

Thus $\mathcal{I}$ is flasque. $\square$

Remark 17.15 See [18] Exercise II 1.19 for a full treatment of extending a sheaf by zero. Extending a sheaf $\mathcal{F}$ by zero *from an open subset* is done by using the technique from Sect. 8.10: Construct a sheaf from its values on a basis for the topology, by assigning $\mathcal{F}(B)$ if the basis open subset B is contained in U, and 0 if B does not meet U. As U is itself open, we may safely discard the basis open subsets meeting both U and its complement.

The concept of objects which are *acyclic* for a (left exact) functor is important in the applications of homological algebra, since such objects allow a painless computation of the functor's (right) derived functors. Nice and understandable notes on this subjects are given in [13]. Flasque sheaves are acyclic for the functor $\Gamma(\mathcal{F})$ of global sections of sheaves:

Theorem 17.16 *If $\mathcal{F}$ is a flasque sheaf on the ringed space X, then*

$$H^i(X, \mathcal{F}) = 0 \quad \text{for all } i > 0.$$

Proof There are enough injectives on X, so we may form the short exact sequence

$$0 \longrightarrow \mathcal{F} \longrightarrow \mathcal{I} \longrightarrow \mathcal{G} \longrightarrow 0$$

where $\mathcal{I}$ is injective. We first show the

Lemma 17.17 *For each open subset $U \subset X$ the following sequence is exact:*

$$0 \longrightarrow \mathcal{F}(U) \longrightarrow \mathcal{I}(U) \longrightarrow \mathcal{G}(U) \longrightarrow 0.$$

Proof of Lemma 17.17.[1] We may assume $U = X$. Let $g \in \mathcal{G}(X)$. We have to show that it is the image of a section of $\mathcal{I}$ on X. Consider the set $\mathcal{S}$ of all pairs (V, s) such that $s \in \mathcal{I}(V)$ is mapped to $\rho_{X,V}(g) \in \mathcal{G}(V)$. These pairs are ordered by $(V, s) \preceq (V's')$ if $V \subseteq V'$ and s' restricts to s. Evidently every chain of such elements have an upper bound. We now use a version of the Axiom of Choice, known as Zorn's Lemma or as the Kuratowski-Zorn lemma:

Lemma 17.18 (Zorn's Lemma) *Let $\mathcal{T}$ be a non empty partially ordered set in which every totally ordered chain has an upper bound. Then $\mathcal{T}$ contains at least one maximal element.*

Now let (V_0, s_0) be the maximal element of $\mathcal{S}$ which exists by Zorn's Lemma. If $V_0 = X$, then $s_0 \in \mathcal{I}(X)$ is mapped to $g \in \mathcal{G}(X)$ and we are done.

If $V_0 \neq X$, then we can find a non empty open $V_1 \neq V_0$ and a section $s_1 \in \mathcal{I}(V_1)$ which is mapped to the appropriate restriction of g in $\mathcal{G}(V_1)$. The restrictions of s_0 and s_1 to $V_1 \cap V_0$ differ by an element in $\mathcal{F}(V_1 \cap V_0)$, which

---

[1] Inspired by [7].

since $\mathcal{F}$ is flasque can be extended to V_1. Thus we can modify s_1 by this extended section to agree with s_0 on $U_0 \cap U_1$. Hence (V_0, s_0) is not maximal, a contradiction. Thus $V_0 = X$, and the proof for Lemma 17.17 is complete. $\square$

We may now prove the

Lemma 17.19 *The sheaf $\mathcal{G}$ is flasque.*

Proof Let $V \subset U$. By Lemma 17.17 any section $s_V \in \mathcal{G}(V)$ comes from a section $t_V \in \mathcal{I}(V)$, which since $\mathcal{I}$ is flasque by Proposition 17.14 extends to a section t_U over U. This is mapped to $s_U \in \mathcal{G}(U)$, extending s_V. $\square$

We proceed with the proof of Theorem 17.16.

The long exact cohomology sequence of

$$0 \longrightarrow \mathcal{F} \longrightarrow \mathcal{I} \longrightarrow \mathcal{G} \longrightarrow 0$$

looks like this:

$$0 \longrightarrow \Gamma(X, \mathcal{F}) \longrightarrow \Gamma(X, \mathcal{I}) \longrightarrow \Gamma(X, \mathcal{G})$$
$$\partial^1$$
$$H^1(X, \mathcal{F}) \longrightarrow 0 \longrightarrow H^1(X, \mathcal{G})$$
$$\partial^2$$
$$H^2(X, \mathcal{F}) \longrightarrow 0 \longrightarrow H^2(X, \mathcal{G})$$
$$\partial^3$$
$$\cdots$$
$$\partial^n$$
$$H^n(X, \mathcal{F}) \longrightarrow 0 \longrightarrow H^n(X, \mathcal{G})$$
$$\partial^{n+1}$$
$$\cdots$$

By Lemma 17.17 applied to $U = X$ we get the exact sequence

$$0 \longrightarrow \Gamma(X, \mathcal{F}) \longrightarrow \Gamma(X, \mathcal{I}) \longrightarrow \Gamma(X, \mathcal{G}) \longrightarrow 0$$

from which we conclude that $H^1(X, \mathcal{F}) = 0$. The same procedure may be applied to other flasque sheaves, in particular to $\mathcal{G}$. Hence $H^1(X, \mathcal{G}) = 0$, thus $H^2(X, \mathcal{F}) = 0$, etc. This completes the proof of Theorem 17.16. $\square$

17.5 Čech Cohomology

An earlier, more explicit and concrete definition of sheaf cohomology follows the lines of Čech theory. We give the definition in a general setting. Let $\mathcal{F}$ be a sheaf of Abelian groups on the topological space X, and $\mathcal{U} = \{U_i\}_{i \in I}$ be an open covering of X. Let $\sigma = (i_0, i_1, \ldots, i_p)$ be a sequence of $p+1$ indices from the indexing set I, in this context usually referred to as a *p-simplex*. Put

$$U_\sigma = U_{i_0} \cap U_{i_1} \cap \cdots \cap U_{i_p}.$$

An alternating p-cochain c of the covering $\mathcal{U}$ with coefficients in $\mathcal{F}$ is a function which to all p-simplexes σ as above assigns an element $c_\sigma \in \mathcal{F}(U_\sigma)$ subject to the conditions that if two indices in σ are equal then $c_\sigma = 0$ and that c_σ changes sign when two indices are interchanged. The set of all such alternating p-cochains form an Abelian group $C^p(\mathcal{U}, \mathcal{F})$. The following observation is immediate:

Lemma 17.20 *There is a canonical isomorphism*

$$C^p(\mathcal{U}, \mathcal{F}) \cong \prod_{\{\sigma \mid i_0 < i_1 < \cdots < i_p\}} \mathcal{F}(U_\sigma).$$

We shall identify the left hand side with the right hand side in the above lemma. Now define a homomorphism

$$d^p : C^p(\mathcal{U}, \mathcal{F}) \longrightarrow C^{p+1}(\mathcal{U}, \mathcal{F})$$

by

$$d^p(c)_{i_0, i_1, \ldots, i_{p+1}}$$

$$= \sum_{k=0}^{p+1} (-1)^k \rho_{U_{(i_0, \ldots, \hat{i}_k, \ldots, i_{p+1})}, U_{(i_0, \ldots, \hat{i}_k, \ldots, i_{p+1})}} (c_{i_0, \ldots, \hat{i}_k, \ldots, i_{p+1}})$$

where as always $\rho_{U,V}$ denotes the restriction map from the open set U to the open subset V, and $\hat{i}_k$ means that this index should be omitted. Applying this homomorphism twice, we see that the composition is the zero map, $d^{p+1} \circ d^p = 0$. Indeed, let $c = (c_{j_0, \ldots, j_p} \mid j_0 < j_1 < \cdots < j_p) \in C^p(\mathcal{U}, \mathcal{F})$. Then $d^{p+1}(d^p((c))_{i_0, i_1, \ldots, i_{p+2}}$ is a double sum of the form

$$\sum_{k=0}^{p+1} (-1)^k \sum_{\ell=0}^{p+2} (-1)^\ell \rho(c_{i_0, \ldots, \hat{i}_k, \ldots, \hat{i}_\ell, \ldots, i_{p+2}})$$

where ρ denotes the appropriate restriction map. But then the same element will appear in the sum twice, with opposite signs. Thus the double sum yields 0.

Hence we get a complex of abelian groups

$$C^0(\mathcal{U},\mathcal{F}) \xrightarrow{d^0} C^1(\mathcal{U},\mathcal{F}) \xrightarrow{d^1} \cdots \xrightarrow{d^{i-1}} C^i(\mathcal{U},\mathcal{F}) \xrightarrow{d^i} \cdots .$$

The group $H^p(C^\bullet(\mathcal{U},\mathcal{F}))$ is denoted by $H^p(\mathcal{U},\mathcal{F})$ and referred to as the pth Čech cohomology of X for the covering $\mathcal{U}$ with coefficients in $\mathcal{F}$.

If the open covering $\mathcal{V}$ is a refinement of $\mathcal{U}$, then there is an obvious canonical group-homomorphism

$$H^p(C^\bullet(\mathcal{U},\mathcal{F})) \xrightarrow{\iota_{\mathcal{U},\mathcal{V}}} H^p(C^\bullet(\mathcal{V},\mathcal{F})).$$

Letting $\Upsilon(X)$ denote the set of open coverings of X, partially ordered by refinement, we immediately verify that the set

$$\{\check{H}^p(\mathcal{U},\mathcal{F})\}_{\mathcal{U}\in\Upsilon(X)}$$

forms an inductive system for the mappings $\iota_{\mathcal{U},\mathcal{V}}$.

Definition 17.3 The pth Čech cohomology group of the topological space X with coefficients in the sheaf $\mathcal{F}$ is

$$\check{H}^p(X,\mathcal{F}) = \varinjlim_{\mathcal{U}\in\Upsilon(X)} \check{H}^p(\mathcal{U},\mathcal{F}).$$

The advantage of Čech cohomology is that it is more readily computable than the Grothendieck cohomology, and of course that the Čech— and Grothendieck versions give the same result for the most important cases in algebraic geometry. Here we shall not pursue this further than to the two theorems stated below, but refer the reader to references like [15, 18] and [35].

We observe that the group $C^i(\mathcal{U},\mathcal{F})$ is the group of global sections of a sheaf $\mathcal{C}^i(\mathcal{U},\mathcal{F})$ which we define as the associated sheaf of the presheaf

$$\mathcal{C}^i_{\mathrm{pre}}(\mathcal{U},\mathcal{F})(V) = C^i(\mathcal{U}|V,\mathcal{F})|V.$$

Clearly

$$\mathcal{C}^i(\mathcal{U},\mathcal{F})(X) = C^i(\mathcal{U},\mathcal{F}),$$

and there is an injective morphisms of Abelian sheaves on X

$$\varepsilon : \mathcal{F} \longrightarrow \mathcal{C}^0(\mathcal{U},\mathcal{F})$$

obtained by applying the canonical

$$\mathcal{F}(X) \longrightarrow \prod_{i\in I} \mathcal{F}(U_i) \quad \text{where } f \mapsto (\rho_{X,U_i}(f))_{i\in I} \in \mathcal{C}^0(\mathcal{U},\mathcal{F})$$

to all open subsets U of X, injectivity then follows from the sheaf property. Moreover, in the obvious way we obtain morphisms of Abelian sheaves

$$\mathcal{C}^i(\mathcal{U}, \mathcal{F}) \xrightarrow{d_{\mathcal{C}}^i} \mathcal{C}^{i+1}(\mathcal{U}, \mathcal{F}).$$

Proposition-Definition 17.21 *We then have that*

$$0 \longrightarrow \mathcal{F} \xrightarrow{\varepsilon} \mathcal{C}^0(\mathcal{U}, \mathcal{F}) \xrightarrow{d_{\mathcal{C}}^0} \mathcal{C}^1(\mathcal{U}, \mathcal{F}) \xrightarrow{d_{\mathcal{C}}^1} \cdots \xrightarrow{d_{\mathcal{C}}^{i-1}} \mathcal{C}^i(\mathcal{U}, \mathcal{F}) \xrightarrow{d_{\mathcal{C}}^i} \cdots$$

is an exact sequence, i.e., we have a resolution of $\mathcal{F}$, it is referred to as the Čech-resolution.

Proof Indeed, we have observed that ε is injective, and the image is equal to the kernel $d_{\mathcal{C}}^0$ by the sheaf property for $\mathcal{F}$.

By applying the global case to any open V we find, at the level of presheaves that

$$\ker(d_{\mathcal{C}}^i) \subseteq \operatorname{im}(d_{\mathcal{C}}^{i-1})$$

hence this is so for the associated sheaves.

To show the converse inclusion, it suffices to do so in each stalk. So let $u \in \mathcal{C}^p(\mathcal{U}, \mathcal{F})_x$ be such that $d_{\mathcal{C}x}^p(u) = 0$, we may assume that $x \in U_i$. There exists an open neighborhood V of x contained in U_i and an element $s \in \mathcal{C}^p(\mathcal{U}, \mathcal{F})(V)$ such that $s_x = u$. If $\sigma = (i_0, \ldots, i_{p-1})$ is a $(p-1)$- simplex, we let $i\sigma$ denote the p-simplex $(i, i_0, \ldots, i_{p-1})$. Making V smaller if necessary we may assume that u is a section of the *presheaf*, so $u \in C^p(\mathcal{U} \cap V, \mathcal{F}|V)$, hence we have that s is a family (s_τ) where τ runs through the p-simplexes and $s_\tau \in \mathcal{F}(V \cap U_\tau)$. Define $t \in C^{p-1}(\mathcal{U} \cap V, \mathcal{F}|V)$ by the rule $t_\sigma = s_{i\sigma} \in \mathcal{F}(v \cap U_i \cap U_\sigma) = \mathcal{F}(V \cap U_\sigma)$. This rule may be taken literally as it stands if $i < i_0$, but otherwise the expression $s_{i\sigma}$ must be modified by rearranging the indices in increasing order: Thus if $i = i_j$ for some i_j, it should be interpreted as 0, and otherwise the indices written in increasing order with the appropriate change of sign for $s_{i\sigma}$. However, for simplicity we argue as if i were less than i_0 in the following, the adjustment to the general case is unproblematic. For simplicity of notation we also denote restriction of a section ξ from some larger open set to the smaller one W by $\xi|W$. Also, for the simplex $\tau = (j_0, j_1, \ldots, j_p)$ we let τ_k denote the face $(j_0, j_1, \ldots, \hat{j}_k, \ldots, j_p)$.

With these preparations we have

$$(d^{p-1}(t))_\tau = \sum_{k=0}^{p} (-1)^k t_{\tau_k} |V \cap U_\tau$$

$$= \sum_{k=0}^{p} (-1)^k s_{i\tau_k} |V \cap U_\tau.$$

Removing $|V \cap U_\tau$ to simplify notation we get

$$(dt)_\tau = s_{(i,j_1,\dots,j_p)} - s_{(i,j_0,j_2,\dots,j_p)} + \dots + (-1)^p s_{(i,j_0,\dots,j_{p-1})}$$

and similarly

$$(ds)_{i\tau} = s_{(j_0,\dots,j_p)} - s_{(i,j_1,\dots,j_p)} + \dots - (-1)^p s_{(i,j_0,\dots,j_{p-1})}$$

and so

$$(ds)_{i\tau} = s_\tau - (dt)_\tau.$$

Since $ds = 0$ we thus have

$$s_\tau = (dt)_\tau$$

and the claim follows.   □

Theorem 17.22 *Let $X = \mathrm{Spec}(A)$ be an affine scheme and let $\mathcal{F} = \widetilde{M}$ be a quasi coherent module on X, and $\mathcal{U}$ be an open affine covering of X by subsets of the form $D(f_i)$. Then*

$$\check{H}^i(\mathcal{U}, \mathcal{F}) = H^i(X, \mathcal{F}) = 0 \quad \text{for all } i > 0.$$

Proof Take an injective resolution of M in the category of A-modules,

$$0 \longrightarrow M \longrightarrow I^\bullet.$$

Whenever I is an injective A-module, $\widetilde{I}$ is a flasque sheaf: Indeed, one needs to show that for all open $U \subset X$ the restriction

$$\widetilde{I}(X) \longrightarrow \widetilde{I}(U)$$

is surjective.[2] It follows from this that $H^i(X, \mathcal{F}) = 0$ for all $i > 0$, in view of Theorem 17.16.   □

We conclude this section by giving complete proofs of the following two results, given in [18] respectively as Lemma 4.4 and Theorem 4.5:

Theorem 17.23 *Let X be a topological space and $\mathcal{U}$ an open covering. Then for all $p \geq 0$ there is a map $c_{X,\mathcal{F},p}$, functorial in $\mathcal{F}$,*

$$c_{X,\mathcal{F},p} : \check{H}^p(\mathcal{U}, \mathcal{F}) \longrightarrow H^p(X, \mathcal{F})$$

---

[2]We give several references for this verification: Hartshorne provides a proof in [18], on pp. 214–215. Grothendieck has provided a proof in [16], and J.M. Campbell has given a very understandable, elementary proof in the spirit of [35], in [5].

Proof Following [13], we let X be a topological space, $\mathcal{U}$ an open covering and $\mathcal{F}$ a sheaf on X. We compare the Check- resolution and an injective resolution of $\mathcal{F}$:

$$0 \longrightarrow \mathcal{F} \longrightarrow \mathcal{C}^0(\mathcal{U}, \mathcal{F}) \longrightarrow \mathcal{C}^1(\mathcal{U}, \mathcal{F}) \longrightarrow \mathcal{C}^2(\mathcal{U}, \mathcal{F}) \longrightarrow \cdots$$

$$\downarrow \text{id}$$

$$0 \longrightarrow \mathcal{F} \longrightarrow \mathcal{J}^0 \longrightarrow \mathcal{J}^1 \longrightarrow \mathcal{J}^2 \longrightarrow \cdots$$

The top line is the Check resolution, the bottom line an injective resolution. We consider the leftmost part of the diagram

$$0 \longrightarrow \mathcal{F} \longrightarrow \mathcal{C}^0(\mathcal{U}, \mathcal{F})$$

$$\downarrow \text{id}$$

$$0 \longrightarrow \mathcal{F} \longrightarrow \mathcal{J}^0$$

and observe that by injectivity of $\mathcal{J}^0$ there exists a morphism

$$\varphi_0 : \mathcal{C}^0(\mathcal{U}, \mathcal{F}) \longrightarrow \mathcal{J}^0$$

making the diagram commutative. Using Proposition 17.4 we may continue, and obtain the diagram with all squares commuting

$$0 \longrightarrow \mathcal{F} \longrightarrow \mathcal{C}^0(\mathcal{U}, \mathcal{F}) \longrightarrow \mathcal{C}^1(\mathcal{U}, \mathcal{F}) \longrightarrow \mathcal{C}^2(\mathcal{U}, \mathcal{F}) \longrightarrow \cdots$$

$$\downarrow \text{id} \qquad \downarrow \varphi_0 \qquad \downarrow \varphi_1 \qquad \downarrow \varphi_2$$

$$0 \longrightarrow \mathcal{F} \longrightarrow \mathcal{J}^0 \longrightarrow \mathcal{J}^1 \longrightarrow \mathcal{J}^2 \longrightarrow \cdots$$

Thus we have a morphism of complexes

$$\varphi : \mathcal{C}(\mathcal{U}, \mathcal{F}) \longrightarrow \mathcal{J}$$

and we obtain

$$c_{X,\mathcal{F},p} = H^p(\varphi)) : \check{H}^p(\mathcal{U}, \mathcal{F}) \longrightarrow H^p(X, \mathcal{F}). \qquad \square$$

Theorem 17.24 *Let $\mathcal{U}$ be an open affine covering of the Noetherian separated scheme X, and let $\mathcal{F}$ be a quasi coherent sheaf on X. Then for all $p \geq 0$ the map $c_{X,\mathcal{F},p}$ is an isomorphism.*

Proof We use the following concept:

Definition 17.4 (Acyclicity Condition) The covering $\mathcal{U}$ of X is said to be acyclic for the sheaf $\mathcal{F}$ if for all $U_0, \ldots, U_n \in \mathcal{U}$ and $i > 0$

$$H^i(U_0 \cap \cdots \cap U_q, \mathcal{F}|U_0 \cap \cdots \cap U_n) = 0$$

Theorem 17.24 will now follow from the

Lemma 17.25 *Let $\mathcal{F}$ be a sheaf on the topological space and $\mathcal{U}$ be an open covering of X which is acyclic for $\mathcal{F}$. Then the maps*

$$c_{X,\mathcal{F},p} : \check{H}^p(\mathcal{U}, \mathcal{F}) \longrightarrow H^p(X, \mathcal{F})$$

are isomorphisms for all $p \geq 0$.

Proof of Lemma 17.25 For $p = 0$ the assertion of the lemma is true even without the acyclicity condition, since in this case both $\check{H}^0(\mathcal{U}, \mathcal{F})$ and $H^0(X, \mathcal{F})$ are the global sections of $\mathcal{F}$. We now proceed by induction on i, and embed $\mathcal{F}$ in an injective sheaf $\mathcal{I}$ and form the exact sequence

$$0 \longrightarrow \mathcal{F} \longrightarrow \mathcal{I} \longrightarrow \mathcal{I}/\mathcal{F} = \mathcal{Q} \longrightarrow 0. \tag{17.1}$$

For each non-empty $q + 1$-fold intersection

$$U = U_0 \cap \cdots \cap U_q$$

of subsets in the cover $\mathcal{U}$, we get the long exact sequence of cohomology for the right derived functors of left exact functor $\Gamma(U, \ )$

$$
\begin{array}{ccccccc}
0 & \longrightarrow & H^0(U, \mathcal{F}) & \longrightarrow & H^0(U, \mathcal{I}|U) & \longrightarrow & H^0(U, \mathcal{Q}|U) \\
 & & & & \partial^1 & & \\
 & H^1(U, \mathcal{F}|U) & \longrightarrow & H^1(U, \mathcal{I}|U) & \longrightarrow & H^1(U, \mathcal{Q}|U) \\
 & & & & \partial^2 & & \\
 & H^2(U, \mathcal{F}|U) & \longrightarrow & H^2(U, \mathcal{I}|U) & \longrightarrow & H^2(U, \mathcal{Q}|U) \\
 & & & & \partial^2 & & \\
 & \cdots & & & & &
\end{array}
$$

By the hypothesis of $\mathcal{F}$ being acyclic for $\mathcal{U}$, $H^i(U, \mathcal{F}|U) = 0$ for all $i > 0$, and $H^i(U, \mathcal{I}|U) = 0$ for all $i > 0$ since $\mathcal{I}|U$ is flasque. Thus of course $H^i(U, \mathcal{Q}|U) = 0$ for all $i > 0$, hence $\mathcal{Q}$ as well as $\mathcal{F}$ are acyclic for $\mathcal{U}$. Moreover the long exact cohomology sequence of 17.1 yields

$$0 \longrightarrow \Gamma(U, \mathcal{F}) \longrightarrow \Gamma(U, \mathcal{I}) \longrightarrow \Gamma(U, \mathcal{Q}) \longrightarrow 0. \tag{17.2}$$

The short exact sequences above yield an exact sequence of complexes

$$0 \longrightarrow \mathcal{C}^\bullet(\mathcal{U}, \mathcal{F}) \longrightarrow \mathcal{C}^\bullet(\mathcal{U}, \mathcal{I}) \longrightarrow \mathcal{C}^\bullet(\mathcal{U}, \mathcal{Q}) \longrightarrow 0. \qquad (17.3)$$

In this case we therefore get a long exact sequence of Čech cohomology, cf. also [18], p. 222. Taking into account the vanishing

$$H^i(U, \mathcal{I}|U) = 0 \quad \text{for all } i > 0$$

we obtain

$$0 \longrightarrow \check{H}^0(\mathcal{U}, \mathcal{F}) \longrightarrow \check{H}^0(\mathcal{U}, \mathcal{I}) \longrightarrow \check{H}^0(\mathcal{U}, \mathcal{Q}) \longrightarrow \check{H}^1(\mathcal{U}, \mathcal{F}) \longrightarrow 0 \qquad (17.4)$$

and isomorphisms

$$0 \longrightarrow \check{H}^i(X, \mathcal{Q}) \longrightarrow \check{H}^{i+1}(\mathcal{U}, \mathcal{F}) \longrightarrow 0 \text{ for all } i > 1. \qquad (17.5)$$

We now take the long exact sequence for the derived functors of $\Gamma(X, \ )$ applied to (17.3). We obtain the analogous sequences of (17.2), (17.4) and (17.5). These morphisms are compatible with the canonical morphisms $c_{X,\mathcal{F},p}$ defined above, and we get the commutative diagrams with exact rows:

$$
\begin{array}{ccccccccc}
0 & \longrightarrow & \check{H}^0(\mathcal{U}, \mathcal{F}) & \longrightarrow & \check{H}^0(\mathcal{U}, \mathcal{I}) & \longrightarrow & \check{H}^0(\mathcal{U}, \mathcal{Q}) & \longrightarrow & \check{H}^1(\mathcal{U}, \mathcal{F}) & \longrightarrow & 0 \\
& & \downarrow{\scriptstyle c_{X,\mathcal{F},0}} & & \downarrow{\scriptstyle c_{X,\mathcal{I},0}} & & \downarrow{\scriptstyle c_{X,\mathcal{Q},0}} & & \downarrow{\scriptstyle c_{X,\mathcal{F},1}} & & \\
0 & \longrightarrow & H^0(X, \mathcal{F}) & \longrightarrow & H^0(X, \mathcal{I}) & \longrightarrow & H^0(X, \mathcal{Q}) & \longrightarrow & H^1(X, \mathcal{F}) & \longrightarrow & 0
\end{array}
$$

Since $c_{X,\mathcal{F},0}$, $c_{X,\mathcal{I},0}$ and $c_{X,\mathcal{Q},0}$ are isomorphisms, being the canonical mappings of global sections, $c_{X,\mathcal{F},1}$ is also an isomorphism.

Next, consider

$$
\begin{array}{ccccccc}
0 & \longrightarrow & \check{H}^{i-1}(\mathcal{U}, \mathcal{Q}) & \longrightarrow & \check{H}^i(\mathcal{U}, \mathcal{F}) & \longrightarrow & 0 \\
& & \downarrow{\scriptstyle c_{X,\mathcal{Q},i-1}} & & \downarrow{\scriptstyle c_{X,\mathcal{F},i}} & & \\
0 & \longrightarrow & H^{i-1}(X, \mathcal{Q}) & \longrightarrow & H^i(X, \mathcal{F}) & \longrightarrow & 0
\end{array}
$$

For $i = 2$ this identifies $c_{X,\mathcal{F},2}$ with $c_{X,\mathcal{Q},1}$. But $\mathcal{Q}$ is also $\mathcal{U}$ acyclic, so we may apply the result we found for $\mathcal{F}$ to $\mathcal{Q}$, and conclude that $c_{X,\mathcal{Q},1}$ is an isomorphism. Thus $c_{X,\mathcal{F},2}$ is an isomorphism. We proceed by induction, and conclude that all $c_{X,\mathcal{F},i}$ are isomorphisms. Complete proof for Lemma 17.25 and hence for Theorem 17.24.   □

Chapter 18
Intersection Theory

In this chapter we first list some basic facts on divisors, restricting the attention to quasi projective schemes over a field for simplicity. The sheaf of quotients $\mathcal{K}_X$ which plays the role of the function field for varieties is introduced, as well as the group of Cartier divisors $\mathrm{Div}(X) = \Gamma(X, \mathcal{K}_X^*/\mathcal{O}_X^*)$. The basic concepts related to them are given, as is relation to the group of *Weil divisors*. Then follows a section on Chow homology and Chow cohomology, leading up to bivariant theories.

18.1 Basic Facts on Divisors

In this chapter we let X be a quasi projective scheme over a field k. But we start out with a construction which works in a more general setting. In face, let X be a scheme and let $U = \mathrm{Spec}(A)$ be an open affine subset. Let $T \subset A$ be the multiplicatively closed set of all non zero divisors $t \in A$, and let $\mathcal{K}_0(U) = T^{-1}A$. With the obvious restriction maps $\mathcal{K}_0$ then becomes a presheaf on X, let $\mathcal{K}_X$ denote the associated sheaf. The sheaf $\mathcal{K}_X$ plays the role which the function field has when X is reduced and irreducible. Deleting the zero element from $\mathcal{K}_X$ we get a sheaf $\mathcal{K}_X^*$ with the sheaf $\mathcal{O}_X^*$ of units in the structure sheaf as a subsheaf. We form the quotient sheaf $\mathcal{K}_X^*/\mathcal{O}_X^*$. Then there is an exact sequence

$$1 \longrightarrow \mathcal{O}_X^* \longrightarrow \mathcal{K}_X^* \longrightarrow \mathcal{K}_X^*/\mathcal{O}_X^* \longrightarrow 1$$

which yields

$$1 \longrightarrow \Gamma(X, \mathcal{O}_X^*) \longrightarrow \Gamma(X, \mathcal{K}_X^*) \longrightarrow \Gamma(X, \mathcal{K}_X^*/\mathcal{O}_X^*) \longrightarrow H^1(X, \mathcal{O}_X^*) \longrightarrow 0$$

since $\mathcal{K}_X^*$ is a flasque sheaf.

We now return to the case when X is a quasi projective scheme over the field k, and make the following definition:

A. Holme, *A Royal Road to Algebraic Geometry,*
DOI 10.1007/978-3-642-19225-8_18, © Springer-Verlag Berlin Heidelberg 2012

Definition 18.1 The group $\mathrm{Div}(X) = \Gamma(X, \mathcal{K}_X^*/\mathcal{O}_X^*)$ is called the group of Cartier Divisors on X. The operation in this group is written as $+$.

An element $f \in \Gamma(X, \mathcal{K}_X^*)$ is called a *meromorphic function on* X, and its image in $\mathrm{Div}(X)$ is denoted by (f), the notation $\mathrm{Div}(f)$ is also used. Such a divisor is called a *principal divisor*, they form the subgroup $\mathrm{DivPrinc}(X)$. Clearly

$$\mathrm{DivPrinc}(X) = \Gamma(X, \mathcal{K}_X^*)/\Gamma(X, \mathcal{O}_X^*).$$

It is also clear that

$$\mathrm{Div}(X)/\mathrm{DivPrinc}(X) \cong H^1(X, \mathcal{O}_X^*).$$

Two Cartier-divisors D_1 and D_2 are said to be linearly equivalent if $D_1 - D_2 \in \mathrm{DivPrinc}(X)$.

Thus a Cartier-divisor is given by a collection $\{f_U\}$ where $f_U \in \Gamma(U, \mathcal{K}_X^*)$, over some open covering $\mathcal{B}$ of X. If we may assume that all $f_U \in \Gamma(U, \mathcal{O}_X^*)$, then we write $D \succeq 0$, and if D is also non-zero we write $D \succ 0$, such a divisor is called *positive*.

Moreover, for each Cartier divisor D we define an invertible sheaf $\mathcal{O}_X(D)$ as follows: Whenever V is an open subset of some $U \in \mathcal{B}$ let f_V denote the restriction of f_U to V, then put $\mathcal{O}_X(D)(V) = \frac{1}{f_V}\mathcal{O}_X(V)$. It is easily seen that this is well defined, and independent of the choice of the covering $\mathcal{B}$. As is easily seen we then get a presheaf, the associated sheaf is denoted by $\mathcal{O}_X(D)$. It is locally free of rank 1, or *invertible*.

The isomorphism classes of invertible sheaves $\mathcal{P}$ form a group, the multiplication being induced by $\otimes_{\mathcal{O}_X}$ and the inverse by taking the dual module $\mathcal{H}om(\mathcal{O}_X, \mathcal{P})$.

It is easily seen that $\mathcal{O}_X(-D)$ is an ideal on X if and only if $D \succeq 0$.[1] Whenever $D \succ 0$ we denote the corresponding closed subscheme of X by $Y(D)$.

To sum up, we have defined a group homomorphism

$$\tau : \mathrm{Div}(X) \longrightarrow \mathrm{Pic}(X) \ \text{ by } \ D \mapsto [\mathcal{O}_X(D)].$$

Clearly $\ker(\tau) = \mathrm{DivPrinc}(X)$.

Let $\mathrm{Div}^+(X)$ denote the semigroup of Cartier divisors which are $\succ 0$. Now let $Z^1(X)$ be the Abelian group generated by codimension 1 subvarieties, the 1-cycles. For every $D \in \mathrm{Div}^+(X)$ the scheme $Y(D)$ has an associated cycle of codimension 1, defined as follows: For an irreducible component Z of $Y(D)$, of codimension 1 in X, write

$$N_{D,Z} = \mathrm{length}(\mathcal{O}_{Y(D),z})$$

---

[1] See e.g. [15] IV (21.2.7.1).

where $Z = \overline{\{z\}}$. Then write

$$\text{Cyc}(D) = \sum N_{D,z} Z$$

extending the sum over all components of $Y(D)$ of codimension 1.

This gives a mapping

$$\text{Div}^+(X) \longrightarrow Z^1(X)$$

which extends to a homomorphism of ordered groups

$$\text{Cyc}: \text{Div}(X) \longrightarrow Z^1(X).$$

This homomorphism is an isomorphism if $\mathcal{O}_{X,x}$ is a UFD for all points $x \in X$, see Proposition 6.11 in [18]. In particular this is the case when X is smooth, *loc. cit.* Remark 6.11.1A. By this observation we can make the following definition:

Definition 18.2 Assume that X is a non singular projective variety, and let $\omega_{X/k}$ be its canonical sheaf. Then the corresponding divisor is denoted by K_X and referred to as the canonical divisor of X.

In general we make the

Definition 18.3 $Z^1(X)$ is called the group of Weil divisors on X.

Frequently a positive, or *effective*, Cartier divisor is identified with the closed subscheme $Y(D)$, and thus one speaks of the *Cartier divisor and its embedding into* X. At other times a Cartier divisor is identified with its associated Weil divisor. This is harmless in the smooth case, but in general the context must be carefully kept in mind.

18.2 Chow Homology and Chow Cohomology

Let X be a quasi projective variety over the field k, and consider the free Abelian group $Z(X)$ generated by all irreducible closed subsets of X. $Z(X)$ is graded by dimension and by codimension. When considered as graded by dimension we write $Z_\bullet(X)$, and when the grading is by codimension we write $Z^\bullet(X)$. Let $\mathcal{F}$ be a coherent $\mathcal{O}_X$-module on X, and assume that $\dim(\text{Supp}(\mathcal{F})) \leq n$. Let $s_1, \ldots, s_m$ be the generic points of the irreducible components of $\text{Supp}(\mathcal{F})$ which are of dimension n.

Then define

$$Z_n(\mathcal{F}) = \sum_{i=1}^m \text{length}_{\mathcal{O}_{X,s_i}} (\mathcal{F}_{s_i}) \overline{\{s_i\}}$$

where $\text{length}_{\mathcal{O}_{X,s_i}}$ denotes length of an Artinian module over $\mathcal{O}_{X,s_i}$.

For a closed subscheme $i : Y \hookrightarrow X$ we put

$$[Y] = Z_{\dim(Y)}(i_*(\mathcal{O}_Y)).$$

The cycle $[X] \in Z_{\dim(X)}(X)$ itself is referred to as the *fundamental cycle* of X.

If $f : X \longrightarrow Y$ is a proper morphism, so that $R^i f_*(\mathcal{F})$ is coherent by Theorem 17.11, then we define

$$f_* : Z(X) \longrightarrow Z(Y)$$

by

$$f_*(Z_k(\mathcal{F})) = \sum_{i \geq 0} (-1)^i Z_k(R^i(f_*(\mathcal{F}))) \tag{18.1}$$

for all coherent $\mathcal{O}_X$-modules $\mathcal{F}$ such that

$$\dim(\mathrm{Supp}(\mathcal{F})) \leq k$$

compare [11], Sect. 1.2. Here there can be a non zero contribution only in the case when equality holds. Indeed, we have the

Lemma 18.1 *Only the term with $i = 0$ yields a non-zero contribution in* (18.1).

Proof First, we need the following result from [6], stated on p. 95 as Exercise 3.22(e) in Hartshorne's textbook [18]:

Theorem 18.2 (Chevalley) *Let $f : X \longrightarrow Y$ be a morphism of reduced and irreducible schemes over the field k. For each integer h, let C_h be the set of points $y \in Y$ such that $\dim(X_y) = h$. Then the subsets C_h are constructible, and if $e = \dim(X) - \dim(Y)$ then C_e contains an open dense subset of Y.*

We proceed with the proof of Lemma 18.1: Replacing X by a smaller closed subscheme if necessary, we may assume that $X = \mathrm{Supp}(\mathcal{F})$ and that $k = \dim(X)$. Moreover, we may assume that X and Y are reduced and irreducible.

Let $i > 0$. If $\dim(Y) < k$ then $Z_k(R^i f_*(\mathcal{F})) = 0$ by definition, and hence we are done. So assume $\dim(Y) = k (= \dim(X))$. Then by Chevalley's Theorem there is an open dense subset U of Y such that f induces a morphism

$$\overline{f} = f_U : f^{-1}(U) \longrightarrow U$$

all of whose fibers are zero-dimensional.

This morphism is quasi finite, and proper by base extension, hence finite and hence affine. Using that formation of higher direct images commute with base extension we may assume that $Y = U$ and $X = f^{-1}(U)$. But as $\overline{f}$ is affine, the functor $\overline{f}_*$ is exact, so

$$R^i f_*(\mathcal{F})|_U = R^i \overline{f}_*(\mathcal{F}|_{f^{-1}(U)}) = 0$$

for $i > 0$. Hence the support of $R^i f_*(\mathcal{F})$ is a proper closed subset, thus of dimension $< k$. This completes the proof. $\qquad\qquad\qquad\qquad\qquad\qquad\square$

For V reduced and irreducible,

$$f_*([V]) = d[f(V)]$$

where

$$d = \begin{cases} [k(V) : k(f(V))] & \text{when } \dim(V) = \dim(f(V)) \\ 0 & \text{otherwise.} \end{cases}$$

This is the approach used for the classical definition of f_*.

We thus have that $f_* : Z_\bullet(X) \longrightarrow Z_\bullet(Y)$ is a homogeneous homomorphism of degree 0, and $X \mapsto Z_\bullet(X)$ is a covariant functor for proper morphisms.

Now let $f : X \longrightarrow Y$ be a flat morphism with fiber dimension d. For all irreducible closed subsets W of Y of dimension n, we put

$$f^*([W]) = Z_{n+d}(f^*(\mathcal{O}_W)) = Z_{n+d}([f^{-1}(W)])$$

where $f^{-1}(W)$ is the scheme theoretic inverse image. One immediately shows the equality to the right. This leads to a homogeneous homomorphism of degree d. One verifies that this makes $X \mapsto Z_\bullet(X)$ into a contravariant functor for flat morphisms.

Now let $D = \text{Div}(t)$ be a principal and effective Cartier divisor, and $i : D \hookrightarrow X$ its embedding. The we define a "wrong way" or *Gysin* homomorphism

$$i^* : Z^k(X) \longrightarrow Z^k(D)$$

by

$$i^*([V]) = \begin{cases} 0 & \text{for } V \subset D \\ [V_t] & \text{for } V \not\subset D \end{cases}$$

where $[v_t]$ is the class of the closed subscheme of V defined by $t = 0$.

The main source for the exposition which follows is Fulton's article [11] as well as his book [12].

Definition 18.4 Let $W \subset X \times_k Y$ be a closed subscheme, flat over Y by the induced morphism $\pi : W \longrightarrow Y$ induced by the projection. Let $y_1, y_2 \in Y(k)$. Put $W_i = \pi^{-1}(y_i)$, they are closed subschemes of $(X \times_k Y)_{y_i}$, canonically identified with X by the morphism induced by the projection onto X. Then $[W_i] = \pi^*([y_i]) \in Z_\bullet(X)$. Now put

$$\delta^Y_{y_1, y_2}(W) = [W_1] - [W_2] \in Z_\bullet(X).$$

We now consider the subgroup $C_\bullet(X)_\text{alg} \subset Z_\bullet(X)$ generated by all elements of this type for all smooth, connected curves in Y.

Definition 18.5 The subgroup $C_\bullet(X)$ is referred to as the group of elements in $Z_\bullet(X)$ which are algebraically equivalent to zero.

If the curve is a copy of $\mathbb{P}^1_k$, then we get the subgroup of cycles rationally equivalent to zero, $C_\text{rat}\bullet(X) \subset Z_\bullet(X)$. Form $A_\text{rat}\bullet(X) = Z_\bullet(X)/C_\text{rat}\bullet(X)$, this Abelian group is referred to as the group of (rational) cycle classes, or the (rational) Chow homology group.

The Chow group of cycle classes $A_\bullet(X)$ is graded by dimension of the cycles, but it may also be graded by the codimension, in which case it is denoted by $A^\bullet(X)$.

From now on we work with the rational case unless otherwise stated.

We note the following:

Lemma 18.3 *For the morphisms where we have defined pushforward f_* and pullback f^* at the level of $Z_\bullet(X)$, they carry over to $A_\bullet(X)$.*

Proof To show is that pushforward f_* and pullback f^* at the level of $Z_\bullet(X)$ respects "rationally equivalent to zero". This is a straightforward verification. □

We note the following guaranteeing compatibility with fiber products, see [11], Sect. 1.6:

Theorem 18.4 (Fiber Products) *Consider the product diagram over Y*

$$
\begin{array}{ccc}
X' & \xrightarrow{g'} & X \\
{\scriptstyle f'}\downarrow & & \downarrow{\scriptstyle f} \\
Y' & \xrightarrow{g} & Y
\end{array}
$$

where g is flat and f proper. Then

$$f'_* g'^*(\alpha) = g^* f_*(\alpha)$$

for all $\alpha \in Z_\bullet(X)$. The same holds for $A_\bullet$.

Proof This is a straightforward verification using the definition of flat pullback and proper pushforward. □

The following result is in some sense an excision-type theorem for Chow homology, see [11], Sect. 1.9:

Theorem 18.5 *Let U be an open subscheme of X and let $Y = X - U$ with some structure as a closed subscheme. Denote the open, respectively closed, embeddings as follows*

$$i : U \hookrightarrow X \quad and \quad j : Y \hookrightarrow X.$$

Then we have an exact sequence of Abelian groups

$$A_{\bullet}(Y) \xrightarrow{j_*} A_{\bullet}(X) \xrightarrow{i^*} A_{\bullet}(U) \longrightarrow 0.$$

Proof We first remark that on the level of cycles we have an exact sequence

$$0 \longrightarrow Z_{\bullet}(Y) \xrightarrow{j_*} Z_{\bullet}(X) \xrightarrow{i^*} Z_{\bullet}(U) \longrightarrow 0.$$

Indeed, an open embedding is flat, so i^* is defined. Moreover, this map is surjective since an irreducible subvarieties Z' of U is the image under i^* of its closure $\overline{Z'}$. It is also clear that j_* is injective, and likewise that the composition of j_* and i^* is zero.

Finally, if a cycle on X restricts to zero on U, it must have support in the complement which is Y. Thus the exact sequence on the level of cycles follows. To prove the assertion for $A_{\bullet}$, we have to show that the constructions for cycles are compatible with rational, respectively algebraic, equivalence. This is true and proved in a straightforward manner, with the exception that a cycle on Y may be rationally equivalent to zero on X without being so on Y. The proof of the compatibility is straightforward, and is omitted here.   □

Example 18.1 Let $X = \mathbb{A}^1_k$. Then the mapping

$$\mathbb{Z} \longrightarrow A(X) \quad \text{defined by } n \mapsto n[X]$$

is an isomorphism.

Proof As usual we regard $\mathbb{A}^1_k$ as $D_+(X_0) \subset \mathbb{P}^1_k$. To show is that all zero cycles on $\mathbb{A}^1_k$ are rationally equivalent to 0, for this it suffices to show that for all points $P \in \mathbb{A}^1_k$, the 0-cycle P is rationally equivalent to 0. Let $\Delta \subset \mathbb{P}^1_k \times_k \mathbb{P}^1_k$ be the diagonal, and let $W = \Delta \cap \mathbb{A}^1_k \times_k \mathbb{P}^1_k = \Delta \cap (\mathbb{P}^1_k - [\infty]) \times_k \mathbb{P}^1_k$ where we, as usual, put $\infty = (0 : 1)$. Then we have the commutative diagram

Then, as cycles, $\pi^*(P) = P$ and $\pi^*(\infty) = 0$. This proves that the cycle P is rationally equivalent to 0 and we are done.   □

Chow homology behaves nicely with respect to products and coproducts of quasi projective varieties. We have the following two facts, which are not particularly difficult:

Proposition 18.6 *There is a canonical graded isomorphism*

$$\sigma_{X,Y} : A_\bullet(X) \oplus A_\bullet(Y) \longrightarrow A_\bullet(X \sqcup Y)$$

functorial in X and Y.

Proof The canonical embeddings into the disjoint unions

$$i_X : X \hookrightarrow X \sqcup Y \quad \text{and} \quad i_Y : Y \hookrightarrow X \sqcup Y$$

immediate yield

$$i_{X*} : Z_\bullet(X) \longrightarrow Z_\bullet(X \sqcup Y)$$

and

$$i_{Y*} : Z_\bullet(Y) \longrightarrow Z_\bullet(X \sqcup Y)$$

and thus

$$i_{X*} \oplus i_{Y*} : Z_\bullet(X) \oplus Z_\bullet(Y) \longrightarrow Z_\bullet(X \sqcup Y)$$

which is clearly an isomorphism. This is also evidently compatible with rational equivalence.   □

Theorem 18.7 *There is a canonical graded homomorphism*

$$\kappa_{X,Y} : A_\bullet(X) \otimes A_\bullet(Y) \longrightarrow A_\bullet(X \times_k Y)$$

i.e.,

$$\kappa_{X,Y\,m} : \bigoplus_{i+j=m} A_i(X) \otimes A_j(Y) \longrightarrow A_m(X \times_k Y)$$

defined by extending by linearity

$$[V] \otimes [W] \mapsto [V \times_k W]$$

such that the following holds:

(i) *For proper morphisms $f : X \longrightarrow X'$ and $g : Y \longrightarrow Y'$*

$$(f \times_k g)_*(\alpha \times \beta) = f_*(\alpha) \times g_*(\beta).$$

(ii) *For flat morphisms f and g as in (i),*

$$(f \times_k g)^*(\alpha \times \beta) = f^*(\alpha) \times g^*(\beta).$$

Proof (ii): Consider the bilinear mapping

$$\Phi : Z_i(X) \times Z_j(Y) \longrightarrow Z_{i+j}(X \times Y)$$

induced by

$$([V], [W]) \mapsto [V \times_k W]$$

for closed subvarieties $V \subset X$ and $W \subset Y$. This yields

$$\kappa_{i,j} : Z_i(X) \otimes Z_j(Y) \longrightarrow Z_{i+j}(X \times_k Y)$$

from which the claim follows by observing that the homomorphism of cycle groups is compatible with rational (and algebraic) equivalence.   □

Definition 18.6 The mapping $\kappa_{X,Y}$ is referred to as the exterior product, or the Künneth relation, of X and Y.

Proposition 18.8 *The Gysin homomorphism*

$$p_X^* : A_\bullet(X) \longrightarrow A_\bullet(X \times_k \mathbb{A}_k^n)$$

associated to the projection

$$p_X : X \times_k \mathbb{A}_k^n \longrightarrow X$$

is an isomorphism.

Proof Since $X \times_k \mathbb{A}_k^n = (X \times_k \mathbb{A}_k^{n-1}) \times_k \mathbb{A}_k^1$ it is enough to prove the claim for $n = 1$. By Theorem 18.6, if X has several connected components it suffices to show the claim for reach component separately. Thus we may assume that X is connected.

We first show that p_X^* is surjective, and proceed by induction on $m = \dim(X)$. If $m = 0$, then $X = \mathrm{Spec}(k)$ and $A_\bullet(X) = \mathbb{Z}$, while $X \times_k \mathbb{A}_k^1 = \mathbb{A}_k^1$, and the claim was shown in Example 18.1. Now assume the claim for $m - 1$ and let Y be a hyperplane section of X, and put $U = X - H$.

We use Theorem 18.5, and get the following diagram, which we denote by $(*)_U$:

$$
\begin{array}{ccccccc}
A_\bullet(Y) & \longrightarrow & A_\bullet(X) & \longrightarrow & A_\bullet(U) & \longrightarrow & 0 \\
\downarrow{\scriptstyle p_Y^*} & & \downarrow{\scriptstyle p_X^*} & & \downarrow{\scriptstyle p_U^*} & & \\
A_\bullet(Y \times_k \mathbb{A}_k^1) & \longrightarrow & A_\bullet(X \times_k \mathbb{A}_k^1) & \longrightarrow & A_\bullet(U \times_k \mathbb{A}_k^1) & \longrightarrow & 0
\end{array}
$$

To prove surjectivity of p_X^* it suffices to show that the classes $[Z]$ of all irreducible closed subsets Z of $X \times_k \mathbb{A}_k^1$ are in $\mathrm{im}(p_X^*)$. If $\overline{p_X(Z)} \neq X$, take $U =$

$X - \overline{p_X(Z)}$. From $(*)_U$ we see the following: The image of $[Z]$ in $A_\bullet(U \times_k \mathbb{A}_k^1)$ is 0, by exactness of the lower sequence the element $[Z]$ thus comes from $A_\bullet(Y \times_k \mathbb{A}_k^1)$, and it is therefore the image of an element $y \in A_\bullet(Y)$ by the induction assumption. Now $p_X^*(y) = [Z]$. On the other hand, if $\overline{p_X(Z)} = X$, then either $Z = X \times_k \mathbb{A}_k^1$, in which case the claim is trivial, or else Z is of codimension 1 in $X \times_k \mathbb{A}_k^1$.

We use the following lemma:

Lemma 18.9 *Let* $U = \mathrm{Spec}(A)$ *be a smooth affine scheme over* k, $A = k[T_1, \ldots, T_N]/\mathfrak{a}$. *Let* D *be a divisor on* $U \times_k \mathbb{A}_k^1 = \mathrm{Spec}(A[T])$. *Then there is a divisor* D' *on* U *such that* D *is rationally equivalent to the pullback of* D'.

The *proof* will be omitted here, but the claim will follow e.g. by Exercise 12.6 on p. 292 in [18].

Now let $U = \mathrm{Spec}(A)$ be an affine open subscheme of X, smooth over k. Let D be the divisor on $U \times_k \mathbb{A}_k^1 = \mathrm{Spec}(A[T])$ which corresponds to the codimension 1 subscheme $Z \cap \mathrm{Spec}(A[T])$.

Using Lemma 18.9 we apply the diagram $(*)$ to this U. We then have the following situation, where dotted arrows indicate "*there exists an element which maps to* . . .":

$$
\begin{array}{ccccc}
y & & x & \longmapsto [D'] & \longmapsto 0 \\
\uparrow & & & \uparrow & \\
\vdots & p_Y^* & & \vdots\ p_U^* & \\
\vdots & & & \downarrow & \\
\vdots & & [Z] & \longmapsto [D] & \longmapsto 0 \\
\downarrow & & & & \\
y' & \longmapsto & \epsilon = [Z] - p_X^*(x) & \longmapsto 0 &
\end{array}
$$

So we proceed as follows: $[D]$ comes from $[D']$ by the lemma. Then $[D']$ comes from an element x. Use this x to define the element $\epsilon = [Z] - p_X^*(x)$. It maps to 0 in $A_\bullet(U \times_k \mathbb{A}_k^1)$. Hence it comes from $y' \in A_\bullet(Y \times_k \mathbb{A}_k^1)$. By a second use of the induction hypothesis y' comes from an element y. Denote its image in $A_\bullet(X)$ by $\overline{x}$. Then we can sum up what we have got as

$$p_X^*(x + \overline{x}) = p_X^*(x) + p_X^*(\overline{x})$$
$$= p_X^*(x) + \epsilon = p_X^*(x) + [Z] - p_X^*(x) = [Z].$$

It remains to show injectivity for p_X^*. For this, consider the zero section of the projection, $\sigma : X \longrightarrow X \times_k \mathbb{A}_k^1$, $\sigma(x) = (x, 0)$. Then $\sigma^* \circ p_X^* = \mathrm{id}_{A_\bullet(X)}$. Here σ^* is the Gysin homomorphism which exists since σ is the canonical embedding of principal and effective Cartier divisor, in the abuse of language we have introduced. With this relation the injectivity of p_X^* is clear, and the proof is complete. The same proof applies to $C_\bullet$, as is immediately verified. $\square$

Now let X and Y be quasi projective schemes over the field k, where we assume Y to be smooth over k. Consider a morphism

$$f : X \longrightarrow Y.$$

Also let there be given two closed and irreducible subsets $S \subset X$, $T \subset Y$.

Definition 18.7 S and T are said to intersect properly along f if for all $x \in S$

$$\mathrm{codim}_x(S \cap f^{-1}T), S) = \mathrm{codim}_{f(x)}(T, Y).$$

The following important concept was introduced by J.P. Serre, see [12] pp. 401–405.

Definition 18.8

$$[S] \bullet_f [T] = \sum_{i \geq 0} (-1)^i Z_n(\mathrm{Tor}_i^{\mathcal{O}_X}(\mathcal{O}_S, \mathcal{O}_T))$$

where $n = \dim(S \cap f^{-1}(T))$. When x and y are cycles on X and Y, respectively, such that all component of x are in proper position to all component of y, then we define $x \bullet_f y$ in the obvious way by forming the $\bullet_f$ for all pairs of components and extending by linearity.

Example 18.2 Let $X = Y$ and $f = \mathrm{id}_X$. Then S and T are in general position with respect to id_X if

$$\mathrm{codim}_y(S \cap T, S) = \mathrm{codim}_y(t, Y)$$

for all $t \in S$, that is to say

$$\mathrm{codim}(W, S) = \mathrm{codim}(T, Y)$$

for all irreducible components W of $S \cap T$. Since

$$\mathrm{codim}(W, S) = \mathrm{codim}(W, Y) - \mathrm{codim}(S, Y)$$

we arrive at the equality

$$\mathrm{codim}(W, Y) = \mathrm{codim}(S, Y) + \mathrm{codim}(T, Y),$$

that is to say, S and T have *proper intersection* in all components. In this case we write simply $x \bullet y$ instead of $x \bullet_{\mathrm{id}_X} y$.

We also note that we have the following result, which is of course essential:

Proposition 18.10 *Algebraic and rational equivalence are congruence relations for* $\bullet_f$.

If in addition to the assumptions made above we now introduce an assumption on smoothness. We have the following result, which is a general form of the so called *Chow's moving lemma*, [11], Sect. 2.3:

Theorem 18.11 (Chow's Moving Lemma) *Let Y be non-singular and quasi projective, and let $f_i : X_i \longrightarrow Y$ be morphisms and x_i be cycles on X_i for $i = 1, \ldots, m$, and let y be a cycle on Y. Then there is a cycle y' on Y, rationally equivalent to y, such that the cycles $x_i \bullet_{f_i} y$ are defined, i.e., all x_i intersect y' properly along the respective $f_i s$.*

By means of the general Chow's Moving Lemma we find the following:

Theorem 18.12 (The Chow Ring) *If X is smooth over k, then $A^\bullet(X)$ is a commutative graded ring with respect to the multiplication $\bullet = \bullet_{\mathrm{id}_X}$.*

From now on we assume X to be smooth and quasi projective, over an algebraically closed field k. We thus stay in the situation of [18], Appendix A. For smooth X the complexity of the notation we started out with above is really not needed. In this case we therefore refer to $A^\bullet(X)$ as *the Chow ring of X*, and in the following we will assume X to be smooth. In accordance with this we write $A(X)$ for $A^\bullet(X)$ and replace $\bullet = \bullet_{\mathrm{id}_X}$ by $\cdot$, the multiplication in the commutative ring $A(X)$, still graded by codimension.

Example 18.3 The Chow ring of the projective space $\mathbb{P}^n_X$ over the quasi projective non-singular variety X is $A(X)[t]/(t^{n+1})$.

Definition 18.9 (Pullback) Let $f : X \longrightarrow X'$ be a morphism of quasi projective non singular varieties over the algebraically closed field k. Let Y' be a subvariety of X', so $y' = [Y'] \in A(X')$. Write

$$X \times X' \xrightarrow{\ \mathrm{pr}_{X'}\ } X'$$
$$\mathrm{pr}_X \downarrow$$
$$X$$

Then put

$$f^*(y') = \mathrm{pr}_{X*}([\Gamma_f \cdot \mathrm{pr}_{X'}^{-1}(Y')]).$$

In this simplified situation, when the varieties are assumed to be non singular and quasi projective over an algebraically closed field, we do not need the full power of the theory explained in [11]. We do need the important *Projection Formula*, however:

Let $f : X \longrightarrow X'$ be a proper morphism. Let $x \in A(X)$ and $y \in A(X')$ Then

$$f_*(x \cdot f^*(y)) = f_*(x) \cdot y. \qquad (18.2)$$

See [18], p. 426.

Chapter 19
Characteristic Classes in Algebraic Geometry

This chapter is on characteristic classes. We first explain some basic facts on $\mathbb{P}(\mathcal{E})$ and then proceed to Chern classes, Chern characters and Todd classes. After comments on the singular case we define homological Segre classes and proceed to a study of the Grothendieck Group $K(X)$. Apart from these comments on the singular case, we assume that all schemes are non singular for the remainder of this book. But using our references, the task of extending the treatment which follows to cover the singular case should be straightforward.

19.1 Basics on $\mathbb{P}(\mathcal{E})$ Revisited

From Sect. 16.3, recall the construction $\pi_{\mathcal{E}} : \mathbb{P}(\mathcal{E}) \longrightarrow Y$ where there is a surjective

$$\psi_{\mathcal{E}} = \alpha_1^{\#} : \pi_{\mathcal{E}}^*(\mathcal{E}) \longrightarrow \mathcal{O}_{\mathbb{P}(\mathcal{E})}(1).$$

In Sect. 16.3 we defined the contravariant functor F of equivalence classes of 1-quotients of $\mathcal{E}$ on Y and showed this functor to be representable by a Y-scheme $\mathbb{P}(\mathcal{E})$.

The following theorem extends and supplements the statement of Theorem 16.11:

Theorem 19.1 1. *The functor F is representable by the Y-scheme $\mathbb{P}(\mathcal{E}) \xrightarrow{\pi} Y$. The universal element is*

$$(\mathcal{O}_{\mathbb{P}(\mathcal{E})}(1), \pi_{\mathbb{P}(\mathcal{E})}).$$

2. *Let $\mathcal{E}$ be locally free of rank e. Then there is an exact sequence*

$$0 \longrightarrow \mathcal{K} \longrightarrow \pi^*(\mathcal{E}) \longrightarrow \mathcal{O}_{\mathbb{P}(\mathcal{E})}(1) \longrightarrow 0$$

where $\mathcal{K}$ is locally free, in fact we have

$$\mathcal{K} = \Omega^1_{\mathbb{P}(\mathcal{E})/Y}(1).$$

A. Holme, *A Royal Road to Algebraic Geometry*,
DOI 10.1007/978-3-642-19225-8_19, © Springer-Verlag Berlin Heidelberg 2012

3. *Assume in addition that the quasi projective variety Y is reduced, but possibly with singularities. In this case the ring homomorphism*

$$\pi^* : A^{\cdot}(Y) \longrightarrow A^{\cdot}(\mathbb{P}(\mathcal{E}))$$

is injective, and the $A^{\cdot}(Y)$-module $A^{\cdot}(\mathbb{P}(\mathcal{E}))$ is generated as a free module over $A^{\cdot}(Y)$ by $1, \xi, \ldots, \xi^{e-1}$ where $\xi = [D]$, the Weil divisor on $\mathbb{P}(\mathcal{E})$ which corresponds to $\mathcal{O}_{\mathbb{P}(\mathcal{E})}(1)$.

Proof 1. Was shown in Theorem 16.11. For 2., the exact sequence was shown in Theorem 16.12.

3. To prove the statement about the Chow ring, we proceed by induction on $n = \dim(Y)$. For $n = 0$ we may assume that it is connected, so Y is just a point with reduced structure. Thus $A^{\cdot}(Y) = \mathbb{Z}$ and $\mathbb{P}(\mathcal{E}) = \mathbb{P}_k^{e-1}$, thus the claim is known by Example 18.3. Now let U be the open dense subset of Y over which $\mathcal{E}$ is free of rank e. Then $\mathbb{P}(\mathcal{E})_U = \mathbb{P}(\mathcal{E}_U) = \mathbb{P}_U^{e-1}$, so the claim holds over U by Example 18.3. By induction it holds over $Y' = Y - U$, and thus the general claim follows by Theorem 18.5.   □

We note the following

Corollary 19.2

$$\pi_*(\xi^i \wedge [\mathbb{P}(\mathcal{E})]) = \begin{cases} 0 & \text{for } i = 0, 1, \ldots, e-2 \\ [Y] & \text{for } i = e-1. \end{cases}$$

Proof Indeed, since $\pi_* : A_j(\mathbb{P}(\mathcal{E})) \longrightarrow A_j(Y)$ and $\xi^i \wedge [\mathbb{P}(\mathcal{E})] \in A_j(\mathbb{P}(\mathcal{E}))$ this element is mapped into $A_{\dim(Y)+e-1-i}(Y)$, and hence the dimension would have to drop, i.e., the image would be zero, unless $i = e-1$. That the image is $[Y]$ in the latter case follows from the same inductive argument that was used in proving Theorem 19.1   □

Let X be a scheme and $\mathcal{E}$ be locally free on X. By putting $\mathcal{E} = \mathcal{E}_0$ and letting $\mathcal{E}_1 = \mathcal{K}$ on $P_1 = \mathbb{P}(\mathcal{E})$ we obtain two smooth morphisms

$$P_2 = \mathbb{P}(\mathcal{E}_1) \xrightarrow{\pi_2} P_1 \xrightarrow{\pi_1} X$$

where the fiber dimension of π_1 is $e - 1$, while that of π_2 is $e - 2$.

Repeating the process we get a sequence of smooth morphisms, the fibers of which are linear spaces of decreasing dimensions from $e - 1$ and down to 1:

$$P = P_{e-1} \xrightarrow{\pi_{e-1}} P_{e-2} \xrightarrow{\pi_{e-2}} \cdots \xrightarrow{\pi_2} P_1 \xrightarrow{\pi_1} P_0 = X.$$

Let $P \xrightarrow{\pi} X$ denote the composition. Then it follows by the construction that $\pi^*(\mathcal{E})$ splits completely in the sense that there are exact sequences on P_i

$$0 \longrightarrow \mathcal{E}_1 \longrightarrow \pi_1^*(\mathcal{E}) \longrightarrow \mathcal{L}_1 \longrightarrow 0$$
$$0 \longrightarrow \mathcal{E}_2 \longrightarrow \pi_2^*(\mathcal{E}_1) \longrightarrow \mathcal{L}_2 \longrightarrow 0$$
$$\vdots$$
$$0 \longrightarrow \mathcal{E}_{e-1} \longrightarrow \pi_{e-1}^*(\mathcal{E}_{e-2}) \longrightarrow \mathcal{L}_{e-1} \longrightarrow 0$$

where $\mathcal{L}_1, \mathcal{L}_2, \ldots, \mathcal{L}_{e-1}$ are the universal 1-quotients on $P_1, P_2, \ldots, P_{e-1}$.

We sum this up as the following:

Theorem 19.3 (Splitting Principle) *For all locally free sheaves $\mathcal{E}$ on X there is a proper morphism $\pi : P \longrightarrow X$ such that*

$$\pi^* : A^\cdot(X) \longrightarrow A^\cdot(P)$$

is injective and $\pi^*(\mathcal{E})$ has a composition series of locally free sheaves with invertible quotients.

19.2   Chern Classes, Chern Characters and Todd Classes

We saw in Part 2 of Theorem 19.1 that the Chow ring $A^\cdot(\mathbb{P}(\mathcal{E}))$ for a locally free sheaf on Y is a free module over the Chow ring $A^\cdot(Y)$ on the generators $1, \xi, \ldots, \xi^{e-1}$. In particular it then follows that we have a relation

$$\xi^e + \pi^*(a_1)\xi^{e-1} + \cdots + \pi^*(a_{e-1})\xi + \pi^*(a_e) = 0$$

for uniquely defined elements $a_i \in A^i(Y)$. We make the following important definition, valid for a locally free sheaf on a quasi projective scheme, possibly with singularities:

Definition 19.1 *The elements*

$$c_i(\mathcal{E}) = (-1)^i a_i$$

are referred to as the Chern classes of the locally free sheaf $\mathcal{E}$.

It follows from this definition that $c_0(\mathcal{E}) = 1$.

We formally define the *Chern polynomial* as

$$c_T(\mathcal{E}) = 1 + c_1(\mathcal{E})T + \cdots + c_i(\mathcal{E})T^i + \cdots + c_r(\mathcal{E})T^r$$

where $r = \text{rk}(\mathcal{E})$, while the *total Chern Class* is defined as

$$c(\mathcal{E}) = 1 + c_1(\mathcal{E}) + \cdots + c_i(\mathcal{E}) + \cdots + c_r(\mathcal{E}).$$

It is some times useful to formally split the Chern polynomial as follows:

$$c_T(\mathcal{E}) = \prod_{i=1}^{r}(1 + \rho_i(\mathcal{E})T).$$

Some times the entities $\rho_i(\mathcal{E})$ are referred to as the (inverse) Chern roots. So far they are not elements in a "Chow ring", but only carry a symbolic meaning.

We now sum up the most basic properties of Chern classes.

Proposition 19.4 (Divisors) *If $\mathcal{E} \cong \mathcal{O}_X(D)$ for the divisor D, then $c_T(\mathcal{E}) = 1 + DT$.*

Proposition 19.5 (Pullback) *If $X \longrightarrow Y$ is a morphism and $\mathcal{F}$ is locally free on then for any morphism $f : X' \longrightarrow X$*

$$c_i(f^*(\mathcal{F})) = f^*(c_i(\mathcal{F})).$$

Proposition 19.6 (Exact Sequence) *If*

$$0 \longrightarrow \mathcal{E}' \longrightarrow \mathcal{E} \longrightarrow \mathcal{E}'' \longrightarrow 0$$

is an exact sequence of locally free sheaves, then

$$c_T(\mathcal{E}) = c_T(\mathcal{E}')c_T(\mathcal{E}'').$$

Proposition 19.7 (Splitting Principle) *If $\mathcal{E}$ splits with filtration by the invertible sheaves $\mathcal{L}_i$, where $i = 1, \ldots, r$, then*

$$c_T(\mathcal{E}) = \prod_{i}^{r} c_T(\mathcal{L}_i)$$

where r is the rank of $\mathcal{E}$.

Proposition 19.8 (Usual Formulas) *Let a_i, $i = 1, \ldots, r$ and b_j, $j = 1, \ldots, s$ denote the Chern roots of $\mathcal{F}$ and $\mathcal{G}$, respectively. Then*

$$c_T(\mathcal{E} \otimes \mathcal{G}) = \prod_{i,j}(1 + (a_i + b_j)T),$$

$$c_T(\wedge^p \mathcal{E}) = \prod_{1 \le i_1 \le \cdots \le i_p \le r}(1 + (a_{i_1} + \cdots + a_{i_p})T),$$

where r is the rank of $\mathcal{E}$, and

$$c_T(\mathcal{E}^\vee) = c_{-T}(\mathcal{E}).$$

We finally compute the canonical sheaf of projective space. We claim the following:

Proposition 19.9 *We consider projective r-space over* $\mathrm{Spec}(k)$, *the scheme* $\mathbb{P}^r_k$. *Then*

$$\omega_{\mathbb{P}^r_k/k} = \mathcal{O}_{\mathbb{P}^r_k}(-r-1).$$

Proof We use Theorem 16.12 with $\mathcal{E} = k^{r+1}$ on $\mathrm{Spec}(k)$. Recall that $\omega_{\mathbb{P}^r_k/k} = \Lambda^r(\Omega^1_{\mathbb{P}^r_k/k})$ is locally free of rank 1, cf. Proposition 16.9. Thus $\omega_{\mathbb{P}^r_k/k} = \mathcal{O}_{\mathbb{P}^r_k/k}(m)$ for some integer m, as $\mathrm{Pic}(\mathbb{P}^r_k) \cong \mathbb{Z}$. Now $\omega_{\mathbb{P}^r_k/k} = \Lambda^r(\Omega^1_{\mathbb{P}^r_k/k})$, so $c_T(\omega_{\mathbb{P}^r_k/k}) = 1 + c_1(\omega_{\mathbb{P}^r_k/k})T$ where the (first and only) Chern number of the invertible sheaf $\omega_{\mathbb{P}^r_k/k}$ is equal to the first Chern number of $\Omega^1_{\mathbb{P}^r_k/k}$) by the second formula in Proposition 19.8. Now recall the exact sequence from Theorem 19.1 with $Y = \mathrm{Spec}(k)$ and $\mathcal{E} = k^{r+1}$

$$0 \longrightarrow \Omega^1_{\mathbb{P}^r_k}(1) \longrightarrow \mathcal{O}^{r+1}_{\mathbb{P}^r_k} \longrightarrow \mathcal{O}_{\mathbb{P}^r_k}(1) \longrightarrow 0$$

or

$$0 \longrightarrow \Omega^1_{\mathbb{P}^r_k} \longrightarrow \mathcal{O}^{r+1}_{\mathbb{P}^r_k}(-1) \longrightarrow \mathcal{O}_{\mathbb{P}^r_k} \longrightarrow 0.$$

Hence $c_1(\Omega^1_{\mathbb{P}^r_k}) = -(r+1)t$, where $t \in A(\mathbb{P}^r_k)$ is a hyperplane class. $\qquad\square$

For an element in $\alpha \in A^\cdot(Y)$ the expression

$$e^\alpha = 1 + \alpha + \frac{1}{2}\alpha^2 + \cdots + \frac{1}{i!}\alpha^i + \cdots \in A^\cdot(Y) \otimes_\mathbb{Z} \mathbb{Q}$$

has meaning, as $\alpha^m = 0$ for all $m \gg 0$. We now define the total Chern character $\mathrm{ch}(\mathcal{E})$ and the total Todd class $\mathrm{td}(\mathcal{E})$ in terms of the Chern roots as

$$\mathrm{ch}(\mathcal{E}) = \sum_{i=1}^r e^{\rho_i(\mathcal{E})} \quad \text{and} \quad \mathrm{td}(\mathcal{E}) = \prod_{i=1}^r \frac{\rho_i(\mathcal{E})}{1 - e^{-\rho_i(\mathcal{E})}}$$

where the expressions are interpreted as truncated power series, the truncation taking place after the entire computations of the power series are completed. The first five terms are

$$\mathrm{ch}(\mathcal{E}) = 1 + c_1 + \frac{1}{2}(c_1^2 - 2c_2) + \frac{1}{6}(c_1^3 - 3c_1c_2 + 3c_3)$$

$$+ \frac{1}{24}(c_1^4 - 4c_1^2c_2 + 4c_1c_3 + 2c_2^2 - 4c_4) + \cdots$$

and

$$\mathrm{td}(\mathcal{E}) = 1 + \frac{1}{2}c_1 + \frac{1}{12}(c_1^2 + c_2) + \frac{1}{24}c_1 c_2$$
$$- \frac{1}{720}(c_1^4 - 4c_1^2 c_2 - 3c_2^2 - c_1 c_3 + c_4) + \cdots$$

where $c_i = c_i(\mathcal{E})$. The Chern character operator ch has the following important properties:

Proposition 19.10 (Chern Character is Additive and Multiplicative) *The Chern Character is additive on a short exact sequence, and multiplicative for tensor product.*

Proof The proof is immediate by the behavior the Chern roots for short exact sequences and for tensor product. □

19.3 Serre Duality

The aim of this section is to give a very brief survey of Serre duality. We start by presenting the following theorem. Recall the definition of the canonical sheaf $\omega_{X/k}$ in Sect. 16.3. The following important theorem has had an enormous influence on algebraic geometry. It was published in 1955 in [38].

Theorem 19.11 (J.-P. Serre 1955) *Let X be a non singular projective variety over the algebraically closed field k. Then for a locally free sheaf $\mathcal{F}$ on X there are functorial isomorphisms*

$$H^i(X, \mathcal{F}) \cong H^{n-i}(X, \mathcal{F}^\vee \otimes \omega_{X/k}).$$

In contemporary sources the theorem is usually proved by first verifying the case $X = P = \mathbb{P}_k^r = \mathrm{Proj}(S)$ where $S = k[X_0, \ldots, X_r]$ over a field k. Note that in this case $\omega_{P/k} \cong \mathcal{O}_P(-r-1)$, by Proposition 19.9. One then proceeds to treat the case of a smooth variety over a field k, and even much more general situations. This falls outside the scope of the present book. However, we shall briefly sketch the first step, namely a proof for the case $X = \mathbb{P}_k^r$.

The verification amounts to a computation of cohomology, which we briefly sketch below.

Theorem 19.12 *Let A be a commutative ring with 1, and let $X = P = \mathbb{P}_A^r = \mathrm{Proj}(S)$ $S = A[X_0, \ldots, X_r]$. Then*

(a) *The natural map*

$$S \longrightarrow \Gamma_*(\mathcal{O}_X) = \bigoplus_{n \in \mathbb{Z}} H^0(X, \mathcal{O}_X(n))$$

is an isomorphism of graded S-modules.

(b) $H^i(X, \mathcal{O}_X(n)) = 0$ for $0 < i < r$ and all $n \in \mathbb{Z}$.
(c) $H^r(X, \mathcal{O}_X(-r-1)) \cong A$.
(d) The natural map

$$H^0(X, \mathcal{O}_X(n)) \times H^r(X, \mathcal{O}_X(-n-r-1)) \longrightarrow H^r(X, \mathcal{O}_X(-r-1)) \cong A$$

is a perfect pairing of finite free A-modules for each $n \in \mathbb{Z}$.

Proof We sketch the proof, following Hartshorne's treatment in [18], pp. 225–227. By Theorem 17.24 we may prove the theorem by computing Čech cohomology for an open affine covering of Proj(S).

(a): Let $\mathcal{F}$ be the quasi-coherent sheaf of modules $\bigoplus_{n \in \mathbb{Z}} \mathcal{O}_X(n)$. Then

$$H^r(X, \mathcal{F}) = \bigoplus_{n \in \mathbb{Z}} H^r(X, \mathcal{O}_X(n))$$

since the operators H and $\oplus$ commute.

We now consider the standard open covering

$$\mathbb{P}_A^r = D_+(X_0) \cup D_+(X_1) \cup \cdots \cup D_+(X_r),$$

then the open sets $U_{i_0, \dots, i_p}$ are $D_+(X_{i_0} \dots X_{i_p})$, thus

$$\mathcal{F}(D_+(X_{i_0})) = S_{X_{i_0}}, \quad \dots, \quad \mathcal{F}(D_+(X_{i_0} \cdots X_{i_p})) = S_{X_{i_0} \cdots X_{i_p}}$$

and so the Čech complex is given by

$$\mathcal{C}^{\cdot}(\mathcal{U}, \mathcal{F})$$
$$= \prod S_{X_{i_0}} \longrightarrow \prod S_{X_{i_0} X_{i_1}} \longrightarrow \cdots \longrightarrow \prod S_{X_0 \dots \hat{X}_k \dots X_r} \longrightarrow S_{X_0 X_1 \cdots X_r}$$

where the gradings on S and on $\mathcal{F}$ correspond. Now a simple algebraic consideration shows that the kernel of the leftmost mapping is S, and as this is the 0th. Čech cohomology of $\mathcal{F}$ by definition, (a) follows.

We then turn to (c). We have a sequence, exact to the right:

$$\prod_k S_{X_0 \dots \hat{X}_k \dots X_r} \longrightarrow S_{X_0 \dots X_r} \xrightarrow{d^{r-1}} H^r(X, \mathcal{F}) \longrightarrow 0.$$

Now $S_{X_0 \dots X_r}$ is a free A-module with basis consisting of all $X_0^{\ell_0} \cdots X_r^{\ell_r}$ with all $\ell_i \in \mathbb{Z}$. The image of d^{r-1} if the free submodule generated by those basis elements $X_0^{\ell_0} \cdots X_r^{\ell_r}$ for which not all exponents are negative. The quotient therefore is generated by the monomials for which all exponents are negative. But there is only one such monomial, namely $x_0^{-1} \cdots x_r^{-1}$. Thus the cokernel is generated by this $x_0^{-1} \cdots x_r^{-1}$, so it is isomorphic to A.

Proof of (d): By (a) we have $H^0(X, \mathcal{O}_X(n)) = 0$ and $H^r(X, \mathcal{O}_X(-n-r-1)) = 0$ since in this case $-n-r-1 > -r-1$, and there are no monomials of this degree in the basis for $H^r(X, \mathcal{F})$. So the statement is trivial for $n < 0$, both modules are 0.

For $n \geq 0$ $H^0(X, \mathcal{O}_X(n))$ has a basis consisting of monomials of degree n

$$\{X_0^{m_0} \cdots X_r^{m_r} \mid m_0 + \cdots + m_r = n\}.$$

The natural pairing of $H^r(X, \mathcal{O}_X(-n-r-1))$ with $H^r(X, \mathcal{O}_X(-r-1))$ where $\sum \ell_i = -n - r - 1$, and the monomial to the right vanishes if any $m_i + \ell_i \geq 0$. So we have a perfect pairing where the dual basis element to $X_0^{m_0} \cdots X_r^{m_r}$ is $X_0^{-m_0-1} \cdots X_r^{-m_r-1}$.

It remains to prove (b), which is done by induction on r. For $r = 1$ there is nothing to prove. So assume $r > 1$. Now localize the complex as graded S-modules in X_r, this gives the Čech-complex for $\mathcal{F}|D_+(X_r)$ with respect to the open affine covering $\{D_+(X_+(X_i X_r))\}$ of $D_+(X_r)$. By Theorem 17.24 the ith. cohomology of this complex vanishes for $i > 0$. Localization being exact, we find $H^i(X, \mathcal{F})_{X_r} = 0$ for $i > 0$. Therefore every element of $H^i(X\mathcal{F})$, $i > 0$ is annihilated by some power of X_r. The proof of (b) is finally completed by showing that for all $0 < i < r$ multiplication by X_r induces a bijective mapping of $H^i(X, \mathcal{F})$ into itself. This evidently implies that this module is 0. For this conclusion of the proof we refer the reader to [18]. $\square$

Using Theorem 19.11, Serre proved the following:

Theorem 19.13 *Let X be a closed subscheme of $\mathbb{P}_k^N$ for a Noetherian ring A. Let $\mathcal{F}$ be a coherent module on X. Then*

(a) *For each $i \geq 0$ $H^i(X, \mathcal{F})$ is a finitely generated A-module.*
(b) *There is an integer n_0 depending on $\mathcal{F}$ such that $H^i(X, \mathcal{F}(n)) = 0$ for all $i > 0$ and all $n \geq n_0$.*

We have the following

Corollary 19.14 *Let $X = \mathbb{P}_k^r$ and let $M = \Gamma_*(\mathcal{F}) = \bigoplus_{n=0}^{\infty} H^0(X, \mathcal{F}(n))$ considered as graded $S = k[X_0, \ldots, X_r]$-module. Then $\chi(\mathcal{F}(n))$ is a polynomial in n for n sufficiently big, and for $\mathcal{F} = \mathcal{O}_X$ this is just the Hilbert polynomial defined in Sect. 5.5. In particular, if X is a closed subscheme and $\mathcal{F}$ is of $\mathcal{O}_X$, then $\chi(\mathcal{F}(n))$ is the Hilbert polynomial of X, and so the arithmetic genus of X is $(-1)^{\dim(X)}(\chi(\mathcal{O}_X) - 1)$.*

Chapter 20
The Riemann-Roch Theorem

This chapter is on the Riemann-Roch Theorem. We start with Hirzebruch's Riemann-Roch Theorem, and deduce from it the Riemann-Roch Theorem for curves, and for surfaces. We state the general Grothendieck's Riemann-Roch theorem, deducing that of Hirzebruch from it.

Some general constructions and concepts used in this chapter have been extended so that schemes with singularities are covered. This theory is developed in the paper [11] and in the book [12] by W. Fulton. One of the most important applications of this marvelous work is the general *Baum-Fulton-McPehrson Riemann-Roch Theorem* for singular varieties, [2]. However, we do not include a survey of this work here, as it reaches beyond the scope of the present book. The non singular case is challenging enough at this stage, but the theorem proved in [2] would be an exiting source for further study!

20.1 Hirzebruch's Riemann-Roch Theorem

In this and the following three sections we assume that X is a non singular, projective variety, defined over an algebraically closed field k. With the concepts we have introduced so far, we are now able to state the following important formula:

Theorem 20.1 (Hirzebruch's Riemann-Roch Theorem) *Let $\mathcal{E}$ be a locally free $\mathcal{O}_X$-module on the non singular projective variety X of dimension n, defined over the algebraically closed field k. Let $\alpha \in A^{\cdot}(X)$, and let*

$$\alpha \otimes_{\mathbb{Z}} 1 = \alpha_0 + \cdots + \alpha_n \in A^n(X) \otimes_{\mathbb{Z}} \mathbb{Q}$$

and $(\alpha)_n$ denote the degree of α_n. As before let $\mathcal{T}_{X/k} = (\Omega^n_{X/k})^{\vee}$ denote the tangent bundle of X. Then

$$\chi(\mathcal{E}) = (\mathrm{ch}(\mathcal{E}) \cdot \mathrm{td}(\mathcal{T}_X))_n.$$

A. Holme, *A Royal Road to Algebraic Geometry*,
DOI 10.1007/978-3-642-19225-8_20, © Springer-Verlag Berlin Heidelberg 2012

Friedrich Hirzebruch proved this marvelous theorem over the field of complex numbers, his proof is given in his path breaking book [20]. Here we shall deduce it as a consequence of a more general theorem, subsequently found by *Alexander Grothendieck*, which we present in Sect. 20.4.

However, in the next two sections we deduce the Riemann-Roch Theorems for curves and surfaces, as special cases of the Hirzebruch Riemann-Roch Theorem.

20.2 The Riemann-Roch Theorem for Curves, Revisited

Let X be a non singular projective curve over the algebraically closed field k. Let D be a divisor on X. Then $c_1(\mathcal{O}_X(D)) = [D]$, so $\mathrm{ch}(\mathcal{O}_X(D)) = 1 + [D]$. Let K_X denote the canonical divisor of X, then $\Omega^1_{X/k} \cong \mathcal{O}_X(K_X)$. Thus $\mathrm{td}(\mathcal{T}_X) = 1 - \frac{1}{2}K_X$. Let $\mathcal{E} = \mathcal{O}_X(D)$. Denote the genus g_X of X by g. Then

$$(\mathrm{ch}(\mathcal{E}) \cdot \mathrm{td}(\mathcal{T}_X)) = (1 + [D])\left(1 - \frac{1}{2}[K_X]\right)$$

the degree 1 part of which is $[D] - \frac{1}{2}[K_X]$. Thus we get the formula

$$\chi(\mathcal{O}_X(D)) = \deg\left(D - \frac{1}{2}K_X\right).$$

Choosing $D = 0$ in this formula, we obtain $\dim_k H^0(X, \mathcal{O}_X) - \dim_k H^1(X, \mathcal{O}_X) = -\frac{1}{2}\deg(K_X)$ as the zero divisor has degree zero, thus

$$1 - g = -\frac{1}{2}\deg(K_X)$$

and hence

$$\chi(\mathcal{O}_X(D)) = \deg(D) + 1 - g.$$

By Serre duality it follows that the vector space $H^1(X, \mathcal{O}_X(D))$ is dual to $H^0(X, \mathcal{O}_X(K_X - D))$, thus

$$\ell(D) - \ell(K_X - D) = \deg(D) + 1 - g$$

which is the usual form of the Riemann-Roch theorem for curves.

20.3 The Riemann-Roch Theorem for a Surface

Let X be a non singular projective surface over the algebraically closed field k. Let $\mathcal{E}$ be a locally free $\mathcal{O}_X$-module on X. Then the Chern character and the

Todd class of $\mathcal{E}$ take the forms

$$\mathrm{ch}(\mathcal{E}) = 1 + c_1 + \frac{1}{2}(c_1^2 - 2c_2)$$

and

$$\mathrm{td}(\mathcal{E}) = 1 + \frac{1}{2}c_1 + \frac{1}{12}(c_1^2 + c_2)$$

where $c_i = c_i(\mathcal{E})$, $i = 1, 2$. The Riemann-Roch Theorem becomes

$$\chi(\mathcal{E}) = (\mathrm{ch}(\mathcal{E}) \cdot \mathrm{td}(\mathcal{T}_X))_2,$$

in other words,

$$\chi(\mathcal{E}) = \left(1 + c_1(\mathcal{E}) + \frac{1}{2}(c_1(\mathcal{E})^2 - 2c_2(\mathcal{E}))\right.$$
$$\times \left.\left(1 + \frac{1}{2}c_1(X) + \frac{1}{12}(c_1(X)^2 + c_2(X))\right)\right)_2$$
$$= \left(1 + c_1(\mathcal{E}) + \frac{1}{2}(c_1(\mathcal{E})^2 - 2c_2(\mathcal{E}))\right.$$
$$\times \left.\left(1 + \frac{1}{2}c_1(X) + \frac{1}{12}(c_1(X)^2 + c_2(X))\right)\right)_2$$
$$= \left(1 + c_1(\mathcal{E}) + \frac{1}{2}(c_1(\mathcal{E})^2 - 2c_2(\mathcal{E}))\right.$$
$$\times \left.\left(1 - \frac{1}{2}K_X + \frac{1}{12}(K_X^2 + c_2(X))\right)\right)_2,$$

i.e. when setting all terms of degree ≥ 3 to zero,

$$\chi(\mathcal{E}) = \frac{1}{12}(K_X^2 + c_2(X)) - c_1(\mathcal{E}) \cdot \frac{1}{2}K_X + \frac{1}{2}(c_1(\mathcal{E})^2 - 2c_2(\mathcal{E})).$$

When applied to $\mathcal{E} = \mathcal{O}_X(D)$ for a divisor D this yields

$$\chi(\mathcal{O}_X(D)) = \frac{1}{12}(K_X^2 + c_2(X)) - \frac{1}{2}D \cdot K_X + \frac{1}{2}D^2$$

or

$$\chi(\mathcal{O}_X(D)) = \frac{1}{2}D \cdot (D - K_X) + \frac{1}{12}(K_X^2 + c_2(X)).$$

By taking $D = 0$ we find by Corollary 19.14 that

$$1 + p_a(X) = \chi(\mathcal{O}_X) = \frac{1}{12}(K_X^2 + c_2(X)),$$

$p_a(X)$ being the arithmetic genus of X, and hence the Riemann-Roch Theorem for surfaces takes the form

$$\chi(\mathcal{O}_X(D)) = \frac{1}{2}D \cdot (D - K_X) + 1 + p_a(X).$$

Using Serre Duality, Theorem 19.11 which gives

$$H^2(X, \mathcal{O}_X(D)) \cong H^0(X, \mathcal{O}_X(D)^\vee \otimes \omega_{X/k})$$

we arrive at the final form of Riemann-Roch for surfaces:

$$\ell(D) - s(D) + \ell(K_X - D) = \frac{1}{2}D \cdot (D - K_X) + 1 + p_a$$

where $s(D) = \dim_k H^1(X, \mathcal{O}_X(D))$ is referred to as the *superabundance* of the divisor D. As for curves, $\ell(K_X - D)$ is called the *index of speciality* of D.

We note the following

Proposition 20.2 *Assume that D is an effective divisor on the non singular projective surface X. Then we have the following formula involving the arithmetic genus $p_a(D)$ of the curve D:*

$$2p_a(D) - 2 = D \cdot (D + K_X).$$

Proof By the exact sequence

$$0 \longrightarrow \mathcal{O}_X(-D) \longrightarrow \mathcal{O}_X \longrightarrow \mathcal{O}_D$$

we get

$$\chi(\mathcal{O}_X) = \chi(\mathcal{O}_X)(-D) + \chi(\mathcal{O}_D)$$

and hence

$$\chi(\mathcal{O}_D) = \chi(\mathcal{O}_X) - \chi(\mathcal{O}_X(-D))$$
$$= 1 + p_a(X) - \frac{1}{2}(-D) \cdot (-D - K_X) - (1 + p_a(X))$$
$$= -\frac{1}{2}D \cdot (D + K_X).$$

Since by Corollary 19.14 $p_a(D) = 1 - \chi(\mathcal{O}_D)$, we have

$$2p_a(D) - 2 = -2\left(-\frac{1}{2}(D \cdot (D + K_X))\right) = D \cdot (D + K_X)$$

and the claim is proven. $\square$

For a divisor D on X, effective or not, we define the invariant $\pi_a(D)$ by the formula

$$2\pi_a(D) - 2 = D \cdot (D + K_X)$$

and refer to it as the *virtual arithmetic genus of D*.

20.4 Grothendieck's Riemann-Roch Theorem

Assume that X is a non singular quasi projective variety. We defined the ring homomorphism

$$\mathrm{ch} : K(X) \longrightarrow A(X) \otimes_{\mathbb{Z}} \mathbb{Q}$$

in Sect. 19.2. We now ask how this homomorphism behaves with respect to $f^!$ and $f_!$. First, since $f^!$ is induces in the obvious way by f^* and the universal property of $K^{\cdot}(X)$, the ring homomorphism $f^! : K^{\cdot}(Y) \longrightarrow K^{\cdot}(X)$ makes the following diagram commutative:

$$
\begin{array}{ccc}
K(X) & \xrightarrow{\;\mathrm{ch}\;} & A(X) \otimes_{\mathbb{Z}} \mathbb{Q} \\[4pt]
{\scriptstyle f^!}\big\uparrow & & \big\uparrow{\scriptstyle f^* \otimes_{\mathbb{Z}} \mathbb{Q}} \\[4pt]
K(Y) & \xrightarrow[\;\mathrm{ch}\;]{} & A(Y) \otimes_{\mathbb{Z}} \mathbb{Q}
\end{array}
$$

However, for $f_!$ the situation is more complicated: This is the subject of Grothendieck's Riemann-Roch Theorem. Indeed, we have the

Theorem 20.3 (Grothendieck's Riemann-Roch) *Let $f : X \longrightarrow Y$ be a proper morphism of quasi projective schemes over k. Then the diagram*

$$
\begin{array}{ccc}
K(X) & \xrightarrow{\;\mathrm{td}(X)\mathrm{ch}\;} & A(X) \otimes_{\mathbb{Z}} \mathbb{Q} \\[4pt]
{\scriptstyle f_!}\big\downarrow & & \big\downarrow{\scriptstyle f_* \otimes_{\mathbb{Z}} \mathbb{Q}} \\[4pt]
K(Y) & \xrightarrow[\;\mathrm{td}(Y)\mathrm{ch}\;]{} & A(Y) \otimes_{\mathbb{Z}} \mathbb{Q}
\end{array}
$$

commutes, i.e.

$$f_*(\mathrm{td}(X)\mathrm{ch}(x)) = \mathrm{td}(Y)\mathrm{ch}(f_!(x))$$

for all $x \in A(X)$.

Remark 20.4 This yields Hirzebruch's Riemann-Roch as the special case
when $Y = \mathrm{Spec}(k)$. Indeed, a coherent sheaf on Y is then nothing but a finite
vector space over k, so $K(Y) \cong \mathbb{Z}$ by $V \mapsto \dim_k(V)$. When $\mathcal{F}$ is a coherent $\mathcal{O}_X$-
module on X then $f_!([\mathcal{F}]) = \sum(-1)^q[R^q f_*(\mathcal{F})] = \sum(-1)^q \dim_k(H^q(X, \mathcal{F}))$
and thus $f_*(\mathrm{td}(X)\mathrm{ch}(\mathcal{F})) = (\mathrm{td}(X)\mathrm{ch}(\mathcal{F}))_n$. Finally $\mathrm{td}(Y) = 1$, and hence we
get the formula

$$\chi(X, \mathcal{F}) = (\mathrm{ch}(\mathcal{F}) \cdot \mathrm{td}(X))_n$$

which is Hirzebruch's Riemann-Roch Theorem.

Proof We give one of the basic ideas behind the proof of Grothendieck's
Riemann Roch, for the full proof we refer to the paper by Borel and Serre [3],
which has the important appendix by Grothendieck [14]. Let $i : X \hookrightarrow \mathbb{P}_k^N$ be
an embedding of the quasi projective scheme X. Then f factors through the
graph embedding and the canonical projection π_Y as follows:

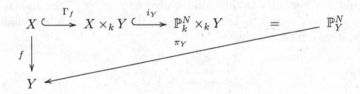

Thus f factors into an embedding $f_1 : X \longrightarrow Z = \mathbb{P}_Y^N$ followed by the canonical projection $f_2 : Z = \mathbb{P}_Y^N \longrightarrow Y$. I claim that if the Grothendieck Riemann-
Roch Theorem holds for f_1 and f_2, then it holds for f. Indeed, the two
assumptions are

$$f_{1*}(\mathrm{td}(X) \cdot \mathrm{ch}(x)) = \mathrm{td}(Z) \cdot \mathrm{ch}(f_{1!}(x))$$

and

$$f_{2*}(\mathrm{td}(Z) \cdot \mathrm{ch}(z)) = \mathrm{td}(Y) \cdot \mathrm{ch}(f_{2!}(z)).$$

We now get

$$f_*(\mathrm{td}(X) \cdot \mathrm{ch}(x)) = f_{2*}(f_{1*}(\mathrm{td}(X) \cdot \mathrm{ch}(x))) = f_{2*}(\mathrm{td}(Z) \cdot \mathrm{ch}(f_{1!}(x)))$$

$$= \mathrm{td}(Y)\mathrm{ch}(f_{2!}(f_{1!}(x))) = \mathrm{td}(Y) \cdot \mathrm{ch}(f_!(x)).$$

Thus the proof of the general theorem has been reduced to verifying the
claim for embeddings and for projections. There still remains serious work,
but we leave the proof here, referring to [3] for this verification.   □

Chapter 21
Some Basic Constructions in the Category of Projective k-Varieties

In this chapter we assume, basically for simplicity only, that all schemes be projective varieties over a field (of any characteristic unless otherwise stated), which without significant loss of generality may be assumed algebraically closed.

The chapter gives some basic constructions in the category of projective k-varieties: the blowing-up of a closed subscheme and of subbundles. We introduce the Grassmann bundles, and the related construction of a parameter variety for the *joining lines* for a projective, embedded scheme. The secant variety and the join are given as applications of these constructions.

21.1 The Blowing-up of a Closed Subscheme

Let P be a scheme and $Y \hookrightarrow P$ a closed subscheme given by the ideal $\mathfrak{J}_Y = \mathfrak{J}$. Let

$$\pi_Y : \tilde{P} = \mathbb{B}_{\mathfrak{J}}(P) \longrightarrow P$$

be the blowing up of P with center Y. We recall some of the basic facts on the blowing-up construction.

First of all, we may define the blowing-up of P with center in the subscheme Y corresponding to the ideal $\mathfrak{J}$ on X as

$$\mathrm{Proj}(\mathrm{Pow}_P(\mathfrak{J})) \quad \text{where } \mathrm{Pow}_P(\mathfrak{J}) = \mathcal{O}_P \oplus \mathfrak{J} \oplus \mathfrak{J}^2 \oplus \cdots.$$

The morphism from this scheme onto P is denoted by $\pi = \pi_Y = \pi_{\mathfrak{J}}$. It is then easily verified that the ideal $\pi^{-1}(\mathfrak{J})$ on $\mathbb{B}_{\mathfrak{J}}(P)$ generated by $\mathfrak{J}$ is an invertible ideal. Moreover, the blowing-up is universal with respect to this property in the sense that whenever $p : Z \longrightarrow P$ is such that $p^{-1}(\mathfrak{J})$ is invertible, then there is a unique morphism $f : Z \longrightarrow \mathbb{B}_{\mathfrak{J}}(P)$ such that $p = \pi_Y f$. In another language, this amounts to saying that the blowing-up represents a certain

A. Holme, *A Royal Road to Algebraic Geometry*,
DOI 10.1007/978-3-642-19225-8_21, © Springer-Verlag Berlin Heidelberg 2012

contravariant functor of P-schemes, which we leave to the reader to make explicit.

We have the following simple but useful observation:

Proposition 21.1 *In the situation above, let $\mathcal{S}$ denote a coherent graded $\mathcal{O}_P$-algebra on P, such that $\mathcal{S}_0 = \mathcal{O}_P$ and such that $\mathcal{S}$ is generated as an $\mathcal{O}_P$-algebra by $\mathcal{S}_1$. Assume that there is a surjective homomorphism*

$$\varphi : \mathcal{S}_1 \longrightarrow\!\!\!\!\!\rightarrow \mathfrak{J}.$$

Then there is a closed embedding

$$i = i(\varphi) : \mathbb{B}_{\mathfrak{J}}(P) \hookrightarrow \mathrm{Proj}(\mathcal{S})$$

such that the following diagram commutes:

$$
\begin{array}{ccc}
\mathbb{B}_{\mathfrak{J}}(P) & \overset{i(\varphi)}{\hookrightarrow} & \mathrm{Proj}(\mathcal{S}) \\
{\scriptstyle \pi_Y} \downarrow & \swarrow {\scriptstyle p_{\mathrm{Proj}(\mathcal{S})}} & \\
P & &
\end{array}
$$

Proof Under the given assumptions φ induces a surjective homomorphism of graded $\mathcal{O}_P$-algebras on P,

$$\Phi : \mathcal{S} \longrightarrow\!\!\!\!\!\rightarrow \mathrm{Pow}_P(\mathfrak{J}) = \mathcal{O}_P \oplus \mathfrak{J} \oplus \mathfrak{J}^2 \oplus \cdots$$

which yields the closed embedding $i(\varphi)$ with the property stated in the proposition. □

We note the following

Corollary 21.2 *If $\mathcal{G}$ is a coherent $\mathcal{O}_P$-module with a surjective*

$$\varphi : \mathcal{G} \longrightarrow\!\!\!\!\!\rightarrow \mathfrak{J},$$

then we get a closed embedding $i(\varphi)$ such that the following diagram commutes:

$$
\begin{array}{ccc}
\mathbb{B}_{\mathfrak{J}}(P) & \overset{i(\varphi)}{\hookrightarrow} & \mathbb{P}(\mathcal{G}) \\
{\scriptstyle \pi_Y} \downarrow & \swarrow {\scriptstyle p_{\mathbb{P}(\mathcal{G})}} & \\
P & &
\end{array}
$$

Moreover, if $\mathfrak{J}$ embeds its zero subscheme regularly in P, and we take $\mathcal{G} = \mathfrak{J}$ and let φ be the identity, then $i(\varphi)$ is an isomorphism.

Proof The first assertion is immediate from the proposition, and the second follows by the following lemma, given as Exercise 17.14a in Eisenbud [9], p. 441:

Lemma 21.3 *Let $x_1, \ldots, x_r$ be a regular sequence in the commutative ring R, and let $I = (x_1, \ldots, x_r)R$. Then the natural surjective map from the symmetric algebra $S_R(I)$ to the so-called blowup algebra $\mathrm{Pow}_R(\mathfrak{I})$*

$$S_R(I) \longrightarrow \mathrm{Pow}_R(\mathfrak{I})$$

is an isomorphism of graded algebras over R. $\square$

21.2 Blowing Up of Subbundles

Let S be a scheme, which we will assume to be a smooth projective variety (although this assumption is not essential). Let

$$0 \longrightarrow \mathcal{E} \longrightarrow \mathcal{F} \longrightarrow \mathcal{G} \longrightarrow 0$$

be an exact sequence of locally free sheaves of ranks e, f and g, respectively. Write

$$W = \mathbb{P}(\mathcal{E}), \qquad X = \mathbb{P}(\mathcal{F}), \qquad Y = \mathbb{P}(\mathcal{G}).$$

We have the canonical exact sequence

$$0 \longrightarrow \mathcal{K} \longrightarrow \mathcal{E}_W \longrightarrow \mathcal{O}_W(1) \longrightarrow 0,$$

where $\mathcal{K} = \Omega^1_{W/S}(1)$. This yields a diagram of locally free sheaves with exact rows and columns

$$
\begin{array}{ccccccccc}
 & & 0 & & 0 & & & & \\
 & & \downarrow & & \downarrow & & & & \\
0 & \to & \mathcal{K} & \to & \mathcal{E}_W & \to & \mathcal{O}_W(1) & \to & 0 \\
 & & \| & & \downarrow & & \downarrow & & \\
0 & \to & \mathcal{K} & \to & \mathcal{F}_W & \to & \mathcal{M} & \to & 0 \\
 & & \downarrow & & \downarrow & & & & \\
 & & \mathcal{G}_W & = & \mathcal{G}_W & & & & \\
 & & \downarrow & & \downarrow & & & & \\
 & & 0 & & 0 & & & &
\end{array}
$$

We obtain the following fundamental diagram:

$$
\begin{array}{ccccccc}
Y \times_S W & & & & & & X \times_S W \\
\| & & & & & & \| \\
\mathbb{P}(\mathcal{E}_Y) = \ \mathbb{P}(\mathcal{G}_W) & \overset{j}{\hookrightarrow} & \mathbb{P}(\mathcal{M}) & \overset{h}{\hookleftarrow} & & & \mathbb{P}(\mathcal{F}_W) = \mathbb{P}(\mathcal{E}_X) \\
\end{array}
$$

(SD) with vertical maps pr_1, π, λ, pr_1, pr_2:

$$
\begin{array}{ccccccc}
Y & \overset{i}{\hookrightarrow} & X & & & & W \\
\| & & \| & & & & \| \\
\mathbb{P}(\mathcal{G}) & & \mathbb{P}(\mathcal{F}) & & & & \mathbb{P}(\mathcal{E})
\end{array}
$$

where the canonically isomorphic S-schemes along the top row are identified, i, j and h are the canonical closed embeddings given by the corresponding surjective maps of locally free sheaves. $\lambda = p_{\mathcal{M}}$ is the canonical projection for the $\mathbb{P}$-construction on locally free sheaves on W, and π is the morphism which corresponds to the surjective map of locally free sheaves on $\mathbb{P}(\mathcal{M})$,

$$
\mathcal{F}_{\mathbb{P}(\mathcal{M})} \longrightarrow \mathcal{O}_{\mathbb{P}(\mathcal{M})}(1) \longrightarrow 0
$$

obtained as the composition of the two canonical surjections

$$
(\mathcal{F}_W)_{\mathbb{P}(\mathcal{M})} \longrightarrow \mathcal{M}_{\mathbb{P}(\mathcal{M})} \longrightarrow \mathcal{O}_{\mathbb{P}(\mathcal{M})}(1).
$$

As is immediately checked, all the subdiagrams are commutative. We shall refer to the diagram (SD) as *the standard diagram* associated to the given short exact sequence of locally free sheaves on S.

Proposition 21.4 *The left square in the standard diagram* (SD) *is the blowing up diagram for* $X = \mathbb{P}(\mathcal{F})$ *with center* $Y = \mathbb{P}(\mathcal{G})$. *$\mathbb{P}(\mathcal{G}_W)$ is the exceptional divisor, in fact the conormal sheaf of Y in X is* $\mathcal{E}_Y(-1)$. *h is the canonical closed embedding related to the description of the blowing up as a monoidal transformation.*

The statement about the conormal bundle follows from the

Lemma 21.5 *The closed embedding i in the standard diagram* (SD) *identifies $\mathbb{P}(\mathcal{G})$ with the scheme of zeroes of the mapping of locally free sheaves on X given by the composition*

$$
\mathcal{E}_X \longrightarrow \mathcal{F}_X \longrightarrow \mathcal{O}_X(1).
$$

21.3 Grassmann Bundles

We consider *Grassmann schemes* $p_{\mathcal{F}} : \mathbf{Grass}_r(\mathcal{F}) \longrightarrow X$ over X, representing the functor of isomorphism classes of locally free rank r quotients of pullbacks of the locally free rank f sheaf $\mathcal{F}$ on X, $\mathbb{P}(\mathcal{F})$ being the case $r = 1$:

$$\mathrm{Grass}_r(\mathcal{F}) : \mathrm{Sch}_X \longrightarrow \mathrm{Sets}$$

$$\mathrm{Grass}_r(\mathcal{F})(T \to X) = \left\{ (\tau, \mathcal{Q}) \,\middle|\, \begin{array}{l} \tau : \mathcal{F}_T \to \mathcal{Q} \text{ onto with } \mathcal{Q} \\ \text{locally free of rank } r \text{ on } T \end{array} \right\} \Big/ \cong$$

$\mathbf{Grass}_r(\mathcal{F})$ carries the universal locally free rank r quotient $\mathrm{Quot}_r(\mathcal{F})$ as well as the universal subbundle $\mathrm{Sub}_r(\mathcal{F})$, which fit into the canonical exact sequence on $\mathbf{Grass}_r(\mathcal{F})$:

$$0 \longrightarrow \mathrm{Sub}_r(\mathcal{F}) \longrightarrow p_{\mathcal{F}}^*(\mathcal{F}) \longrightarrow \mathrm{Quot}_r(\mathcal{F}) \longrightarrow 0. \qquad (21.1)$$

Applying the above to the case $X = \mathrm{Spec}(k)$ and $\mathcal{F} = k^f$, we obtain the k-scheme $\mathbb{G}_r(f)$, which is the grassmanian parameterizing $\mathbb{P}^{r-1}$'s in $\mathbb{P}^{f-1}$. In line with this notation the universal subbundle and the universal quotient is denoted by $\mathrm{Sub}_r(f)$ and $\mathrm{Quot}_r(f)$, respectively.

The Chow ring of a Grassmann scheme $\mathbf{Grass}_r(\mathcal{F})$ over X is generated as an $A(X)$-algebra by the Chern classes, equivalently by the Segre classes, of the locally free rank r sheaf $\mathrm{Quot}_r(\mathcal{F})$. The *Basis Theorem* (cf. [12], Theorem 14.6.5) asserts that a basis for $A(\mathbf{Grass}_r(\mathcal{F}))$ as an $A(X)$-module is given by the Schur polynomials in the Segre classes of $\mathrm{Quot}_r(\mathcal{F})$.

21.4 The Parameter Variety for the Joining Lines

In the construction of the variety of secant lines to an embedded projective variety, we consider the standard diagram from (21.2) for the following exact sequence on $\mathbb{P}^N$:

$$0 \longrightarrow \Omega^1_{\mathbb{P}^N}(1) \longrightarrow \mathcal{O}^{N+1}_{\mathbb{P}^N} \longrightarrow \mathcal{O}_{\mathbb{P}^N}(1) \longrightarrow 0.$$

This yields the diagram below, where $f : \mathbb{P}(\Omega^1_{\mathbb{P}^N}(1)) \longrightarrow \mathbb{P}^N$ denotes the canonical morphism, $f = p_{\mathcal{E}}$.

$$
\begin{array}{ccc}
0 & & 0 \\
\downarrow & & \downarrow \\
0 \to \mathcal{K} \to f^*(\Omega^1_{\mathbb{P}^N}(1)) \to \mathcal{O}_{\mathbb{P}(\Omega^1_{\mathbb{P}^N}(1))}(1) \to 0 \\
\| \qquad\qquad \downarrow \qquad\qquad\qquad \downarrow \\
0 \to \mathcal{K} \to f^*(\mathcal{O}^{N+1}_{\mathbb{P}^N}) \to \mathcal{M} \qquad \to 0 \\
\downarrow \qquad\qquad\qquad \downarrow \\
f^*(\mathcal{O}_{\mathbb{P}^N}(1)) = f^*(\mathcal{O}_{\mathbb{P}^N}(1)) \\
\downarrow \qquad\qquad\qquad \downarrow \\
0 \qquad\qquad\qquad 0
\end{array}
\qquad (21.2)
$$

and we obtain the following version of the standard diagram:

$$\mathbb{J}(N) = \quad \mathbb{P}^N \times_{\mathbb{P}^N} \mathbb{J}(N) \qquad\qquad \widetilde{\mathbb{P}^N \times \mathbb{P}^N} \qquad \mathbb{P}^N \times \mathbb{J}(N) = \quad \mathbb{P}^N \times \mathbb{P}^N \times_{\mathbb{P}^N} \mathbb{J}(N)$$

$$\| \qquad\qquad\qquad\qquad \| \qquad\qquad\qquad\qquad \|$$

$$\mathbb{P}(\mathcal{E}_{\mathbb{P}^N}) = \quad \mathbb{P}(\mathcal{G}_{\mathbb{J}(N)}) \overset{j}{\hookrightarrow} \qquad \mathbb{P}(\mathcal{M}) \overset{h}{\hookrightarrow} \qquad \mathbb{P}(\mathcal{F}_{\mathbb{J}(N)}) = \mathbb{P}(\mathcal{E}_X)$$

(SD 1) $\quad \mathrm{pr}_1 \qquad\qquad\qquad\qquad\qquad \pi_\Delta \qquad\qquad \lambda \qquad \mathrm{pr}_1 \qquad\qquad \mathrm{pr}_2$

$$\mathbb{P}^N \qquad \overset{i=\Delta_{\mathbb{P}^N}}{\hookrightarrow} \qquad \mathbb{P}^N \times \mathbb{P}^N \qquad\qquad \mathbb{J}(N)$$

$$\| \qquad\qquad\qquad\qquad \| \qquad\qquad\qquad\qquad \|$$

$$\mathbb{P}(\mathcal{G}) \qquad\qquad\qquad \mathbb{P}(\mathcal{F}) \qquad\qquad\qquad \mathbb{P}(\mathcal{E})$$

$$\mathcal{G} = \mathcal{O}_{\mathbb{P}^N}(1) \qquad\qquad \mathcal{F} = (\mathcal{O}_{\mathbb{P}^N})^{N+1} \qquad \mathcal{E} = \Omega^1_{\mathbb{P}^N}(1)$$

$\mathbb{P}(\mathcal{O}_{\mathbb{P}^N}^{N+1})$ is identified with $\mathbb{P}^N \times \mathbb{P}^N$ in such a way that the canonical morphism

$$p_{\mathcal{F}} : \mathbb{P}(\mathcal{O}_{\mathbb{P}^N}^{N+1}) \longrightarrow \mathbb{P}^N$$

is the second projection pr_2. Here π_Δ denotes the blowing up of $\mathbb{P}^N \times \mathbb{P}^N$ with center in the diagonal $\Delta = i(\mathbb{P}^N)$. λ is a $\mathbb{P}^1$-bundle, and $\mathbb{J}(N)$ denotes the incidence correspondence.

We extract the following diagram from the standard diagram above, where we include the structural morphisms f and pr_2 of the bundles $\mathbb{P}(\mathcal{F}) = \mathbb{P}^N \times \mathbb{P}^N$ and $\mathbb{P}(\mathcal{E}) = \mathbb{J}(N)$:

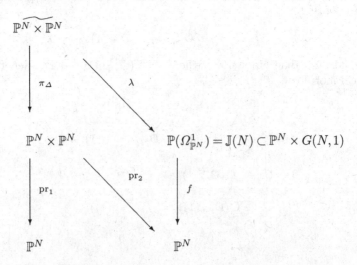

21.5 The Secant Variety and the Join

Now let $X \hookrightarrow \mathbb{P}^N$ be a reduced, projective subscheme, which we assume to be of positive dimension, so $X \times X \not\subseteq \Delta$. We denote by $\widetilde{X \times X}$ the strict transform of $X \times X$. The *secant bundle* of X is defined as

$$\mathrm{Sb}(X) = \lambda(\widetilde{X \times X}).$$

$\mathrm{Sb}(X)$ is not a bundle, so the name is somewhat misleading. Rather, it is a parameter variety for the secant lines to X through smooth points of X.

Moreover, we define the subscheme $S(X) \hookrightarrow \mathbb{P}^N \times \mathbb{P}^N$ by

$$S(X) = \pi_\Delta(\lambda^{-1}(\mathrm{Sb}(X))),$$

and we then have that

$$\mathrm{Sec}(X) = \mathrm{pr}_1(S(X))$$

is the closure of the union of all secant lines to the subvariety X, passing through smooth points of X.

The construction of the "secant bundle" $S(X)$ and the secant subscheme $\mathrm{Sec}(X)$ is generalized to a "join bundle" and a join subscheme by working with $X \times Y$ where X and Y are two projective subschemes in $\mathbb{P}^N$. The join bundle and scheme then are direct generalizations of the corresponding concepts for secants. We omit the details here.

Chapter 22
More on Duality

This final chapter is devoted to duality, the dual variety and the conormal scheme of an embedded projective variety are given as applications. Reflexivity and biduality are studied, in particular duality of hyperplane sections and projections. An application we present here is a very nice theorem of Hefez and Kleiman. Finally we give a brief presentation of some further results on duality and reflexivity.

22.1 The Dual Variety and the Conormal Scheme

Let $\mathbb{P}^N = \mathbb{P}(V)$ denote projective N-space, V being an $N + 1$-dimensional vector space over k. $\mathbb{P}^{N\vee} = \mathbb{P}(V^*)$ denotes the *dual* projective space, whose k-points are identified with the hyperplanes in $\mathbb{P}^N$:

$$\mathbb{P}^{N\vee} = \left\{ H \mid H \subset \mathbb{P}^N \text{ hyperplane} \right\}.$$

Let $X \hookrightarrow \mathbb{P}^N$ be a projective, closed subscheme of $\mathbb{P}^N$. By definition the dual variety is the (reduced) closure of the set of hyperplanes which are tangent to X at some smooth point:

$$X^\vee = \overline{\left\{ H \in \mathbb{P}^{N\vee} \,\middle|\, \begin{array}{l} H \text{ tangent to } X \\ \text{at a smooth point } x \in X \end{array} \right\}}.$$

The condition in the above definition can be expressed as

$$H \supset T_{X,x} \quad \text{for some } x \in X_{sm}.$$

By definition $X^\vee$ is reduced, but it need neither be irreducible nor of pure dimension. While the definition does make sense for non reduced subvarieties X, the nilpotent components do not contribute to $X^\vee$, and in particular $X^\vee = \emptyset$ if and only if X has no reduced components.

A. Holme, *A Royal Road to Algebraic Geometry*, 343
DOI 10.1007/978-3-642-19225-8_22, © Springer-Verlag Berlin Heidelberg 2012

To understand $X^\vee$ better, we look at the following diagram:

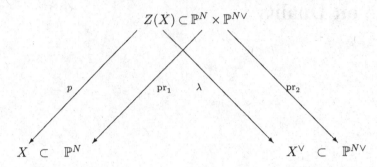

Here

$$Z(X) = \left\{ (x, H) \in X_{sm} \times \mathbb{P}^{N\vee} \,\middle|\, \begin{array}{l} H \text{ tangent to } X \\ \text{at the smooth point } x \in X \end{array} \right\}.$$

By definition $X^\vee$ is the image of $Z(X)$ under the projection pr_2, thus the morphism λ is induced. Also, pr_1 induces a morphism p.

Assume for the moment that X is smooth. Then we have the following key diagram, where

$$\mathcal{N}_{X/\mathbb{P}^N} = (\mathcal{J}_{X/\mathbb{P}^N}/\mathcal{J}^2_{X/\mathbb{P}^N})^\vee$$

and where $i : X \hookrightarrow \mathbb{P}^N$ is the embedding of X into $\mathbb{P}^N$. Moreover, $\mathcal{P}^1(X)$ denotes the locally free sheaf of *principal parts* of X:

$$
\begin{array}{ccccccccc}
& & 0 & & 0 & & & & \\
& & \downarrow & & \downarrow & & & & \\
0 \to & \mathcal{N}^\vee_{X/\mathbb{P}^N}(1) & \to & i^*\Omega^1_{\mathbb{P}^N}(1) & \to & \Omega^1_X(1) & \to 0 & & \\
& \| & & \downarrow & & \downarrow & & & \\
0 \to & \mathcal{N}^\vee_{X/\mathbb{P}^N}(1) & \to & \mathcal{O}^{N+1}_X & \to & \mathcal{P}^1(X) & \to 0 & & (22.1) \\
& & & \downarrow & & \downarrow & & & \\
& & \mathcal{O}_X(1) & = & \mathcal{O}_X(1) & & & & \\
& & \downarrow & & \downarrow & & & & \\
& & 0 & & 0 & & & &
\end{array}
$$

The left injective map in the lower exact sequence induces a surjective

$$\mathcal{O}^{N+1}_X \longrightarrow\!\!\!\!\!\rightarrow \mathcal{N}_{X/\mathbb{P}^N}(-1)$$

which gives the closed embedding

$$Z(X) = \mathbb{P}(\mathcal{N}_{X/\mathbb{P}^N}(-1)) \hookrightarrow \mathbb{P}(\mathcal{O}^{N+1}_X) = X \times \mathbb{P}^N.$$

The twist by -1 ensures that the tautological line bundle (i.e., the invertible sheaf) $\mathcal{O}_{\mathbb{P}(\mathcal{N}_{X/\mathbb{P}^N}(-1))}(1)$ is the restriction of the invertible sheaf $\mathrm{pr}_2^*(\mathcal{O}_{\mathbb{P}^{N\vee}}(1))$.

We now assume only that X is a reduced, projective scheme of pure dimension n. Letting X_{sm} denote the open subset of X consisting of smooth points, we have

$$Z(X) = \overline{\mathbb{P}(\mathcal{N}_{X_{sm}/\mathbb{P}^N - \text{Sing}(X)}(-1))}.$$

Much information about $X^\vee$ is contained in the *Chow cohomology class*

$$[Z(X)] \in A^*(\mathbb{P}^N \times \mathbb{P}^{N\vee}) = \mathbf{Z}[s,t]$$

where $A^*(S)$ denotes the Chow (cohomology) ring of the smooth, projective variety S, $s = \text{pr}_1^*([H])$, $t = \text{pr}_2^*([H'])$ and H, H' denote hyperplanes in $\mathbb{P}^N$ and $\mathbb{P}^{N\vee}$, respectively.

If $n = \dim(X)$, then clearly

$$\dim(Z(X)) = n + (N - n - 1) = N - 1$$

and hence, since $Z(X)$ maps onto $X^\vee$ by λ,

$$\dim(X^\vee) \le N - 1.$$

We have the following expression, where the $\delta_j(X)$ are integers for all j:

$$[Z(X)] = \delta_0(X)s^N t + \delta_1(X)s^{N-1}t^2 + \cdots + \delta_n(X)s^{N-n}t^{n+1}$$
$$+ \delta_{n+1}(X)s^{N-n-1}t^{n+2} + \cdots + \delta_{N-1}(X)st^N. \qquad (22.2)$$

It follows easily, and will be shown in the next sections, that

$$\Delta = \delta_{n+1}(X)s^{N-n-1}t^{n+2} + \cdots + \delta_{N-1}(X)st^N$$

is actually equal to *zero*.

22.2 Reflexivity and Biduality

Suppose that we construct the basic diagram for X, $Z(X)$ and $X^\vee$, but this time with $X^\vee$ instead of X:

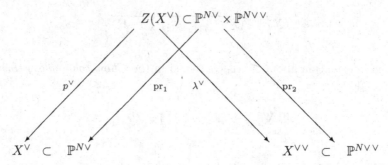

$$Z(X^\vee) \subset \mathbb{P}^{N\vee} \times \mathbb{P}^{N\vee\vee}$$

$$p^\vee \qquad \mathrm{pr}_1 \quad \lambda^\vee \qquad \mathrm{pr}_2$$

$$X^\vee \subset \mathbb{P}^{N\vee} \qquad\qquad X^{\vee\vee} \subset \mathbb{P}^{N\vee\vee}$$

If we make the canonical identification $\mathbb{P}^N = \mathbb{P}^{N\vee\vee}$, then we have $\mathbb{P}^{N\vee} \times \mathbb{P}^{N\vee\vee} \cong \mathbb{P}^N \times \mathbb{P}^{N\vee}$, where the isomorphism is the canonical one interchanging the two copies of projective space $\mathbb{P}^N$ and $\mathbb{P}^{N\vee}$. Thus we can compare the subschemes $Z(X^\vee)$ and $Z(X)$. In the "good" cases these two subschemes are equal:

Definition 22.1 The embedded, reduced projective variety $X \hookrightarrow \mathbb{P}^N$ is said to be *reflexive* if

$$Z(X^\vee) = Z(X).$$

If $X = X^{\vee\vee}$, then we say that *biduality* holds for X.

It is clear that reflexivity implies biduality. But the converse is false, for this there are counterexamples. The following classical result is an important fact, it implies in particular that reflexivity and hence biduality always holds in characteristic zero:

Theorem 22.1 *The embedded, reduced projective variety $X \hookrightarrow \mathbb{P}^N$ is reflexive if and only if the morphism $\lambda : Z(X) \longrightarrow X^\vee$ is generically smooth.*

For the history of this theorem, as well as modern proof using Lagrangian geometry, we refer to Kleiman's article [32]. As a corollary we obtain the following geometric criterion:

Corollary 22.2 X *is reflexive if and only if the contact locus* $\mathrm{Cont}(H,X)$ *with* X *of the generic tangent hyperplane H to X is a linear subspace of $\mathbb{P}^N$.*

Proof of the Corollary Evidently $\mathrm{Cont}(H,X) = Z(X)_h = \lambda^{-1}(h)$, where h is the point of $X^\vee$ which corresponds to the tangent hyperplane H. Thus reflexivity implies that $\mathrm{Cont}(H,X)$ equals the fiber $\lambda^{-1}(h) =$

$\mathbb{P}(\mathcal{N}_{X^\vee{}_{sm}/\mathbb{P}^{N^\vee}-\mathrm{Sing}(X^\vee)}(-1))_h$, provided that H corresponds to a smooth point h of $X^\vee$. Conversely, assume that $\mathrm{Cont}(H, X) = Z(X)_h = \lambda^{-1}(h)$ is linear, hence in particular equidimensional and geometrically regular, for all $h \in U$ where $U \subset X^\vee$ is an open dense subset. Then making U smaller if necessary we may assume that λ is flat over U. Hence λ is smooth over U by standard facts on smooth morphisms, say [18], Chap. III Theorem 10.2. □

In the next section we shall give a *numerical criterion* for reflexivity. At this point we note the following fact, which was noted in [39]:

Theorem 22.3 (T. Urabe) *If X is reflexive, then*

$$\delta_j(X^\vee) = \delta_{N-1-j}(X).$$

Proof By the identifications above and the assumption of reflexivity it is clear that

$$\delta_0(X)s^N t + \cdots + \delta_i(X)s^{N-i}t^{i+1} + \cdots$$
$$= \delta_0(X^\vee)st^N + \cdots + \delta_i(X^\vee)s^{j+1}t^{N-j} + \cdots$$

The claim is immediate from this. □

22.3 Duality of Projective Varieties

If X is smooth, then it is easily seen from diagram (22.1) that we have the following formulas for the numerical invariants occurring in formula (22.2):

$$\text{For all } i \geq 0, \quad \delta_i(X) = \sum_{j=i}^{n}(-1)^{n-j}\binom{j+1}{i+1}\deg(c_{n-j}(X)). \tag{22.3}$$

To prove this, we use a general fact which is referred to as *Scott's Formula*: Let

$$0 \longrightarrow \mathcal{E} \longrightarrow \mathcal{F} \longrightarrow \mathcal{G} \longrightarrow 0$$

be an exact sequence of locally free sheaves of the finite ranks e, f and g, respectively, on the (smooth projective) scheme S. Then there is a canonical closed embedding

$$\mathbb{P}(\mathcal{G}) \hookrightarrow \mathbb{P}(\mathcal{F})$$

so that we get a class

$$[\mathbb{P}(\mathcal{G})] \in A(\mathbb{P}(\mathcal{F})) = A(S)[\xi_F]$$

where $\xi_F \in A(\mathbb{P}(\mathcal{F}))$ denotes $c_1(\mathcal{O}_{\mathbb{P}(\mathcal{F})}(1))$. Letting $p : \mathbb{P}(\mathcal{F}) \longrightarrow S$ denote the canonical projection, we then have Scott's Formula

$$[\mathbb{P}(\mathcal{G})] = \sum_{i \geq 0} p^*(c_{e-i}(\mathcal{E}^\vee))\xi_F^i.$$

A proof of this can be found in [12], p. 61 or in Sect. 2 of the Appendix to [22]. Now a straightforward computation applying this formula to the exact sequence on X

$$0 \longrightarrow \mathcal{P}^1(X)^\vee \longrightarrow \mathcal{O}_X^{N+1} \longrightarrow \mathcal{N}_{X/\mathbb{P}^N}(-1) \longrightarrow 0$$

where the Chern classes of $\mathcal{P}^1(X)$ are computed by means of the exact sequence

$$0 \longrightarrow \Omega_X(1) \longrightarrow \mathcal{P}^1(X) \longrightarrow \mathcal{O}_X(1) \longrightarrow 0$$

yields (22.3).

Formula (22.3) has been generalized to the *singular case* by R. Piene in [37]. In fact this paper was an important contribution to the theory of duality in the presence of singularities, making it possible to understand the relation between singular Chern- or Segre classes on one hand and the *Polar classes* which have the *delta–invariants* as their degrees, on the other.

Here we confine ourselves to some simple observations, which in certain situations can be quite useful as we shall see later.

Before we state the result, we recall some background:

Nevertheless, we have some information in the singular case as well, and in certain situations this can be quite useful, as we shall see later. Before we state the result, we recall some background:

Let Z denote a *smooth, quasi-projective scheme* of pure dimension n, embedded as a (locally closed) subscheme of a projective space $i : Z \hookrightarrow \mathbb{P}^N$. Let $\overline{Z} \subset \mathbb{P}^N$ denote the closure of Z in $\mathbb{P}^N$, so $Z \subset \overline{Z}$ is an open subset, and denote the codimension of $S = \overline{Z} - Z$ in $\overline{Z}$ by r. Finally let $A^*(Y)$ and $A_*(Y)$ be the Chow cohomology ring, respectively the Chow homology module, of the singular quasi-projective scheme Y, in the sense of Chap. 18 or Fulton [12]. Recall that $A^*(Y)$ is a commutative, graded ring with 1, and $A_*(Y)$ is a graded module over $A^*(Y)$. Both multiplications will be denoted by $\cdot$ or by $\wedge$. We then have the graded homomorphism

$$\Psi_Y : A^*(Y) \longrightarrow A_*(Y)$$

which sends the element $\alpha \in A^*(Y)$ to the element $\alpha \wedge [Y] \in A_*(Y)$. Since $A^*(Y)$ is graded by codimension and $A_*(Y)$ by dimension, we have

$$\Psi_{Yj} : A^j(Y) \longrightarrow A_{\dim(Y)-j}(Y).$$

This homomorphism is compatible with the standard properties of the *covariant pair*

$$(A^*(\ ), A_*(\ )),$$

in particular it is functorial, and whenever Y is smooth, it is an isomorphism which only introduces a shift in the grading. Finally, the open embedding i above induces a *Gysin map* $i^* : A_*(Z) \longrightarrow A_*(\overline{Z})$, which fits into the exact sequence below, where $j : S \hookrightarrow \overline{Z}$ is the canonical closed embedding:

$$A_*(S) \xrightarrow{j_*} A_*(\overline{Z}) \xrightarrow{i^*} A_*(Z) \longrightarrow 0.$$

It follows that for all integers $\alpha \leq n - r - 1$ the above maps will induce isomorphisms

$$A^\alpha(\overline{Z}) \xrightarrow{\approx} A^\alpha(Z) \xrightarrow{\approx} A_{n-\alpha}(Z)).$$

We now take $X = X - S$, where $S = \mathrm{Sing}(X)$, thus $X = \overline{Z}$. We make the following

Definition 22.2 Let X be a projective scheme, of pure dimension n, and with singular locus $S = \mathrm{Sing}(X)$ which is of dimension m. Then for all $\alpha \leq n - m - 1$, the Chern class $c_\alpha(X) \in A^\alpha(X)$ is defined as the Chern class $c_\alpha(X - S) \in A^\alpha(X - S)$ where $A^\alpha(X - S)$ is identified with $A^\alpha(X)$ by the isomorphism given above.

We are now ready for a partial extension of (22.3) to the singular case:

Theorem 22.4 *Assume that the singular locus of X is of dimension m. Then the formula (22.3) above holds for all $i \geq m + 1$.*

Remark 22.5 It follows from this that the formula (22.3) holds for all $i \geq m+1$ for any functorial theory of singular Chern-classes which coincides with the usual Chern-classes in the smooth case.

Proof Let $U = X - \mathrm{Sing}(X)$, and let $\mathbf{V} = \mathbb{P}^N - \mathrm{Sing}(X)$. Then U is a closed subscheme of $\mathbf{V}$, and

$$p^{-1}(U) = \mathbb{P}(\mathcal{N}_{U/\mathbf{V}}(-1)).$$

Moreover, we have the exact sequence

$$0 \longrightarrow \mathcal{P}^1(X)^\vee|_U \longrightarrow \mathcal{O}_U^{N+1} \longrightarrow \mathcal{N}_{U/\mathbf{V}}(-1) \longrightarrow 0.$$

As in the case when X is smooth, we now proceed by Scott's Formula, where as before the Chern classes of $\mathcal{P}^1(X)|_U = \mathcal{P}^1(U)$ are computed by means of the exact sequence

$$0 \longrightarrow \Omega^1_U(1) \longrightarrow \mathcal{P}^1(U) \longrightarrow \mathcal{O}_U(1) \longrightarrow 0.$$

The claim follows from this. $\square$

Corollary 22.6 $\delta_n(X)$ *is equal to the degree of X, and for $j \geq n+1$, $\delta_j(X) = 0$.*

Proof Immediate from (22.3).   □

The invariants δ_j determine the dimension of the dual scheme $X^\vee$, and if X is reflexive, then the degree of $X^\vee$ is also given by these invariants: This was proved by *R. Piene* in [37].

In the following theorem we give a stronger version of this result, which gives a *numerical criterion* for reflexivity.

Theorem 22.7 (i) $X^\vee$ *is of dimension* $N - 1 - r$ *if*

$$\delta_0(X) = \cdots = \delta_{r-1}(X) = 0, \quad \delta_r(X) \neq 0.$$

(ii) *With* r *as in* (i), X *is reflexive if and only if*

$$\delta_r(X) = \deg(X^\vee).$$

Proof (i): We have the following sequence of biimplications, where P^r_{generic} denotes a linear r-space in $\mathbb{P}^{N\vee}$ in general position:

$$\dim(X^\vee) \leq N - 1 - r$$
$$\Updownarrow$$
$$X^\vee \cap P^r_{\text{generic}} = \emptyset$$
$$\Updownarrow$$
$$Z(X) \cap \mathrm{pr}_2^{-1}(P^r_{\text{generic}}) = \emptyset$$
$$\Updownarrow$$
$$[Z(X)] \cdot t^{N-r} = \delta_0 s^N t^{N-r+1} + \cdots + \delta_{r-1} s^{N-r} t^N = 0$$
$$\Updownarrow$$
$$\delta_0 = \cdots = \delta_{r-1} = 0.$$

(ii): Let P^{r+1}_{generic} be a linear subspace of $\mathbb{P}^{N\vee}$ in general position, as above. It intersects $X^\vee$ in exactly $\deg(X^\vee)$ smooth points. Letting ϵ denote the degree of the general fiber of λ, we then have the formula

$$\epsilon \cdot \deg(X^\vee) = \delta_r(X).$$

See the definition of $f_*([V])$ in Sect. 18.2. In the proof of Theorem 22.9 we show $\varepsilon = 1$. The claim follows from this, together with Corollary 22.2.   □

22.4 Duality of Hyperplane Sections and Projections

A basic observation in the study of projective duality is that the operation of embedding a projective space as a linear subspace of another, is dual to the operation of projecting a larger space onto a smaller projective space with a linear center: In order to make this correspondence precise, we have to resolve a conflict of notation: Namely, if $L \hookrightarrow \mathbb{P}^N$ is a linear subspace, then the notation $L^\vee$ could mean either the dual variety of L considered as a subvariety of $\mathbb{P}^N$, or the dual space of the linear space L itself. Normally there is no need to distinguish between these two concepts in the notation, since the situation will be clear from the context. But whenever there is a possibility of confusion, *we shall denote the dual of the projective space L itself by L^**. We then have the following elementary observation, the proof of which is immediate:

Proposition 22.8 (i) *Let $L \hookrightarrow \mathbb{P}^N$ be a linear subspace of dimension r. Then $L^\vee \hookrightarrow \mathbb{P}^{N\vee}$ consists of those hyperplanes in $\mathbb{P}^N$ which are tangent to L, i.e. they contain L.*

(ii) *Let $\mathrm{pr}_L : \mathbb{P}^N -\!-\!\rightarrow \mathbb{P}^{N-r-1}$ be the projection with center in the linear r-dimensional subspace L. The pullback of a hyperplane in $\mathbb{P}^{N-r+1}$ yields a hyperplane in $\mathbb{P}^N$ which contains L, thus a point in $L^\vee$. This correspondence is bijective, and establishes an embedding*

$$\mathbb{P}^{N-r-1\vee} \hookrightarrow \mathbb{P}^{N\vee}$$

which identifies $\mathbb{P}^{N-r-1}$ with $L^\vee$.

We next give a simple proof of the theorem below, which is shown in [37].

Theorem 22.9 (R. Piene) *Let X be a reduced scheme of pure dimension n.*

(1) *Assume that X is not a hypersurface in $\mathbb{P}^N$. If $\mathrm{pr}_P : \mathbb{P}^N -\!-\!\rightarrow \mathbb{P}^{N-1}$ is a generic projection with center in the point P, then*

$$\delta_i(X) = \delta_i(\mathrm{pr}_P(X))$$

for all $i = 0, \ldots, n$.

(2) *Let $H \subset \mathbb{P}^N$ be a generic hyperplane. Then*

$$\delta_i(X \cap H) = \delta_{i+1}(X)$$

for all $i = 0, \ldots, n-1$.

Proof We first show (1). By the proposition we have that in the set up

$$\mathrm{pr}_P : \mathbb{P}^N -\!-\!\rightarrow \mathbb{P}^{N-1}$$

$P^\vee$ is identified with $\mathbb{P}^{N-1\vee}$. Under this identification, we find

$$Z(\mathrm{pr}_P(X)) = (\mathrm{pr}_P \times \mathrm{id}_{\mathbb{P}^{N\vee}})(Z(X)) \cap (\mathbb{P}^{N-1} \times \mathbb{P}^{N-1\vee}).$$

A proof of this observation is given in Sect. 6 of [26] as Proposition 6.3, (2). Letting

$$j : \mathbb{P}^{N-1} \times \mathbb{P}^{N-1\vee} \hookrightarrow \mathbb{P}^{N-1} \times \mathbb{P}^{N\vee}$$

be the canonical embedding, we thus find

$$\varepsilon[Z(\mathrm{pr}_P(X))] = j^*((\mathrm{pr}_P \times \mathrm{id}_{\mathbb{P}^{N\vee}})_*([Z(X)]))$$

where ε is an integer, we will show shortly that $\varepsilon = 1$. The symbol $(\mathrm{pr}_P \times \mathrm{id}_{\mathbb{P}^{N\vee}})_*$ does of course represent an abuse of notation, since pr_P is not defined at P. The pushdown is defined by the standard diagram extended to $\mathbb{P}^{N\vee}$:

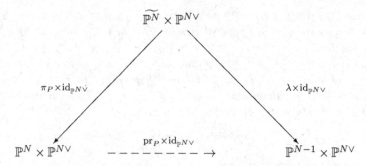

Here π_P denotes the blowing up with center P, and λ is the corresponding bundle map on to $\mathbb{P}^{N-1}$.

As usual we write

$$A(\mathbb{P}^N \times \mathbb{P}^{N\vee}) = \mathbf{Z}[s,t]$$

where s and t are the pullbacks of the hyperplane classes from $\mathbb{P}^N$ and $\mathbb{P}^{N\vee}$, respectively. Similarly

$$A(\mathbb{P}^N \times \mathbb{P}^{N-1\vee}) = \mathbf{Z}[s,\bar{t}]$$
$$A(\mathbb{P}^{N-1} \times \mathbb{P}^{N\vee}) = \mathbf{Z}[\bar{s},t]$$

and

$$A(\mathbb{P}^{N-1} \times \mathbb{P}^{N-1\vee}) = \mathbf{Z}[\bar{s},\bar{t}].$$

For $i \geq 1$ we have $\mathrm{pr}_{P*}(s^i) = \bar{s}^{i-1}$ (where pr_{P*} is defined in the obvious way via the blowing up with center P), and for all i and j

$$j^*(\bar{s}^i t^j) = \bar{s}^i \bar{t}^j.$$

Thus the expressions

$$[Z(X)] = \delta_0(X)s^N t + \delta_1(X)s^{N-1}t^2 + \cdots + \delta_n(X)s^{N-n}t^{n+1}$$

and

$$[Z(\mathrm{pr}_P(X))] = \delta_0(\mathrm{pr}_P(X))\bar{s}^{N-1}\bar{t} + \delta_1(\mathrm{pr}_P(X))\bar{s}^{N-2}\bar{t}^2$$
$$+ \cdots + \delta_n(\mathrm{pr}_P(X))\bar{s}^{N-1-n}\bar{t}^{n+1}$$

immediately yield the claim, since $\delta_n(X) = \delta_n(\mathrm{pr}_P(X)) = \deg(X)$ so that $\varepsilon = 1$.

(2) is shown in an analogous way: Let

$$i_H : \mathbb{P}^{N-1} \hookrightarrow \mathbb{P}^N$$

$$\mathrm{pr}_H : \mathbb{P}^{N\vee} \dashrightarrow \mathbb{P}^{N-1\vee}$$

denote the embedding with image H, and dually the projection with center H regarded as a point of $\mathbb{P}^{N\vee}$. We then have the relation

$$Z(X \cap H) = (\mathrm{id}_{\mathbb{P}^{N-1}} \times \mathrm{pr}_H)(Z(X) \cap (\mathbb{P}^{N-1} \times \mathbb{P}^{N\vee}))$$

which again is shown in Sect. 6. Here $\mathbb{P}^{N-1}$ is identified with H via i_H. Thus

$$[Z(X \cap H)] = (\mathrm{id}_{\mathbb{P}^{N-1}} \times \mathrm{pr}_H)_*((i_H \times \mathrm{id}_{\mathbb{P}^{N\vee}})^*([Z(X)])).$$

Now we have

$$[Z(X \cap H)] = \delta_0(X \cap H)\overline{s}^{N-1}\overline{t} + \delta_1(X \cap H)\overline{s}^{N-2}\overline{t}^2$$
$$+ \cdots + \delta_{n-1}(X \cap H)\overline{s}^{N-1-(n-1)}\overline{t}^n$$

and since

$$(\mathrm{id}_{\mathbb{P}^{N-1}} \times \mathrm{pr}_H)_*((i_H \times \mathrm{id}_{\mathbb{P}^{N\vee}})^*(s^u t^v) = \overline{s}^u \overline{t}^{v-1}$$

we also have

$$(\mathrm{id}_{\mathbb{P}^{N-1}} \times \mathrm{pr}_H)_*((i_H \times \mathrm{id}_{\mathbb{P}^{N\vee}})^*([Z(X)]))$$
$$= \delta_1(X)\overline{s}^{N-1}\overline{t} + \delta_2(X)\overline{s}^{N-2}\overline{t}^2 + \cdots + \delta_n(X)\overline{s}^{N-1-(n-1)}\overline{t}^n$$

where again a multiplicity ε turns out to be 1 since $\delta_n(X) = \delta_{n-1}(X \cap H) = \deg(X)$. This gives the claim.   □

22.5   A Theorem of Hefez-Kleiman

In this section we prove a theorem by Hefez and Kleiman [32] in the following form:

Theorem 22.10 *Let X be a reduced, projective scheme of pure dimension n, and let r be the integer such that*

$$\delta_0(X) = \cdots = \delta_{r-1}(X) = 0, \quad \delta_r(X) \neq 0.$$

Then for all $i \in [r, n]$,

$$\delta_i(X) \geq 1.$$

In characteristic zero, this can be strengthened to

$$\delta_i(X) \geq 2$$

provided that X is not a linear subspace of $\mathbb{P}^N$.

Proof Replacing X by the intersection with an appropriate generic linear subspace if necessary, we may assume that $r = 0$, i.e., that $X^\vee$ is a hypersurface. The claim then amounts to showing the theorem below, which is classical in spirit, but was first discovered by *A. Wallace* in [40] (Theorem 1 on p. 7, as well as the lemmas d and e):

Theorem 22.11 (Wallace) *If $X^\vee$ is a hypersurface in $\mathbb{P}^N$, then $(X \cap H)^\vee$ is a hypersurface in $\mathbb{P}^{N-1\vee}$, provided H is a hyperplane in general position. More generally, if X is cut by an r-dimensional linear subspace L in general position, then the dual of the linear section is a hypersurface in $L^* \cong \mathbb{P}^{r\vee}$.*

Proof We give Wallace's proof, slightly reformulated: The dual of $X \cap H$ in $\mathbb{P}^{N\vee}$ is the closure of all points which correspond to hyperplanes H' which contain $T_{X,x} \cap H$ for some smooth point $x \in X$. Thus letting $T(X, X \cap H)$ denote the subvariety of $\mathbb{P}^{N\vee}$ which corresponds to (the closure of the set of) hyperplanes tangent to X at some (smooth) point of $X \cap H$, we find that the dual of the hyperplane section in $\mathbb{P}^{N\vee}$ is the *cone* over $T(X, X \cap H)$. Since $X^\vee$ is a hypersurface by assumption, the map λ is generically finite, thus $T(X, X \cap H)$ is of codimension 1 in $X^\vee$ whenever H is sufficiently general, hence $(X \cap H)^\vee$ in $\mathbb{P}^{N\vee}$ is a hypersurface as well, and a cone with vertex H. But since the correspondence between points in $\mathbb{P}^{N\vee}$ and points in $H^* = \mathbb{P}^{N-1\vee}$ is given by projection with center $H \in \mathbb{P}^{N\vee}$, this means that the dual of $(X \cap H)^\vee$ is a hypersurface in $\mathbb{P}^{N-1\vee}$.

Next, assume that the characteristic is zero. Replacing X by its intersection with a generic linear subspace of $\mathbb{P}^N$ if necessary, we may assume that $X^\vee$ is a hypersurface. Then cutting $X^\vee$ with a generic linear 2-space $\mathbb{P}2 \subset \mathbb{P}^N$ and using the duality of the delta-invariants, we have from Piene's theorem in the previous section

$$[Z(X^\vee \cap \mathbb{P}^2)] = \delta_0(X)\bar{s}^2\bar{t} + \delta_1(X)\bar{s}\bar{t}^2$$

where $\bar{s}$ and $\bar{t} \in A(\mathbb{P}^2 \times \mathbb{P}^{2\vee})$ are the pullbacks of line-classes from $\mathbb{P}^2$ and $\mathbb{P}^{2\vee}$, respectively. In particular, if $X^\vee$ is a hypersurface, then $X^\vee \cap \mathbb{P}^2$ must be a planar curve, of degree $\delta_0(X) \geq 2$: If it were of degree 1, then it would be a line, thus $X^\vee$ would be linear, hence X would be linear by biduality. Moreover, $\delta_1(X)$ is the degree of the curve $(X^\vee \cap \mathbb{P}^2)^\vee$. Hence we also have $\delta_1(X) \geq 2$. Thus we have shown the claim for $i = 0$ and 1. To proceed, we cut X with one hyperplane more and repeat the argument.   □

A further sharpening is given by the

Proposition 22.12 *Assume that the characteristic of the ground field is zero, and that $X \subset \mathbb{P}^N$ is irreducible and is not a hypersurface in some linear subspace. Then the inequality in the theorem can be strengthened to*

$$\delta_i(X) \geq 3.$$

Proof By the same argument as in the last part of the proof of the theorem: Assume that $\delta_i(X) = 2$ for some i. Then we may assume that $i = r = 0$, if necessary after cutting X with an appropriate generic linear subspace of $\mathbb{P}^N$. Thus if $\mathbb{P}^2$ is a generic linear subspace of dimension 2, then $X^\vee \cap \mathbb{P}^2$ is an irreducible curve of degree 2, hence a smooth conic, thus $X^\vee$ is either smooth or a cone of degree 2. But it can not be a cone, as biduality holds and X is not contained in a hypersurface. Hence $X^\vee$ is smooth of degree 2, so X is also of degree 2. We are thus finished by observing that a variety of degree 2 is necessarily a hypersurface in some linear subspace of the ambient space. □

The number r referred to in the theorem above is called the *duality defect* of the embedded variety X. The concept is important for the classification of embedded projective varieties.

But here, with a view into a vast and very interesting field of research, is where our Royal Road ends.

References

1. Altman, A.B., Kleiman, S.L.: Joins of schemes, linear projections. Compos. Math. **31**(3), 309–343 (1975)
2. Baum, P., Fulton, W., MacPherson, R.: Riemann-Roch for singular varieties. Publ. Math. l'HIES **45**, 101–145 (1975)
3. Borel, A., Serre, J.P., Le Théorème de Riemann-Roch. Bull. Soc. Math. Fr. **86**, 97–136 (1958)
4. Bucur, I., Deleanu, A.: Categories and Functors. Interscience Publication. Wiley, New York (1968)
5. Campbell, J.M.: A note on flasque sheaves. Bull. Aust. Math. Soc. **2**, 229–232 (1970)
6. Cartan, H. Chevalley, C.: Géometrie Algébrique. Séminaire Cartan-Chevalley. Secrétariat Math., Paris (1955/56)
7. Chenevert, G., Kassaei, P.: Sheaf Cohomology. Paper on the web (2003). http://www. math.mcgill.ca/goren/SeminarOnCohomology/SheafCohomology.pdf
8. Diao, H.: Chow's Lemma. This paper, as well as a number of other interesting ones, is available at www.math.harvard.edu/ on the home page of Professor Dennis Gaitsgory of Harvard University
9. Eisenbud, D.: Introduction to Commutative Algebra with a View Towards Algebraic Geometry. Graduate Texts in Mathematics. Springer, Berlin (1997)
10. Fulton, W.: Algebraic Curves. Mathematical Lecture Notes Series. W.A. Benjamin, Amsterdam (1969)
11. Fulton, W.: Rational equivalence on singular varieties. Publ. Math. l'HIES **45**, 147–167 (1975)
12. Fulton, W.: Intersection Theory, 1st edn. Ergebnisse der Mathematik und ihrer Grenzgebiete. 3. Folge, vol. 2. Springer, Berlin (1984), 2nd edn. (1998)
13. Garrett, P.: Sheaf Cohomology. Note available at the authors home page www. math.umn.edu/garret
14. Grothendieck, A.: La Théorie des classas de Chern. Bull. Soc. Math. Fr. **86**, 137–154 (1958)
15. Grothendieck, A.: Éléments de géométrie algébrique. i–iv. Publ. Math. I.H.E.S., 4, 8, 11, 17, 20, 24, 28, 32
16. Grothendieck, A.: Local Cohomology. Lecture Notes in Mathematics, vol. 41. Springer, Berlin (1967)
17. Grothendieck, A., Dieudonné, J.A.: Éléments de géométrie algébrique. I. Die Grundlehren der Mathematischen Wissenshaften, vol. 166. Springer, Berlin (1971)
18. Hartshorne, R.: Algebraic Geometry. Graduate Texts in Mathematics. Springer, Berlin (1977)
19. Herodotus: The Histories. A Norton Critical Edition. Translated by Walter Blanco. W.W. Norton, New York (1992)

A. Holme, *A Royal Road to Algebraic Geometry*,
DOI 10.1007/978-3-642-19225-8, © Springer-Verlag Berlin Heidelberg 2012

20. Hirzebruch, F.: Topological Methods in Algebraic Geometry, 1956. Grundlehren der Math. Wissenshaften, vol. 131. Springer, Berlin (1966). 3rd enlarged edn.
21. Holme, A.: Introduccion a la Teoria de las Clases Caracteristicas en la Geometria Algebraica. Monografias del Instituto de Matematicas, vol. 6. Universidad Nacional Autonoma de Mexico, Mexico (1978)
22. Holme, A.: Embeddings, Projective Invariants and Classifications. Monografias del Instituto de Matematicas, vol. 7. Universidad Nacional Autonoma de Mexico, Mexico (1979)
23. Holme, A.: Embedding obstruction for smooth, projective varieties I. In: Rota, G.C. (ed.) Studies in Algebraic Topology. Advances in Mathematics Supplementary Series, vol. 4, pp. 39–67. Addison-Wesley, Reading (1979)
24. Holme, A.: Embedding obstruction for singular algebraic varieties in $\mathbf{P}^n$. Acta Math. **135**, 155–185 (1975)
25. Holme, A.: On the dual of a smooth variety. In: Algebraic Geometry, Copenhagen 1978. Lecture Notes in Mathematics, vol. 732, pp. 144–156. Springer, Berlin (1979)
26. Holme, A.: The geometric and numerical properties of duality in projective algebraic geometry. Manuscr. Math. **61**, 145–162 (1988)
27. Holme, A.: Geometry. Our Cultural Heritage. Springer, Berlin (2001)
28. Holme, A.: Geometry. Our Cultural Heritage, 2nd edn. Springer, Berlin (2010)
29. Holme, A., Roberts, J.: Pinch points and multiple locus for generic projections of singular varieties. Adv. Math. **33**, 212–256 (1979)
30. Holme, A., Roberts, J.: On the embeddings of projective varieties. In: Proceedings of the Sundance Symposium. Springer Lecture Notes in Mathematics. Springer, Berlin (1987)
31. Holme, A., Roberts, J.: Zak's superadditivity theorem. Arkiv Mat. (1995)
32. Kleiman, S.L.: Tangency and duality. In: Proceedings of the 1984 Vancouver Conference in Algebraic Geometry. CMS Conference Proceedings, pp. 163–226. The American Mathematical Society, Providence (1986)
33. Kosters, M.: Injective modules and the injective hull of a module. Notes from Leiden University available at www.math.leidenuniv.nl
34. Lam, T.Y.: Lectures on Modules and Rings. Springer, Berlin (1999)
35. Macdonald, I.G.: Algebraid Geometry. Introduction to Schemes. Mathematics Lecture Notes Series. W.A. Benjamin, New York (1968)
36. Mumford, D.: Introduction to Algebraic Geometry. Preliminary version of the first 3 chapters. Notes, Harvard University (1965)
37. Piene, R.: Polar classes of singular varieties. Ann. Sci. Éc. Norm. Super. **11**, 247–276 (1978)
38. Serre, J.-P.: Cohomologie et géometrie algébrique. In: Proc. of the International Congress of Mathematicians 1954, vol. III, pp. 515–520. Noordhoff, Groningen (1956). North Holland, Amsterdam
39. Urabe, T.: Duality of numerical characters of polar loci. Publ. RIMS, Kyoto Univ. **17**, 331–345 (1981)
40. Wallace, A.H.: Tangency and duality over arbitrary fields. Proc. Lond. Math. Soc. **6**, 321–342 (1956)
41. Weil, A.: Foundations of Algebraic Geometry (1946). Revised and enlarged edition, Amer. Math. Soc. Colloq. Publ. 29 (1962)
42. Zariski, O., Samuel, P.: Commutative Algebra, Volumes I and II. The University Series in Higher Mathematics. D. van Nostrand, Princeton (1960)

Index

A. Holme, *A Royal Road to Algebraic Geometry*,
DOI 10.1007/978-3-642-19225-8, © Springer-Verlag Berlin Heidelberg 2012